The HPLC Expert II

The HPLC Expert II

Find and Optimize the Benefits of your HPLC/UHPLC

Edited by Stavros Kromidas

WILEY-VCH

Editor

Dr. Stavros Kromidas
Consultant, Saarbrücken
Breslauer Str. 3
66440 Blieskastel
Germany

Library of Congress Card No.: applied for

British Library Cataloguing-in-Publication Data
A catalogue record for this book is available from the British Library.

Bibliographic information published by theDeutsche Nationalbibliothek
Die Deutsche Nationalbibliothek lists this publication in the Deutsche Nationalbibliografie; detailed bibliographic data are available on the Internet at http://dnb.d-nb.de.

Print ISBN: 978-3-527-33972-3
ePDF ISBN: 978-3-527-69497-6
ePub ISBN: 978-3-527-69495-2
Mobi ISBN: 978-3-527-69496-9
oBook ISBN: 978-3-527-69494-5

Cover Design Formgeber, Mannheim, Germany
Typesetting SPi Global, Chennai, India
Printing and Binding Markono Print Media Pte Ltd, Singapore

Printed on acid-free paper

Contents

List of Contributors

Siegfried Chroustovsky
Techno Trade and Service
Lichtenklinger Str. 13
Siedelsbrunn, 69483
Wald-Michelbach
Germany

Monika Dittmann
Agilent Technologies
Hewlett-Packard-Straße 8
76337 Waldbronn
Germany

Björn-Thoralf Erxleben
Shimadzu Europa GmbH
Albert-Hahn-Str. 6-10
47269 Duisburg
Germany

Tobias Fehrenbach
Schraudolphstr. 38
80799 Munich
Germany

Michael Heidorn
Thermo Fisher Scientific
Dornierstr. 4
82110 Germering/Munich
Germany

Terence Hetzel
Institut für Energie- und
Umwelttechnik e.V., IUTA
Bliersheimer Str. 58-60
47229 Duisburg
Germany

Stavros Kromidas
Consultant
Breslauer Str. 3
66440 Blieskastel
Germany

Stefan Lamotte
BASF SE, Global Comp. Center
Analysis
GMC/AC-E210
67056 Ludwigshafen
Germany

Juri Leonhardt
Institut für Energie- und
Umwelttechnik e.V., IUTA
Bliersheimer Str. 58-60
47229 Duisburg
Germany

Christoph Portner
Tauw GmbH
Richard-Löchel-Straße 9
47441 Moers
Germany

Arno Simon
Beyontics GmbH
Altonaer Straße 79-81
13581 Berlin
Germany

Frank Steiner
Thermo Fisher Scientific
Dornierstr. 4
82110 Germering/Munich
Germany

Thorsten Teutenberg
Institut für Energie- und
Umwelttechnik e.V., IUTA
Bliersheimer Str. 58-60
47229 Duisburg
Germany

Jens Trafkowski
Agilent Technologies
Hewlett-Packard-Straße 8
76337 Waldbronn
Germany

Jochen Tuerk
Institut für Energie- und
Umwelttechnik e.V., IUTA
Bliersheimer Str. 58-60
47229 Duisburg
Germany

Steffen Wiese
Institut für Energie- und
Umwelttechnik e.V.
IUTA, Bliersheimer Str. 58-60
47229 Duisburg
Germany

Foreword

In "The HPLC-Expert," we discussed several topics of the modern HPLC and illustrated new developments. In this book, "The HPLC-Expert 2" we focus on the modern HPLC/UHPLC device.

Our objective is to give detailed information about the modern HPLC/UHPLC equipment, so that our HPLC Colleagues can use *their* device optimal depending on their requirement. On the one hand, we present in 12 chapters how HPLC-Hardware and also particular modules can be run at the maximal resolution and peak capacity and, on the other hand, the procedure, if robustness is the main focus.

Practice is put forward, and theoretical background information is only given to an extent that we considered absolutely necessary. I hope that practice-oriented laboratory supervisors and experienced operators will find inspiration and hints about chances and constraints of the modern HPLC/UHPLC devices.

My special thanks go to Klaus Illig for his critical hints to the manuscript and to my author colleagues who contributed their experience and knowledge. Also, I want to thank WILEY-VCH and in particular Reinhold Weber for the good and close collaboration.

Stavros Kromidas
Blieskastel, February 2016

The Structure of "The HPLC-Expert 2"

This book contains 12 chapters that are divided into two parts:
1) specifics of the hardware, separation techniques, materials
2) software, experience reports, trends.

In the first introductory chapter, Stavros Kromidas illustrates potentials and limitations of UHPLC devices (**When Should I Use My UHPLC as a UHPLC**) and demonstrates in which laboratory situation and which analytical question the use of either HPLC or UHPLC might be more favorable.

The topic of Chapter 2 refers to **The Modern HPLC/UHPLC Device**. In Section 2.1.1 (**Today's Demands on the Individual Modules**), Stefan Wiese goes systematically into the different modules of a HPLC-device and explains important requirements from a user's viewpoint. Subsequently, Michael Heidorn and Frank Steiner show in Chapter 2.2 (**The Thermostate of Columns – A Minor Matter**) the importance of proper thermostatic control and demonstrate by means of several examples that simply setting the temperature is not enough.

Chapter 3 is devoted to the often discussed topic "band broadening" (**The Issue of External Band Broadening in HPLC/UHPLC Devices**); Monika Dittman shows with the help of numerical examples in which case the actual band broadening is eminent or even not relevant at all.

In Chapter 4 (**The Gradient; Requirements, Optimal Use, Hints, and Pitfalls**) Frank Steiner and Michael Heidorn give attention to gradient elution. The influence of several (pump) systems on quality and reliability is demonstrated and options for improvement provided for the user of present devices.

Thorsten Teutenberg and coauthors provide a great many suggestions in Chapter 5 (**Requirements of LC-Hardware for the Coupling of Different Mass Spectrometers**), as to arrange LC/MS coupling as optimal as possible. Among other things, complex samples and miniaturization play an important role.

In Chapter 6 (**2D chromatography – Opportunities and limitations**) Thorsten Teutenberg discusses how different modi of the 2D chromatography can influence the peak capacity and what has to be considered for a successful use in practice.

The first part of this books ends with Chapter 7 (**Materials in HPLC and UHPLC – What to Use for Which Purpose**). Tobias Fehrenbach and Steffen Wiese talk about a less noticed but important subject: the diversity of materials used in a HPLC/UHPLC device is discussed and the impairment of the analytical results is presented.

The second part of this book starts with Chapter 8 by Arno Simon (**What a Software has to Possess in Order to Use the Hardware Optimally**). The author elaborates the development of evaluation software, shows different philosophies, and ventures an outlook in the future of data systems for chromatography.

In Chapter 9 (**Aspects of the Modern HPLC Device – Experience Report of an Operator**), Steffen Wiese goes into special applications that are meaningful in everyday life and gives tips on how to handle these challenges, for example, high temperature, large injection volumes, transfer of methods.

In Chapter 10 (**Experiences of an Independent Service Engineer – Hints and Recommendations for an Optimal Operation of Agilent and Waters-Devices**), Siegfried Chroustovsky looks at the device from an engineer's point of view and offers a great many of tips on fault finding and troubleshooting.

The core topic of the introductory chapter is taken up in Chapter 11 (**The Analyte, the Question, and the UHPLC – The Use of UHPLC in Practice**) by Stefan Lamotte, who brings out when and how a UHPLC device is used optimally.

Finally in Chapter 12 (**Report of Device Manufacturers – Article by Agilent, Shimadzu, and ThermoScientific**), three manufacturers introduce briefly their newest products and evaluate the future of HPLC.

We think the style and structure of "The HPLC-expert" have proven themselves, so those were kept the same in the subsequent book: the book need not to be read linearly. All chapters present self-contained modules – "jumping" between chapters is always possible.

That way, we try to keep the nature of the book as a reference book. The reader may benefit therefrom.

1

When Should I Use My UHPLC as a UHPLC?

Stavros Kromidas

1.1 Introduction

Modern analytical LC systems are designed without exception as ultrahigh-performance liquid chromatography (UHPLC) systems. However, outside of pure research laboratories, a maximum of about 20–30% of the separations are performed under UHPLC conditions. By this, pressures above approximately 800 bar are meant. In which cases does it make sense, or is even necessary, to use the existing UHPLC system under truly UHPLC conditions? On the other hand, when should the UHPLC system perhaps be used as a fast, but "classical," high-performance liquid chromatography (HPLC)? This chapter deals with exactly this point. To this end, the answers to two questions can help, both of which we will deal with. The first is "What do I really need?" Here it is necessary to define which characteristics of an HPLC method in exactly *this* situation are in the foreground, among others, for example, short retention times, a robust method, maximum resolution/peak capacity, and low detection limit. The second question is much simpler: "Why is the UHPLC more capable than the HPLC?" Afterward, we will discuss the key question: "How do I reasonably combine my requirements on the method and the potential of UHPLC – taking into consideration the real laboratory situation?"

Note Familiarity with the theoretical background is assumed, and the principles of HPLC optimization are therefore only mentioned but not derived. For this, reference is made to the relevant literature (for example, [1–5]).

The HPLC Expert II: Find and Optimize the Benefits of your HPLC/UHPLC, First Edition.
Edited by Stavros Kromidas.

1.2 What Do I Want to Achieve and What Is a UHPLC Capable of?

1.2.1 What Do I Want to Achieve?

One often wants more than only one attribute from an HPLC method, for example, "good" *and* "fast" separation. However, before deciding on the method design – which indeed includes the question of the necessity of UHPLC conditions – two points urgently have to be clarified. *Firstly*, what are the peculiarities of the method, and how is its environment? Here we are interested in, among other things, the following crucial features of the proposed analysis: matrix, time required for sample preparation and manual reintegration, experience of the user, changing or constant chromatographic conditions, research or routine laboratory, and so on. *Secondly*, what is the primary requirement for *this* method in the specific case? The main objective should be clearly identified, a second (or third?) regarded merely as a wish, for example: "In this case we need, for this reason the maximum possible sensitivity – if the method is also precise, that would be good too … " Four typical requirements for an HPLC method, which we will subsequently consider in more detail, are listed as follows:

- *Good separation*: this can mean, firstly, sufficient resolution – separation between two critical peaks or possibly between 2 and 3 relevant peak pairs. Or, secondly, sufficient peak capacity – separation of many (or all?) – possibly chemically similar components, see Section 1.3.1
- *Fast separation*: short retention times; this often goes hand in hand with a low solvent consumption, see Section 1.3.2
- *Sensitive measurement*: decrease in the detection limit, which means an improvement in the relative mass sensitivity, see Section 1.3.3
- *Robust conditions*: reliable methods, which lead to the avoidance of repeat measurements and minimization of equipment downtime, see Section 1.3.4.

1.2.2 What Is a UHPLC Capable of?

Put simply, a UHPLC system is an instrument that, first of all, compared to an HPLC system, has about 10 times lower dead volume (dispersion volume or "Extracolumn Volume": the volume from the autosampler to the detector without a column) and also dwell/delay volume (the volume from the mixing valve/mixing chamber to the head of the column).The dead volume of a modern UHPLC system is nowadays about $\leq 7-10\,\mu l$, with the aid of special kits even about $\leq 4\,\mu l$, the dwell volumes are about $100-200\,\mu l$ with low-pressure gradient (LPG), and about $25-35\,\mu l$ with high-pressure gradient (HPG) systems.

Note Nowadays, we talk less of "Extracolumn Volume" but rather of "Extracolumn Dispersion." This takes into account the fact that the geometry of, for example, connections and mixing valves, and thus the flow profile, has more influence on the peak broadening than the absolute dead volume, see also Chapter 3. Secondly, a modern UHPLC system allows working pressures up to around 1500 bar.

1.3 What Is Required from an HPLC Method?

1.3.1 Separate Well

First of all, we show briefly how the separation in HPLC can be improved in principle, and then we will have a closer look at the contribution UHPLC can make toward a better separation.

In chromatography, we distinguish with respect to the quality of a separation between two cases:

1) I am really only interested in one or a few components. It is therefore a question of– according to my individual criteria – sufficient separation between the component of interest and an "interfering" component – in other words, ultimately on the separation of two peaks. The focus can be on the critical pair (e.g., main and secondary components), possibly on two to three more peak pairs. The criterion here is the resolution, and when simplified, it describes the distance between the peaks at the baseline.

$$R_S = \frac{1}{4}\cdot\sqrt{N}\cdot\frac{k_2}{1+k_2}\cdot\frac{\alpha - 1}{\alpha} \tag{1.1}$$

   where R = resolution, N = plate number (fundamentally defined for isocratic conditions), α = separation factor (formerly selectivity factor), and k = retention factor (formerly capacity factor k′).
2) I want to or have to separate "all" existing peaks sufficiently well, that is, when possible with baseline separation. In this case, the peak capacity comes into play. This is the total number of peaks that I can separate in a certain time with a sufficiently good resolution (commonly R = 1). The sum of all resolutions is often stated as a measure of the peak capacity. In the literature, one finds several formulas for the peak capacity, we consider here the two simplest:

$${}_{c}n_c = \frac{t_{Rl} - t_{\mathrm{Re}}}{w} \tag{1.2a}$$

   or

$$n_c = \frac{t_G}{w} \tag{1.2b}$$

   where n_c = peak capacity, t_{Rl} = retention time of the last peak, t_{Re} = retention time of the first peak, w = peak width, and t_G = gradient duration.

Note About 70–80% of separations, nowadays, are gradient separations. Consequently, today's HPLC/UHPLC system is a high- or low-pressure gradient with DAD and/or MS/MS, and furthermore, aerosol detectors are becoming more common. The thoughts presented here apply in principle for both isocratic and gradient separations, but for the aforementioned reason, I will lay the focus a little more on gradient separations.

Let us look first at the resolution.

Equation 1.1 shows that the resolution can be improved by increasing efficiency, selectivity, and retention. The requirement for the retention term is strong interactions, the optimum value lies around $k \approx 3-5$, and this means that the peaks of interest should elute by or after approximately three to five times the dead or mobile time. From Equation 1.1, it can be seen that the term for selectivity, and thus the separation factor α, is by far the most sensitive function of the resolution: $\alpha - 1/\alpha$! On the other hand, the plate number is under the root, a doubling of N improves the resolution by a factor of "only" 1.4. Two numerical examples illustrate this; for a detailed discussion, see [6]:

1) Assume that two peaks elute with an α-value of 1.01. To achieve baseline separation of these two peaks, one would need about 160 000 plates. If the α-value could be increased from 1.01 to 1.10, for the same resolution, just less than 2000 plates would be required. Even a seemingly small improvement in the α-value from 1.01 to 1.05 means that instead of 160 000 plates, only about 6000 plates are necessary.
2) Further assume that we have a separation with the following values: $k = 2$, $\alpha = 1.05$, and $N = 9000$. This results in a resolution of $R = 0.76$. This is not enough, and the resolution should be improved. To start with, the interactions can be increased, for example, through a more hydrophobic stationary phase or more water in the mobile phase. Assuming that the stronger interactions affect the two components equally, then the selectivity remains constant. The k-value increases from $k = 2$ to, for example, $k = 6$, and the resolution increases to $R = 0.97$. Alternatively, one could use a column with 15 000 plates, and the resolution improves to $R = 0.98$. Both measures are therefore correct; however, they are not particularly effective when it comes to significantly improving the resolution. If the α-value could be increased from 1.05 to 1.10, this would result in a resolution of $R = 1.45$. Let us finish the second example with the following observation: when two peaks are of different sizes (e.g., drug and impurity) and/or tailing is present, the resolution must be about $R \geq 2$ if the error in integration is to remain below 1% [7]. In the present case, to improve the resolution to $R = 2$, there are two alternatives available: increase the α-value from 1.10 just to 1.15 or double the plate number – at a constant α-value of 1.10 – from 9000 to 18 000 plates The last case would be possible with, for example, a 150 mm, 2.5 μm column.

As a rule of thumb, for a baseline separation, one could remember the following:

- If, in a real sample, the α-value of the critical pair is about 1.05, then for a baseline separation, about 20 000 plates would be necessary.
- If the α-value is approximately 1.02, you would need about 100 000 plates.
- With an α-value of 1.01, you will hardly be successful without 2D chromatography (see Chapter 6).

These numerical examples are valid for both isocratic and gradient separations.

Note The plate number is defined for isocratic separations. There are several formulas, of which the simplest is

$$N = \frac{L}{H} = \frac{16 \cdot t_R^2}{w^2} \tag{1.3}$$

where N = plate number, L = length of the column, H = height of one theoretical plate, t_R = retention time, and w = width of the peak at its base.

Let us look first at the question "what significance does the plate number have in isocratic and gradient separations?" When a peak elutes later in an isocratic run, it becomes wider, the ratio t_R/w, however, remains constant and therefore also the plate number. Note that the plate number for a component in the isocratic mode is – at least theoretically – a constant. This means it is independent of the retention time as long as alterations in this are due to a change in the stationary or mobile phases and/or temperature – but not in the flow! Once again, the peaks elute later or earlier and thus are wider or narrower – the ratio retention time/peak width remains constant and therefore also the plate number. Here, it is assumed that the mechanism of the interaction with the stationary phase remains constant.

How are the relationships in gradient mode? In the literature, it is often stated that, strictly speaking, the plate number can only be determined for isocratic separations. In connection with gradients, the terminology "separation efficiency" is often used. Nevertheless, also with gradients, there is no reason in principle against talking about a "plate number" N_{Gr} – at least as an idea.

Consider Equation 1.3 and assume that, due to any measure, a peak elutes in a gradient method later. In this case, the "plate number" N_{Gr} increases, because t_R increases but w remains constant. When simplified, the following applies: isocratic: ratio t_R/w constant and plate number constant; gradient: w constant and the "plate number" N_{Gr} increases.

What significance does this have for the separation? In isocratic separations, I can push the critical pair into the optimal retention area and, due to the optimal selectivity then existing, achieve the maximum resolution – but only for *this* critical pair, it is possible that other peak pairs may be less well separated. In a gradient run, because the peaks move closer together, the result is lower selectivity, but as we have seen that the plate number is higher, the peaks are narrow. As described earlier, the selectivity influences the resolution considerably more than does the plate number, and therefore, for this pair, we have better resolution under isocratic conditions as with a gradient run. Unless there is a particular case where the kinetics of desorption of one or more components from the stationary phase are very slow, e.g. high enthalpy of adsorption, multiple mechanisms, and large molecules. Here the advantage of a higher plate number with a gradient predominates, and under gradient conditions, we have better resolution. With a gradient, compared to isocratic separations, additional possibilities for altering the selectivity in the "front" and "rear" areas of the chromatogram are available.

In comparison to isocratic separations , increased "plate number" N_{Gr} is the reason for an improved average resolution (often reported as the sum of the resolutions), which ultimately means a better peak capacity; see the following discussion. From a practical point of view, this means that, especially for gradient separations, the high plate number of a column is not as important as often suggested in the brochures of column suppliers. Example: "Our new column XYZ has 450 000 plates/m." Common chromatographic conditions for such runs are acetonitrile/water, low flow, simple, neutral aromatics, and 1 μl injection volume. Users can now so understand the specification "450 000 plates/m" to mean that during the use of the column, this plate number (with a 100 mm column, about 40 000–45 000 plates) would actually be available. Note, however, that the plate number is affected not only by the quality of the packing and the particle size. Far more than 10–15 factors play a role, such as eluent composition and temperature (viscosity), the dead volume of the system (more precisely, the dispersion of the substance bands), particle size distribution, retention time, flow, injection volume and sample concentration, constitution and pH of the sample solution, chemical structure and diffusion coefficient of the analyte, and, last but not least, the parameter settings affect the appearance of the peaks.

For example, broad, tailing peaks indicate a slow kinetic (e.g., additional ionic interactions, large molecules) or a significant dead volume in *this* system with *this* column – despite a "good" plate number. In conclusion, note the following: for improvement of the resolution, an increase in selectivity principally "brings" the most, an increase in plate number is secondary, the van Deemter H/u curves are much overrated by the marketing of the manufacturers.

How can I improve the selectivity? A change of pH, the addition of modifiers, and the use of alternative stationary phases are important factors and independent of the hardware. Let us look now at UHPLC. What can it actually accomplish? Of the two advantages of UHPLC – the small dead/delay volume and the ability to work at higher pressures – the second advantage can be used here. In the following cases, the efforts to improve the selectivity are accompanied by an increase in pressure, without doubt, a situation for which a UHPLC system is designed.

- *Methanol as organic solvent*: this often results in better selectivity than with acetonitrile by the separation of polar molecules.
- *Lower temperature*: by the separation of certain substances (enantiomers, α-β-/double-bond isomers), an improvement in selectivity is often seen at lower temperatures.
- *Pressure*: at pressures above around 600–700 bar, the polarizability of certain molecules (e.g., prednisone/prednisolone, conformational isomers, tocopherols, etc.) changes. The selectivity also changes (improvement?), and in combination with certain stationary phases (C_{30}, "Mixed Mode Phases" and

other “shape selectivity phases”), interesting possibilities arise. For example, immediately after the column, a restrictor with a minimal volume could be added. However, the robustness is to be seen critically by small pressure fluctuations in this region.
- *Flow*: increase in the flow leads to an increase in the gradient volume (gradient volume = gradient duration × flow).

You can, of course, change two factors simultaneously and make the best use of the possibility of UHPLC to work at higher pressures. For example, with gradient runs, lower the temperature to 10 °C and at the same time increase the flow. In one case, we have increased the pressure to 1000 bar, lowered the temperature to 15 °C, and, at the same time, increased the gradient volume, once by means of the gradient duration (in this case, necessary due to the slow kinetics) and once by means of the flow. The number of peaks, which then appeared, had increased, compared with the original validated method, by about 30%.

Now we come to peak capacity.

There are cases in which an improvement in selectivity is hardly possible, for example:

- A large number of possibly even similar components, in addition, may be in a complex matrix.
- When hydrophobic interactions dominate, these are not particularly specific, and there is hardly any noticeably different selectivity. When, for example, basic compounds are neutralized by pH, they are present as neutral molecules, and the interactions with the stationary phase are hydrophobic in nature and thus rather unspecific. In such cases, in the course of optimization experiments and when using different stationary phases, interactions of varying strength do occur, resulting in differing retention times and k-values, but the selectivity is often comparable, see Figure 1.1: differing k-values (see bars), but very similar α-values (see lines) are found.

In such cases, a noticeable improvement of the selectivity is hardly feasible. Even if it were possible to improve the selectivity at one specific point in the chromatogram, it could become worse elsewhere. In a case such as this, the peak capacity comes into focus: peaks as narrow as possible (i.e., maximum achievable “plate number” N_{Gr}/separation efficiency), ideally evenly distributed over the entire chromatogram, see, for example, Figure 1.2 (taken from [6]). Here, a separation with a (theoretical) peak capacity of 925 peaks on four 250 mm columns connected in series is shown.

Before we look at how the UHPLC can profitably be used, let us note with reference to Equations 1.2a and 1.2b how in principle the peak capacity can be increased:

1) I need a long gradient or rather a large difference in retention time between the first and the last peak. This requires a large gradient volume, possibly also a long column.
2) I need a small peak width; in other words, I aim for narrow peaks. I can achieve this through a steep gradient, a high start and also end % B, small particles, low viscosity, and high temperature.

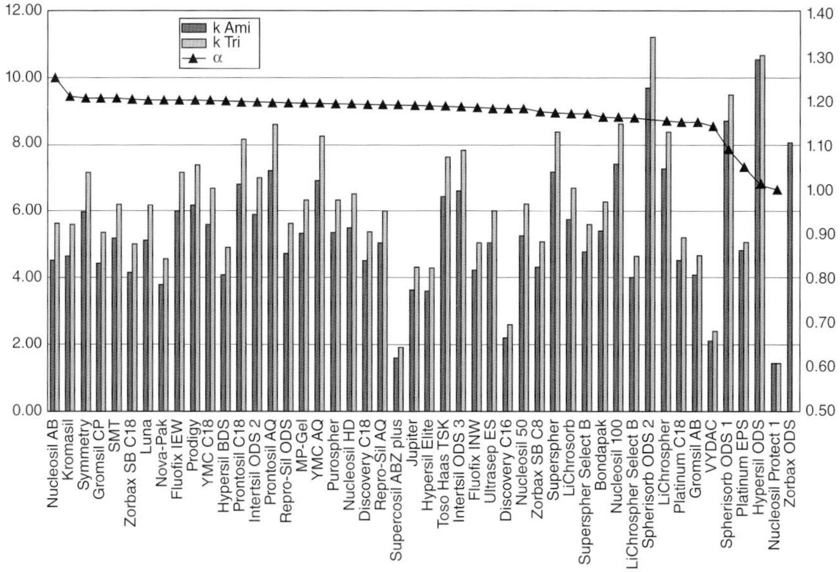

Figure 1.1 Retention (bar) and separation factors (line) of tricyclic antidepressants in acidic acetonitrile/phosphate buffer on differing RP phases; for details, see text. (From "HPLC richtig optimiert," Figure 7, S. 176.)

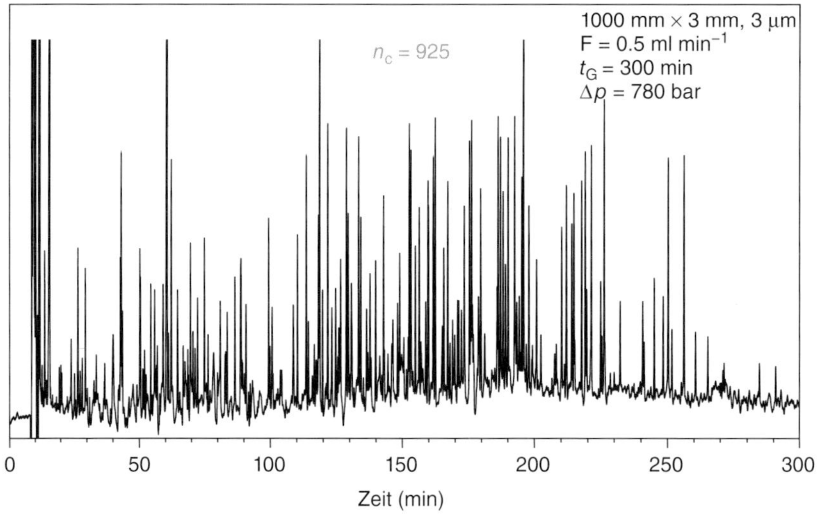

Figure 1.2 High-resolution 1D-UHPLC separation of a tryptic digestion of five proteins. A chain of four 250 mm columns was constructed using dead volume couplings based on Viper fittings (Thermo Scientific). Stationary phase: Acclaim 120 C18 (Thermo Scientific), temperature: 30 °C. Theoretical peak capacity calculated from the peak width of individual well-resolved peaks. (From "HPLC-Experte," Figure 3.25, S. 164.)

As known, the UHPLC gives us small dead volumes and allows high pressures. With reference to the last named advantage, the following would be possible: a long column or several columns connected in series plus perhaps small particles. Note that when columns are connected in series, the negative influence of the dead volume ("Extra Column Effects") is minimized. From experience, this becomes quite apparent in the case of very small column volumes in spite of the most modern UHPLC design, see further discussion and Chapters 3 and 4. Because with gradient separations, the particle size is not so crucial, one possibility would be as follows:

Three 150 mm × 3 mm columns, 2.5–3.5 μm particles, in series plus 40–50 °C.

A survey of the literature showed that separations with higher peak capacity under UHPLC conditions are published more and more frequently. To start with, the following examples seem to be realizable for a "Real-Life" laboratory.

2 × 150 mm, 1.9 μm, 1200 bar at 45 °C: 480 peaks in 40 min (12 peaks/min)
3 × 150 mm, 2.6 μm fused core, 1200 bar at 45 °C: 600 peaks in 50 min (12 peaks/min)
4 × 250 mm, 3.0 μm, 1200 bar at 30 °C: 1000 peaks in 300 min (3–4 peaks/min).

If time is not a significantly limiting factor and the matrix not extremely difficult (polymers, foods, fermented cultures), with UHPLC, about 600–1000 peaks can theoretically be separated, see also Chapter 12. For such cases, in the mid-term long columns with 2.1 mm internal diameter and 1.5–2.6 μm, fused-core material could represent one of the most interesting possibilities. Under optimal conditions and with the most modern UHPLC hardware, the target is "100/100": 100 peaks/100 s. To date, separations with a theoretical peak capacity of 730 peaks in 30 min or 530 peaks in 13 min have been reported. The higher the peak capacity – made possible through an optimal combination of UHPLC system and column – the less necessary a good selectivity becomes, the improvement of which in any case is not exactly a trivial task, especially when pressed for time.

Now let us look at everyday use. In a real chromatogram – except perhaps with a sample containing only homologs – the peaks are rarely evenly distributed. Especially when the peaks also have to be quantified – that is, a resolution of 1.5 or at least 1 is necessary – in practice, only a much smaller capacity can be achieved. According to statistical calculations from Giddings, with a theoretical peak capacity of 1000, 184 peaks could be separated. Taking into consideration a difficult matrix and/or possibly suboptimal equipment, a good rule of thumb is considered to be about 1/10 of the theoretical peak capacity, for the last given example in reality, 100 peaks. Put simply, for really demanding problems (multicomponent samples and/or a complex matrix), the best method is 2D chromatography with orthogonal separation mechanisms, the next best is the modern UHPLC, which nevertheless can provide one-dimensionally a theoretical peak capacity of around 1000.

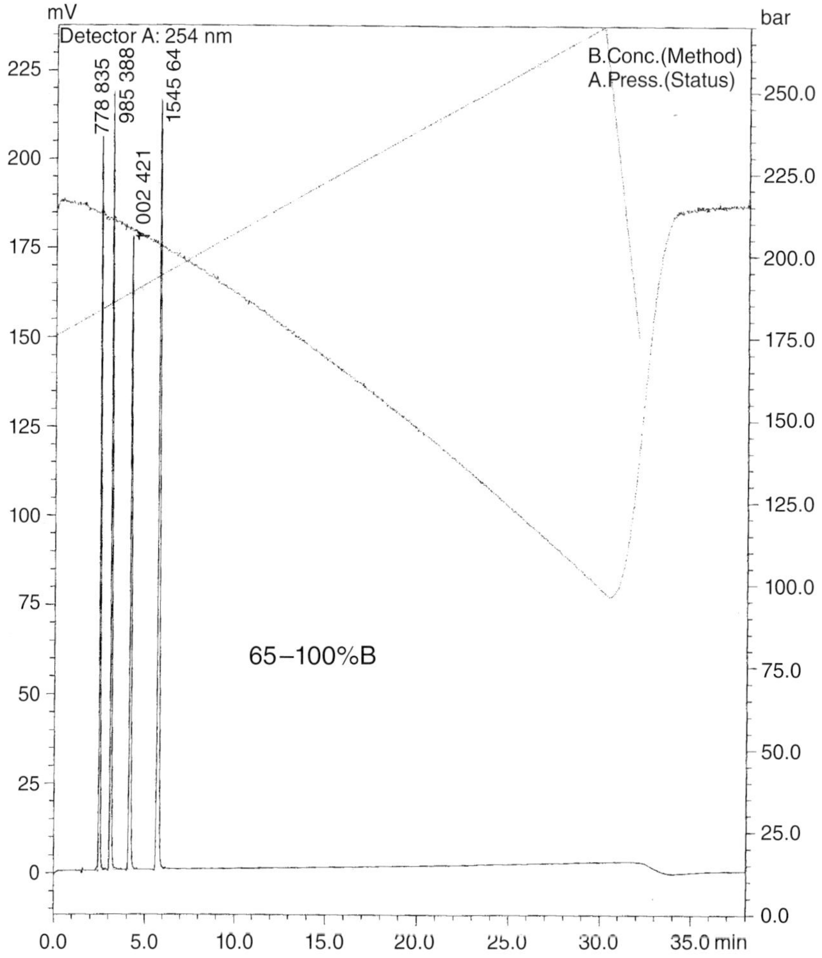

Figure 1.3 Separation of polystyrene; a good peak capacity is obtained starting with 55% B and using a flat gradient. (Source: Waters.)

Here too, you could try to use simultaneously as many parameters as possible, which can contribute to a good peak capacity. The following variations correspond to an "optimal" combination.

A long column (or multiple columns in series), 2–3 μm particles, a high flow, 40–50 °C, acetonitrile as the organic solvent, gradient starting at about 40% B. With ionic components, one could try to achieve good peak symmetry by altering the pH. Depending on the mechanism, a steep gradient, occasionally also a flat gradient, can be beneficial, see Figure 1.3.

Column length and gradient duration have one thing in common: both have less influence on the peak capacity than is generally believed. For example, gradients longer than 20–25 min only make sense in the case of very complex mixtures. With respect to column length and gradient duration, note the following simplified rules of thumb for an optimal peak capacity:

- 50 mm ≤ 5 min
- 100 mm ≈ 10–20 min
- 150 mm ≥ 20 min.

In this context, the aforementioned statements should be recalled. Firstly, increasing the plate number is not really the most effective way to improve the resolution. Secondly, gradient separations produce narrow peaks anyway, and a high "plate number" N_{Gr} is relevant only in difficult cases – when a high peak capacity is needed. Furthermore, a decrease in the particle size by a factor of 2 led (by a fourfold increase in pressure) to an improvement in resolution by a factor of 1.4. An increase in column length by a factor of 2 (the pressure increased only by a factor of 2, the analysis time was longer) also led to an improvement in resolution by a factor of 1.4. If I continue to increase the column length (or use multiple columns), up to a given/critical pressure, I achieve significantly more plates. Thus, if the ultimate aim is "maximum plate number" (question this critically!), then I should connect several columns in series, filled with 2.5–3.5 μm particles – in this case, I would have to accept the long retention time. In other words, I should use the pressure allowed by UHPLC for long columns rather than for small particles; see also comments in Chapter 12. Some numerical examples follow to illustrate this:

- A typical UHPLC column, 100 mm × 2.1 mm, 1.7 μm, and at a pressure of about 1000 bar has approximately 20–25 000 plates. If a pressure of about 1000 bar should be regarded as a limit for prolonged routine use, one soon sees the limits of small particles as a source of maximum efficiency – despite UHPLC. Thus, it becomes clear that the UHPLC, under normal conditions, cannot provide the resolution that would be necessary for difficult separations
- Four 250 mm × 4.6 mm, 5 μm columns in series have approximately 100 000 plates – at a pressure of about 600 bar (interesting: additionally set the temperature to 80 °C; result: excellent peak capacity at moderate pressure).

Note For separations under "Ultrahigh-Resolution Separation" conditions (plate number >150 000 plates), the temperature should not be increased – this would have a negative effect on the B-term of the van Deemter equation, and the result would be increased diffusion.

- One 250 mm × 2.1 mm, 1.9 μm column provides approximately 55 000 plates – at a pressure of about 960 bar. This plate number should be about the maximum, which can nowadays be achieved with a column under real conditions.

In summary, the conclusion regarding column length, number of plates, and pressure is as follows.

Doubling the column length leads to twice the pressure and an improvement in resolution/peak capacity by a factor of 1.4. Half of large particles lead to fourfold pressure and also to an improvement in the resolution/peak capacity by a factor of 1.4. From this it follows that at a given pressure, the plate number can be increased more effectively with a longer column (or columns in series) than by using smaller particles. Once again, for gradient separations, both longer column(s) and small particles do not necessarily have really essential advantages – with the following exceptions: very many, very similar components (long column), very slow kinetics, and low mass sensitivity (small particles).

In this context, the experience that is still valid even after 50 years of HPLC is recalled: a 250 mm × 4.6 mm column can be filled more easily and better than a short and especially a thin column. Another advantage of the long column is the long lifetime. Disadvantages are long runs and high solvent consumption – both not to be underestimated in the long term.

Note

1) As we have seen earlier, neither good packing quality nor small particles guarantee a good peak shape, and among other factors, the dead volume can play an important role: with particles ≤1.7 μm, the plate number and also the peak symmetry often increase with later eluting peaks. This shows that the dead volume of today's UHPLC equipment is too large to fully exploit the efficiency of these particles. A further indication of this fact is that the resolution of early eluting peaks on 5 μm columns is often better than on ≤2 μm columns.
2) As just stated, in gradient separations, neither the length of the column nor the gradient duration or the particle size is of crucial importance – rather the gradient volume, initial and end % B, and the slope. Especially for gradient separations, "real" UHPLC conditions are only needed for very challenging separation problems (e.g., complex mixture, high peak capacity required) and/or in cases where several parameters need to be altered simultaneously, thereby increasing the pressure. If, for example, in the case of polar components, both selectivity and peak capacity are to be improved, one could proceed as follows: long column (or several columns in series), 2–3 μm particles, high flow, 40–50 °C, plus methanol, and as an alternative to increasing temperature, try a run at 10–15 °C. Especially with such experiments, carried out in a short time, the UHPLC can show its strengths.

1.3.2 Separate Fast

First of all: if the selectivity very good, then one could use a 3 mm, 5 μm column with relatively high flow – in any case, one achieves this way a faster separation than under UHPLC conditions. Note that due to the good selectivity, the low plate number resulting from the high flow is of no great importance. The strength par excellence of UHPLC lies in the following situation: when the column being used has almost the optimal selectivity, then a sufficiently good resolution in a short time is obtained. In other words, under UHPLC conditions, we get the best plate number/time ratio – that means the lowest retention time at a given efficiency and also low solvent consumption. A decrease in column length and, at the same time, a decrease in particle size gives us an "identical" separation in a shorter time at a lower solvent consumption.

Requirements for an "identical" separation are equally well-packed columns, no noticeable decrease in the plate number due to the wider size distribution for particles ≤1.7 μm, and no deterioration in peak shape (tailing) due to dead volume, especially with the early eluting peaks. Put differently, by keeping the column length constant and using smaller particles, I can achieve, through an increase in the plate number, better resolution in the same retention time. Here I could even at the same time reduce the retention time, because according to the H/u curve, with the now smaller particles, I can increase the flow without appreciable loss of efficiency. Note, however, that with concentration-sensitive detectors, the area decreases.

Of course, an increase in flow only makes sense when by the interactions of the analyte with the stationary phase, fast kinetics and thus a small C term in the van Deemter equation result. To these relationships, the following remark is valid: they are neither new findings, nor is therefore a UHPLC necessarily required. Such improvements are – at least partially – up to a pressure of 400 bar with a classical HPLC system quite realizable. Only since the introduction of UHPLC technology in the mid-2000s does one risk working at higher pressures.

Conclusions for "fast separation" and UHPLC:

1) If the aim is "fast separation," then UHPLC is outstandingly suited if, firstly, the separation problem is not very demanding (≤15–20 peaks); secondly, constant, simple, robust chromatographic conditions dominate; and thirdly, there is an automated and constant sequence. As a typical application, the IPC (In-Process Control) could be named here.
2) The UHPLC is also suitable when in a short time I have to develop methods and/or identify trends or optimize existing methods by varying various parameters. The UHPLC allows many parameters to be tested rapidly, since short gradient runs using short columns with sufficiently good resolution/peak capacity are possible. The UHPLC is predestined for development departments, which, pressed for time, have to develop or optimize methods with varying parameters or also with the help of generic runs.

1.3.3 Improve Mass Sensitivity

This must be the simplest case, while my goal is clear – the quantification of small peaks. Equally clear is the contribution UHPLC can make with its characteristics. The following is relatively easy to implement:

- *Reduction of the internal diameter*: when one is not able to or allowed to inject more volume, a thinner column in combination with small particles leads to an improvement in the relative mass sensitivity.
- A shorter column with smaller particles leads to a smaller peak volume.

In this context, however, one should consider the following practical aspects.

A reduction of the column inner diameter from, for example, 2 to 1 mm did indeed lead – at a pressure increased by a factor of 4 – to an improvement in the relative mass sensitivity by a factor of about 20. However, to pack a 1 mm capillary well and reproducibly is not an easy task – both with porous and with – in particular! – fused-core materials. In addition, by the analysis of secondary components on thin columns or capillaries, there is the danger of the columns being overloaded by the main peak or the matrix. Finally, when one wants to advance into these regions, the dead volumes of even the most modern UHPLC systems need to be optimized. This means that when the maximum possible relative mass sensitivity does not necessarily have to be reached, under everyday conditions, a 2.1 mm, 1.5–1.7 µm column would be well suitable, in matrix-free sample solutions and with optimal hardware perhaps 1.3 µm. If the relative mass sensitivity has to be further improved, then the UHPLC reaches its limit, this is where nano or capillary LC comes into play.

Although the following points are not UHPLC-specific, for good sensitivity, they are important and are therefore briefly mentioned here:

- The sample solution should be weaker – in RP-HPLC more polar – than the initial eluent, which means dilute the sample solution with water; this leads to a concentration at the head of the column ("On-Column Concentration")
- *With very early eluting, chemically similar peaks*: start the gradient with a high proportion of water/buffer and maybe use a short isocratic step; here too, a concentration is possible
- *For simple separations and with not too polar components*: start with 50–70% B, and run a steep gradient, see Figure 1.4

We conclude this section with the following tip: optimal parameters such as data recording rate (sample rate), and time constant/response time are important at high flows and in general with early eluting, narrow, small peaks. This is increasingly so under UHPLC conditions and even more so under UHPLC conditions and when striving for good mass sensitivity. For suitable numerical values, see Section 1.4.

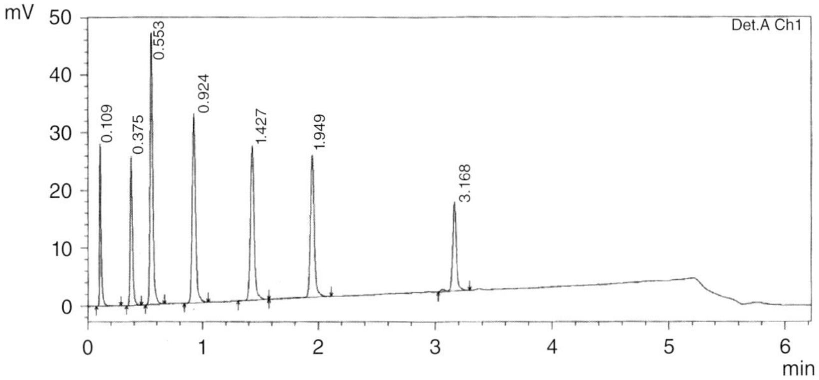

Figure 1.4 Initial gradient conditions with the focus on improvement in peak form; Column GeminiNX, 50 mm × 4 mm, 3 µm, 65–100% B (acetonitrile/water).

Note In connection with the advantages of UHPLC, it is repeatedly claimed that the "sensitivity" in UHPLC is better in comparison to HPLC. But if here the sensitivity of detection is meant, then the opposite is the case! The explanation is that if one uses a conventional column and a DAD as a concentration-sensitive detector, a cell with a long path (Lambert–Beer "longs for" a long light path) can be used. The risk that the sensitivity could be neutralized by dead volumes hardly exists, as the column volume is large in comparison to these. Furthermore, as already known, the injection volume should be at the most 10% of the column volume, with low retention factors, band broadening makes itself noticeable from 1% injection volume. With a conventional column, using larger injection volumes is no problem. With the low-volume columns used in UHPLC, the injection volume should not exceed 1–2 µl. For the same reason (small column volume), in the UHPLC, the "Extracolumn Volume" and consequently the detector cell volume must be small. Despite significant recent progress in cell design, with the cell volume of much less than 1 µl required in the UHPLC, the light path in a detector cell cannot be significantly lengthened. For these reasons, detection sensitivity in UHPLC is intrinsically less sensitive than in HPLC. However, if from the outset, the injection volume is small or cannot/may not be increased, the UHPLC undoubtedly has an advantage: the mass sensitivity in an (optimized) UHPLC system due to the small peak volume is better by magnitudes than in the HPLC.

1.3.4 Robust Separations in Routine Use

In a routine laboratory, the robustness of the method has top priority and the downtime should be reduced to a minimum. In the case of simple chromatographic methods, large numbers of samples, robust conditions, clear sample solutions, no/minimal sample preparation, automatic integration, and so on – see earlier discussion – UHPLC would no doubt come into question. In the following cases, the use of a UHPLC system should, however, be critically questioned:

- *Difficult matrix*: the sample preparation does not lead to homogeneous, clear sample solutions, and they contain possibly substances from the matrix (plant extracts, contaminated soils, coated tablets, ointments, polymers, biological matrices such as tissue, blood, etc.). Before we come to further critical points regarding an appropriate use of the UHPLC, here is an example of the nonsuitability of UHPLC. Assume that the components of interest are only soluble in or extractable with acetonitrile, alcohols, or tetrahydrofuran, and the sample solution is then stronger (more organic) than the eluent/initial gradient. In particular areas – such as pharmaceuticals and environmental analysis – this is unfortunately unavoidable. A typical example would be the extraction/dissolving of active ingredients into ethanolic solutions (analysis of ointment). In this case, we have to deal with fronting, at the worst, even double peaks occur. Due to the small column volumes in the UHPLC (e.g., for a 5 mm × 2.1 mm column about 200 µl), the poor peak shape persists – in the UHPLC, the "lost" plate number remains permanently lost! Somewhat simplified, one can say: a matrix-contaminated sample and/or a strong solvent let every UHPLC fail. With conventional columns (longer/thicker and thus with larger column volume), the problem usually occurs only with the early eluting peaks and injection volumes greater than about 15–20 µl.

- Varying chromatographic conditions in routine use, poor robustness of the method(s).
- Sample preparation, manual integration, and other necessary steps such as documentation, and filing take much longer than the separation time.
- The number of samples is relatively limited.
- Frequent method transfers with several laboratories and anticipated differences in procedure, in know-how, and in the "culture." Here are a few examples: a less experienced HPLC user simply takes a different capillary or cuts it off improperly or connects a PEEK fitting with a steel capillary or uses a column switch or a connector between column and detector, in order to fit columns from different suppliers. The UHPLC does not "forgive" such things, whereas the HPLC is more likely to.

Let us imagine a fictitious laboratory in the quality control of a pharmaceutical company or generic drug manufacturer, where tablets, capsules, or ointments are analyzed. Admittedly simplified, the situation often looks something like this: a user has possibly two HPLC systems to look after, and/or, on top of that, also has other jobs to do (among others, documentation). The methods are old: LiChrospher/Nucleosil 100/Hypersil ODS/Spherisorb ODS, and so on, phosphate buffer, possibly triethylamine or ion-pair reagents and so forth, things do not always run smoothly. Often, time is short, after a system failure – for whatever reason – the system must be requalified or at least, using system suitability tests and possibly repeat injections, the fault-free condition of the equipment must be verified. One already has an uncomfortable feeling before the impending method transfer, since from experience, something like that is seldom completed without complications.

When considering the possible introduction of a UHPLC, the objections could read something like "does it really help much when we reduce the retention time from 20 to 6 min, when the sample preparation takes half an hour and the frequently necessary manual integration just as long, not to mention screening the individual chromatograms? And what use is it, when on the one hand the sequence is already finished at 11 p.m. instead of 5 a.m. the next morning, and on the other hand the system more often shuts itself down? And as far as solvent consumption is concerned, in the laboratory we are talking about liters, in the production of hundreds of liters.... And anyway, we were unsatisfied with the life of the UHPLC columns during the test phase with the UHPLC." One can certainly more or less understand these arguments – up to perhaps the last point: relevant is not the absolute column life, rather the number of injections per unit time or the number of column volumes run before the column becomes unusable. Based on the final criterion, and excluding matrix problems, and so on, no significant difference between HPLC and UHPLC columns can be seen. When a UHPLC is now to be used in this environment without changing

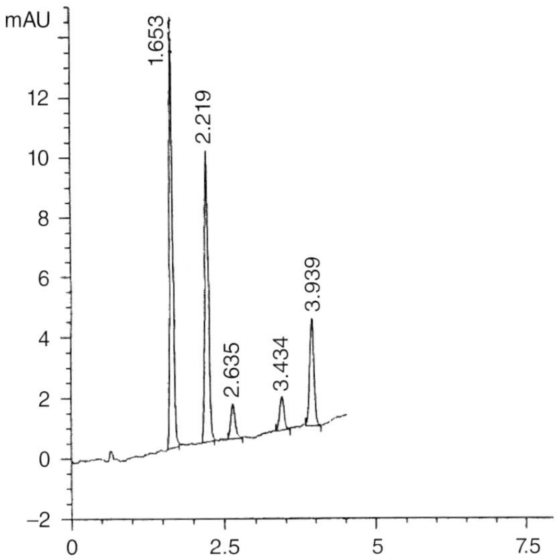

Figure 1.5 Separation of seven components on a Synergi MAX RP 20 mm × 4 mm, 2 μm column in less than 2 min on a high-pressure gradient system from the beginning of the 1990s – the system had considerable dead volume.

attitudes, working habits, and also the expectations on a UHPLC, then many annoying things can happen. Following are just three examples:

- A small salt crystal from the nonfiltered buffer or a part of the matrix can block the thin capillaries used in the UHPLC, resulting in leaks or, at the worst, the system shuts itself down.
- There is no UHPLC column available with similar characteristics to the column specified in the method.
- Method transfer: when SST criteria, for example, in addition to resolution, the symmetry factor, and relative retention time instead of relative retention [8] are required, then this is often quite difficult to achieve – in spite of the existing software tools for scaling method parameters to be found in the literature or available from the equipment manufacturer. The problems of method transfer from UHPLC to HPLC will not be discussed here in detail, reference is rather made to Chapter 4.

With any functioning, classical HPLC system, improvements are possible in any direction. Following are two examples that show that with older HPLC instruments, significant time savings can be achieved without noticeable loss of resolution and in compliance with the SST criteria (Figure 1.5).

1.4 The UHPLC in Routine Use – A Brief Report

This section provides a brief, simplified description of the routine use of UHPLC. As mentioned at the beginning, since the mid-2000s, more and more UHPLC systems have come into use – with both satisfied and less satisfied users.

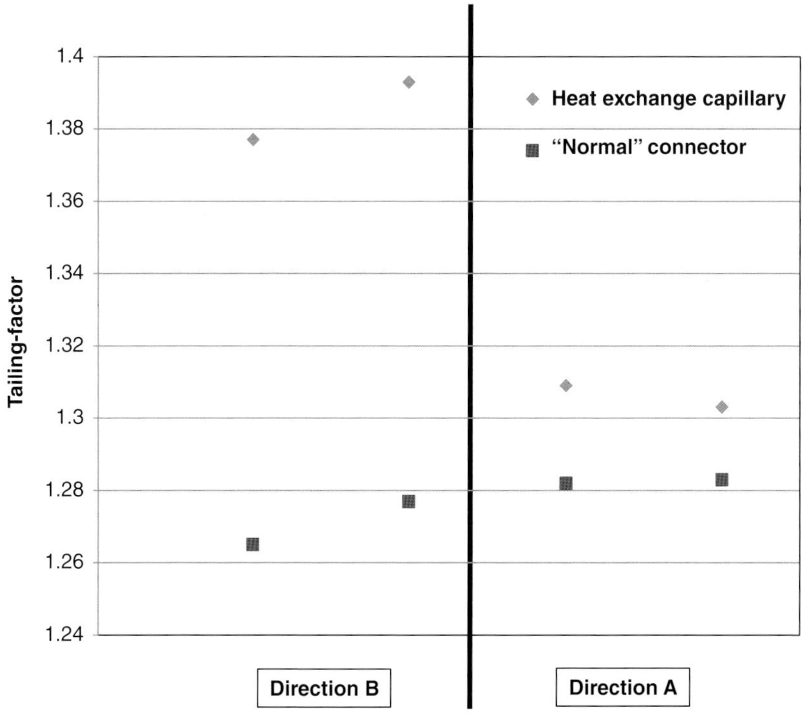

Figure 1.6 Tailing factor depending on the direction of the used capillary. For details, see text.

Satisfied users in routine laboratories are firstly those who, under UHPLC conditions, run chromatographically relatively simple analyses (≤20–25 components with quite different properties) under constant, stable conditions, or, secondly, those who operate their UHPLC system more or less as a fast HPLC, because in the past they have not – for whatever reasons – used the capabilities of their HPLC for rapid separations. The last group of UHPLC owners are not really UHPLC users in the true sense, but they are satisfied – so what? Both user groups are happy with the fast separations, the narrow peaks, and the reduced solvent consumption. Users in development laboratories are satisfied because in a short time they can perform many optimization experiments that does indeed make lot of fun!

Dissatisfied UHPLC users are primarily those who over the years were spoilt due to the relatively minor problems with their HPLC systems and are now faced with a UHPLC instrument with such problems in routine use: solutions and chemicals must be particularly pure or maybe even extra filtered, leaks in the autosampler are a nuisance, "good" columns suddenly show double peaks in the UHPLC, in the case of early eluting peaks, the parameters (sample rate, time constant, etc.) must be adapted, and so on. Even a capillary fitted the other way around can affect the peak symmetry, see Figures 1.6 and 1.7. In short: if, in a real environment, I do indeed want to use my UHPLC as a UHPLC, then I have to increase my care and effort compared to my previous HPLC habits – or I stay below about 600–650 bar and definitely have less problems in routine use.

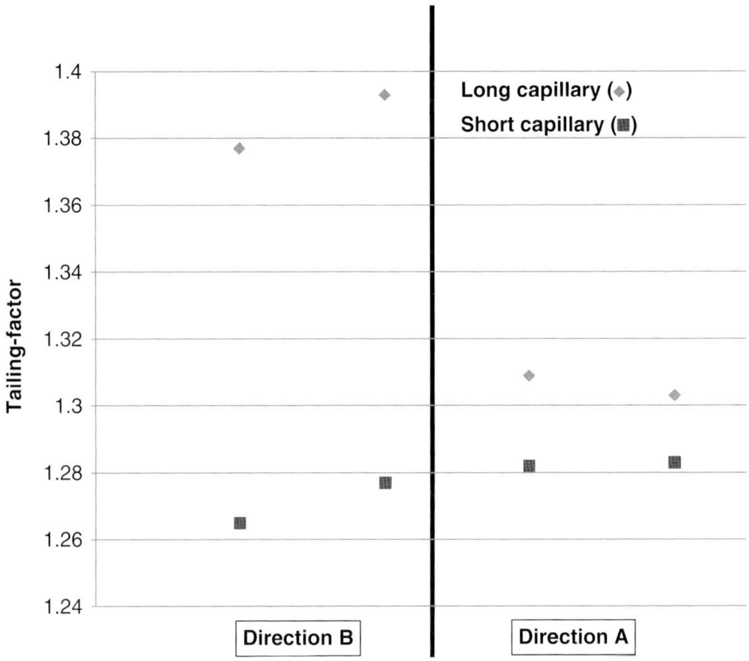

Figure 1.7 Influence of the direction of capillaries on the asymmetry factor; for details, see text.

Note In a highly regulated environment, absolute priority has, without doubt, the confirmation of the expected/desired results with minimization of trouble, repeated measurements, and system downtime. Time and solvent savings are desirable, in fact; however, these are second- or even third-ranking goals. The requirements for a routine method are therefore robustness, time, and only then cost savings. All this can be achieved to a large extent without UHPLC: the U.S. Pharmacopoeia allows a good many adaptations without having to revalidate the method. I could, for example, use a different C_{18}, a 70% shorter column, one 25% thinner, one filled with 50% smaller particles. Furthermore, according to the USP and also European Pharmacopoeias, it is permitted to alter the gradient, only the final % B must be maintained (for details, see [8–10]).

When, after an adjustment, the system suitability criteria – well thought out (!) – are still met, then a revalidation is not necessary. This topic is complex and alone could fill a book. We end this little excursion with the following personal appraisal: it is certainly not easy, but one could at least try to overcome one's own trepidation a little and make use of the flexibility allowed by the authorities and organizations. In the process I could maybe, in compliance with the requirements, make my method more robust or faster, thereby saving my employers money, sparing my nerves, doing something useful for the environment (solvent consumption) – and the adaptations are formally in order. If something similar to this is possible in such a highly regulated area as the pharmaceutical industry, then it could be implemented analogously even more so in other industries.

The message for a routine laboratory is, in my view, as follows: first, make full use of regulatory flexibility and the technical possibilities of your current HPLC system. Even this may be an important step that brings a noticeable improvement toward your aim. When you want to or are allowed to, with reasonable effort and small investment, you can achieve "UHPLC-like" separations on your HPLC system:

- Dilute the sample solution – concentration at the column head will be possibly resulting in an improvement in the peak shape, especially for the early eluting peaks.
- Use a smaller loop in the autosampler.
- Think of the parameters (Sample Rate, Time Constant/Response Time, Dwell Time, Bandwidth, Slit, see Section 1.5).
- Use a smaller UV cell in your DAD (2–4 μl volume is OK, – optimal would, of course, be 0.25–1 μl, length 60–85 mm).

When subsequently, the method works reliably for a while and you want to further improve it (does everyone involved really want this?), then of course you should, with an appropriate budget, now think of UHPLC.

1.5 How Can the Potential of UHPLC Effectively Be Fully Exploited? (See Also Chapters 2, 3, and 9)

A look at the special features of UHPLC is also helpful in answering this question: small dead and delay volumes, high pressures possible. Furthermore, the current analytical problem should always be in the foreground. We can state the following: for not too difficult analytical problems, all commercial UHPLC systems are good enough – especially in gradient mode. The more demanding the separation problem is, the more important the parameter settings become and the more likely hardware optimization could be attained. To this end, the subsequent remarks should provide some clues. Using dead volume as an example, it will first be made clear that, depending on the current situation – which can only be assessed individually – the entire bandwidth is possible, from "nothing needs to be done" to "urgent need for action."

1.5.1 Dead Volumes

The fact is that the dead volume of every UHPLC system on the market is much too large for a 50 mm × 2.1 mm, ≤2 μm column or smaller. Depending on the manufacturer, the loss of efficiency is around about 20–40%, and this is openly admitted by most manufacturers. If you notice a conspicuous tailing in the early peaks under isocratic conditions, then you know that this is currently the case.

The question now is: does it bother me that, due to the dead volume, instead of, for example, the possible peak width of 4 s, it is now 6 s? Or not really, since I have a relatively simple separation problem and fortunately the required symmetry factor of 1.3 is not exceeded? With gradient separations, the dead volume (not the dwell volume!) is even less relevant, because here narrow peaks are almost guaranteed. When, on the other hand, this loss in efficiency interferes with the separation of my early eluting peaks, then something must be done. In this case, optimization kits with particularly thin capillaries, dead-volume-"free" fittings, specially designed UV-cells, and so on, can help. These are now available from the manufacturers.

Conclusion: according to the situation, either "business as usual" or urgent action.

Numerical values for settings and chromatographic parameters as well as hardware requirements, which are necessary or useful for UHPLC separations:

- *Data recording rate (sample rate)*: usually 20–40 Hz, with runs of about ≤1–2 min and about ≤1–2 s peak width, more than 50 data points per peak are necessary to avoid loss of resolution.
- *Time constant*: ≤50 ms. In some software programs, the time constant is, however, coupled to the data recording rate.
- *Bandwidth and slit with a DAD*: for good detection sensitivity, 16 nm, for a good spectral resolution (e.g., for Peak Purity tests), 1–4 nm.
- *Injection*: injection cycles as short as possible. Here, injectors with fixed loops have an advantage, but the problem of carryover should not be lost sight of. Injection volume about 1–2 µl (for a 50 mm × 2.1 mm column with a column volume of approximately 200 µl), in any case <5 µl. As we have seen earlier, the 1% rule is valid – the injection volume must not exceed 1% of the column volume; otherwise, the volume overload is noticeable. Regarding the possibility of larger injection volumes, see Chapter 9.
- *Cell volume*: 1–2 µl, 0.25 µl is worth considering (regarding UHPLC/MS coupling, see Chapter 5).
- *Dead and delay volumes*: at the moment there exists here, so to speak, a sporting competition among the manufacturers, as to who can save how many microliters first. Note that 10 µl, at the most 15 µl, dead volume would be acceptable for a column with 100–300 µl column volume. Here the 10% rule applies; the volume outside the column may not exceed 10% of the column volume; otherwise, the plate number falls considerably for details, see Chapters 2 and 3. As far as dwell volumes are concerned, low-pressure UHPLC instruments with a dwell volume of 100 µl are indeed already available, but 150–200 µl is certainly not bad. As is known, the dwell volume affects the separation; however, it is not possible to make a general statement about whether, as a matter of principle, a large or small dwell volume is good or bad for the separation.
- *In general*: at high pressures and at the same time high flow rates, heat is generated due to friction ("Frictional Heating"). This effect can significantly influence the separation, and one should bear this in mind (see Chapter 2.2). Furthermore, a preheater with a volume of 1–2 µl (for example, a 300 mm, 100 µm capillary prior to the column oven) is part of every modern UHPLC. Commercial UHPLC eluents are usually filtered through a 0.22 µm filter. When, nevertheless, problems with blocked capillaries and frits occur, an additional filtering of the eluent and also of the sample solutions and other chemicals would be appropriate.

1.6 Summary and Outlook

Advantages of UHPLC:

- Increased flexibility in comparison to HPLC and increased capability for improving selectivity and resolution: methanol as the organic solvent, lowering of the temperature, increase in flow rate, and thereby increase of the gradient volume – also when using thin/long columns. The consequence of all these interesting possibilities is increased pressure.
- Peak capacity can be increased: connection of columns containing 2.5–3.5 μm porous or 1.5–5 μm fused-core material.
- For ≤2 μm material, a very good plate number/retention time ratio and thereby sufficiently good resolution in a short time for less demanding separations, and as a result, increased throughput and reduced solvent consumption. The main practical advantage lies less in the rapid analysis of sequences – with autosamplers often possible overnight, rather in the fast control. I see much sooner if something is going wrong and can "rescue" my sequence before leaving work, instead of first noticing problems in the morning when everything is too late and I – also maybe the production – have already lost a lot of time again.
- The small dead volumes in combination with thin columns and small particles result in a small peak volume and thereby an improvement in the relative mass sensitivity/lowering of the detection limit.

As disadvantages, the following can be mentioned:

- Greater effort in sample preparation. In addition, matrix-contaminated samples and strong sample solvents present seemingly unsolvable problems. Especially in the last case, the small particles cannot compensate for the deterioration in the peak shape.
- The reproducibility under changing conditions is not always satisfactory.
- Above about 800 bar noticeable frictional heat, above approximately 600–700 bar, a change in selectivity is possible. The last two named points are one reason why the transfer of a method from HPLC to UHPLC and vice versa, it is not as simple as often presented.
- Difficulties in method transfer. Due to the small column volumes, small differences in dwell volumes with gradient systems cause considerably more problems than in HPLC.
- In general, more care is necessary in the overall handling.

As with almost every technique, depending on the case, typical characteristics of the technique prove to be advantageous as well as disadvantageous. We will finish this discussion with an example to support this. Consider the small column volume typical in UHPLC, the small column and dispersion volumes result in a small peak volume. This is the reason for the excellent mass sensitivity in the UHPLC. However, at the same time, the small column volume is also the reason for the strong fronting in the case of a strong sample solvent. Furthermore, especially in the early eluting peaks, this is also the reason for tailing – since the dispersion volume of UHPLC systems is still much too large for small column volumes. Finally, the small column volume means that the injection volume cannot be increased, leading to the poor detection sensitivity of the UHPLC.

In the following, the practical possibilities of UHPLC for some typical situations are presented in a simplified way:

1) *Existing HPLC method, only small changes are allowed.* You can easily run the method with your classical HPLC column on a UHPLC system, only the injection volume and the gradient need to be adapted. This assumes that as a result of miniaturization, due to the thin capillaries, and so on, no blockage occurs (for details, see earlier discussion). The result is narrow peaks due to the smaller dead volume and thus probably better resolution at higher pressure.
2) *Existing HPLC method, column length, internal diameter, and particle size may be varied and the flow adjusted accordingly.* With the same stationary phase, you use a column shorter by a factor of 2 with particles also smaller by a factor of 2. Result: the "same" separation in half the time at a pressure higher by a factor of 2 and a detection limit lower by a factor of 2. An interesting side effect is solvent consumption reduced by a factor of 2. When instead of a 4 mm column, you use a 3 mm one (corresponding to a decrease by 25%, allowed according to USP and EP as this would be an adaptation and not a change), then this results in a solvent consumption reduced by about 45% (!), at a pressure increased by the same percentage.
3) *The resolution of the method is to be optimized, and you can change what you like.* As explained earlier, you can experiment with methanol as organic solvent, with lower temperatures and long columns/multiple columns. Result: all of this can, at higher pressure, lead to an improvement in selectivity (first two points) or an increase in the plate number/peak capacity (longer column/multiple columns).
4) *The method is fine so far, the resolution could still be a little better.* You may have little time to improve the selectivity by means of the mobile phase, pH, other stationary phases, modifiers, and so on, and just use a column with half as large particles. Result: at a pressure higher by a factor of 4, the plate number is increased by a factor of 2, thus improving the resolution by a factor of 1.4. This could possibly be sufficient and includes an interesting side effect: the detection limit is also reduced by a factor of 1.4.
5) *The resolution of the method is to be optimized. As before, you do not have much time to vary the chromatographic conditions.* You use a column longer by a factor of 2 with particles also smaller by a factor of 2. Result: with a retention time increased by a factor of 2, the resolution is also better by a factor of 2. The problem here is that the pressure is higher by a factor of 8. The original pressure was about 150 bar, it is now about 1200 bar, which a modern UHPLC system should be able to cope with even in routine use.
6) *The detection limit of the method is to be improved,* more precisely, the mass sensitivity, not the detection sensitivity, and that means you cannot/may not increase the injection volume. You use a column with a smaller internal diameter and smaller particles – if necessary, you also reduce the column length. Result: the relative mass sensitivity can be improved by a factor of 5–20, depending on the measures taken.

Note The figures mentioned in the descriptions of example situations relate to isocratic separations. In gradient separations, other pressures can result depending on the gradient, and the retention times in cases 2 and 5 do not change by a factor of 2 but, depending on gradient-specific parameters, only by 10–30%, see also [11].

1.6.1 Outlook

Modern UHPLC systems have more or less the following specifications: 80–100 µl dwell volume with LPG or 25–35 µl with HPG systems, ≤3–5 µl dead volume, 1500 bar back pressure, flow up to 5 ml min^{-1} possible. What can be expected in the future? The pressure on the manufacturers to keep bringing newer products onto the market will probably lead to further developments in the UHPLC hardware. It is quite possible that in the next generation of UHPLCs, the dead and dwell volumes will be further minimized. The focus here is clearly on the dwell volume as such and in particular on the geometry of the gradient mixing chamber, because the mixing quality should not be reduced as a result of miniaturization.

The second critical point is certainly the autosampler: fast injection cycles and prevention/minimization of carryover. The 1–2 µl injection solution remains in the point of the (ceramic) needle until injection, and the loop is not used for the injection process. Instead, it serves as a "buffer" to provide flexibility in dwell volume with different systems. Furthermore, consider the problem of the enormous pressure surges to which the column is subject during injection, their minimization is a further challenge for autosampler design. Finally, pressures of about 2000 bar and flow rates of 1–3 ml min^{-1} should be technically possible. The challenge here is less the pressure to be reached, rather material damage at these pressures.

What would be chromatographically possible on a system with such a design? With a future UHPLC system, 10–20 mm × 2.1 mm columns filled with 1–1.1 µm porous or 0.7–0.8 µm fused-core material could be used. Using the theoretically resulting 20–25 000 plates, challenging separations within a few seconds would be possible – assuming that the three modern detectors DAD, Corona, and MS/Ion mobilization can process the signals without loss of information (see also Chapters 4 and 8). Due to the following limiting factors, these column options will probably, however, not be put into practice: large particle size distribution with the small particles, considerable effort would be required to pack a 10–20 mm column with 0.7–1.0 µm material efficiently and reproducibly, not to mention inhomogeneity in the packing at the column wall and the increased solubility of small particles. Other difficulties would be the – for such a column significant – dead volume of 2–3 µl probably present and the enormous frictional heat, which would then be generated. The future user too will probably be willing to accept "long" runs of a few minutes when under robust conditions.

The more likely scenario for the UHPLC in the future: instead of striving toward extremely short retention times with extremely short columns and extremely small particles, its strengths will more likely be used for the use of long or coupled columns with approximately 1.3–3 μm material, the continuous improvement of peak capacity is likely to stand in the foreground. As far as the matrix of stationary phases is concerned, both porous and fused-core materials have their advantages and weaknesses. A coexistence will probably continue.

References

1 Snyder, L.R., Kirkland, J.J., and Dolan, J.W. (2011) *Introduction to Modern Liquid Chromatography*, John Wiley & Sons, Inc.
2 Meyer, V.R. (2010) *Practical-High-Performance-Liquid-Chromatography*, Wiley-VCH Verlag GmbH.
3 Snyder, L.R. and Dolan, J.W. (2007) *High-Performance Gradient Elution*, John Wiley & Sons, Inc.
4 Kromidas, S. (ed.) (2008) *HPLC Made to Measure*, Wiley-VCH Verlag GmbH.
5 Zuo, Y. (2012) *High-Performance Liquid Chromatography (HPLC): Principles, Practices and Procedures*, Nova Science Publishers Inc.
6 Steiner, F. *et al.* (2015) Optimization strategies in RP-HPLC, in *The HPLC-Expert – Possibillities and Limitations of Modern HPLC* (ed. S. Kromidas), Wiley-VCH Verlag GmbH.
7 Kromidas, S. and Kuss, H.-J. (eds) (2010) *Quantification in LC and GC*, Wiley-VCH Verlag GmbH.
8 www.usppf.com (2016).
9 European Pharmacopoeia, 9.0, Chapter 2.2.46 (Chromatographic separation techniques), 2016.
10 Skriptum Provadis-NOVIA GmbH, HPLC in Pharmaceutical Analysis, 2016.
11 Kromidas, S. *et al.* (2015) Aspects of gradient optimization, in *The HPLC-Expert – Possibillities and Limitations of Modern HPLC* (ed. S. Kromidas), Wiley-VCH Verlag GmbH.

Part I

Hardware and Software, Separation Modes, Materials

2

The Modern HPLC/UHPLC Device

2.1 The Modern HPLC/UHPLC System

Steffen Wiese and Terence Hetzel

2.1.1 Today's Demands on the Individual Modules

2.1.1.1 Overview

One of the most fundamental innovations in the field of liquid chromatography (LC) was the introduction of Ultra-Performance Liquid Chromatography (UPLC™) by Waters in 2004. For the first time, it was possible to perform chromatographic separations with small particles at high pressure (1000 bar). Thereby, very fast, highly efficient, and sensitive separations could be realized. This is achieved, on the one hand, by the use of columns packed with sub 2 μm particles and, on the other hand, by revisions and new developments of the individual components of UPLC systems, in order to minimize the peak dispersion outside the column. Over the past decade, all manufacturers have introduced LC systems, which offer the possibility to work at pressures above 400 bar. At the same time, the commonly accepted nomenclature "Ultrahigh-Performance Liquid Chromatography" (UHPLC) has become established. During this development, manufacturers were faced with different challenges. The use of fully porous sub 2 μm particle columns results in a significant increase in back pressure. At a constant linear velocity, a bisection of the particle diameter leads to a fourth-power higher system pressure. In addition, the minimum of the van Deemter curve shifts to higher linear velocities when the particle diameter is reduced. Therefore, modern UHPLC systems have to be able to deliver the mobile phase very precisely at elevated pressures. Furthermore, the autosampler designs have to be adapted to ensure high reproducibility, especially for the injection of small sample volumes at high pressure. The stability of the column has also to be considered, as during the injection at elevated pressures, pressure drops of several hundred bars are possible. Moreover, with peak widths in the lower second range, which can be achieved by means of sub 2 μm particles, new detectors with sufficient data acquisition rates as well as fast response times need to be developed. In general, all described developments have been made

The HPLC Expert II: Find and Optimize the Benefits of your HPLC/UHPLC, First Edition.
Edited by Stavros Kromidas.

in the context of minimizing the extra-column volumes, which are discussed in Chapter 3 in more detail.

In the following sections, the different assemblies of modern UHPLC systems are described in detail. In addition, the prerequisites for routine operation are discussed.

2.1.2 UHPLC Pump Technology

2.1.2.1 High- and Low-Pressure Pumps

Regarding the applicability of sub 2 μm particles, the pumps of modern UHPLC systems were optimized or newly developed to satisfy the requirements to deliver the mobile phase with the correct flow rate at high pressures. Similar to conventional HPLC systems, modern UHPLC systems have also been differentiated between low- and high-pressure mixing systems. The designs for both mixing types are depicted schematically in Figure 2.1.1 (see also Chapter 4). For High Pressure mixing, which is mostly applied for binary mobile phases, two pumps are usually used.

In contrast, a Low-Pressure mixing system only needs one pump and is often used for quaternary mobile phases. The decisive difference between both systems is the point of mixing of the mobile phases. The mixing of the mobile phases takes place after the pumps for binary High-Pressure mixing systems (Figure 2.1.1a). The mixing of the solvents in the high-pressure part of the UHPLC systems causes a nonnegligible change in volume. For example, with a mobile-phase composition of approximately 60–65/40–35 methanol/water (v/v), there is a change in volume of approximately 4%, whereas at mobile phases consisting of 50–55/50–55 (v/v) acetonitrile/water, a maximum change in volume of

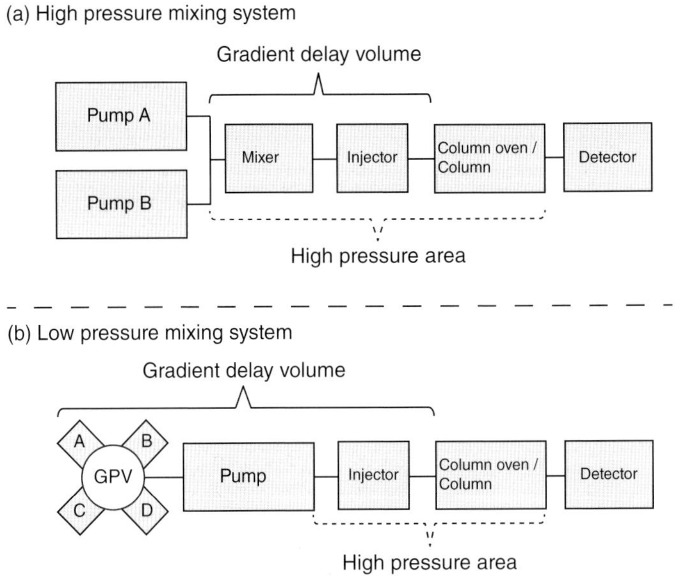

Figure 2.1.1 Schematic design of (a) High- and (b) Low-Pressure mixing systems with gradient proportioning valve (GPV).

approximately 2% can be observed [1]. As a consequence, it is possible that the real flow rate is different in comparison to the programmed one. This topic is discussed in more detail in Chapter 4.

In order to optimize the mixing of the mobile phases (change in volume, compressibility, and viscosity of single solvents), software algorithms have been developed by the manufacturers to control the pumps. In contrast, the mixing of different solvents in a Low-Pressure mixing system is accomplished prior to the pump (Figure 2.1.1b). The different solvents are portioned and mixed using a gradient proportioning valve (solenoid valve). For such systems, the precision of the pumps is responsible for the proper adjustment of the flow rate and not the properties of the different solvents.

The different mixing systems have both advantages and disadvantages. High-Pressure mixing systems are characterized by a low gradient delay volume (see Figure 2.1.1a). This includes the volume from the point of mixing of the mobile phases to the head of the column. A low gradient delay volume is a prerequisite for fast solvent gradient methods (≤5 min) with 2.1-mm-ID columns at relatively slow flow rates, such as those used in the hyphenation to mass spectrometers (see Chapter 5). In contrast, for Low-Pressure mixing systems, the volumes of the pump, the gradient proportioning valve as wells as the volume of the connection tubings are part of the gradient delay volume. As a consequence, the gradient delay volume for Low-Pressure mixing systems is usually two to four times higher when compared to High-Pressure mixing systems [2]. Regarding the increased gradient delay volume of Low-Pressure mixing systems, it should be noted that at higher flow rates (e.g., 1 ml min^{-1}), the influence of the gradient delay volume can be neglected, and therefore, the resulting analysis time is not affected. A particular advantage of Low-Pressure mixing systems is the opportunity to mix and deliver up to four different solvents. This is a useful feature when a chromatographic separation cannot be achieved using binary mobile phases and more than two solvents are required. Therefore, these pumps are very well suited for method scouting systems. Furthermore, these systems provide the possibility to adjust a constant additive or buffer concentration. For example, channels A and B are used for water and an organic cosolvent, whereas channel C continuously delivers a constant concentration of an additive of the mobile phase. In this manner, a constant additive concentration (e.g., buffer, ion-pairing reagents, ionization additives for mass spectrometry (MS), etc.) can be ensured. Moreover, in contrast to High-Pressure mixing systems, Low-Pressure mixing systems are cheaper to purchase because only one pump is needed. A short comparison of the properties of low- and High- Pressure mixing system is given in Table 2.1.1.

Regardless whether a low- or high-pressure mixing system is used, the LC manufacturers have redesigned the appropriate pumps in order to ensure highly efficient separations using sub 2 μm particle packed columns. This includes a revision of the piston design, seals, pressure sensors, inlet and outlet valves, pulsation damper, connection tubings as well as the connection technology (screws and ferrules). The aim of this development was to design a similarly robust and reliable pump similar to a conventional HPLC pump but to be able to work at high pressures. Therefore, the gradient delay volume of the UHPLC pump is dramatically decreased in comparison to the HPLC pump. Compared

Table 2.1.1 Comparison of the individual properties of low- and high-pressure mixing systems, with reference to [3].

Properties	Low-pressure mixing system	High-pressure mixing system
Gradient delay volume	Usually larger	Usually smaller
Volume adjustment of the mixer	Often time-consuming	Usually simply possible
Degassing of the mobile phase	Essential	Recommended
Precision flow rate (compressibility)	Less influenced	Strongly influenced
Complex mobile phases	Usually 3–4 solvents	One pump per solvent
Applicability of low flow rates	More difficult	Usually simple
Small changes of the organic content in the mobile phase	Good response	Usually better response than low-pressure mixing systems
Independent pump use	Not possible, only one pump	Pumps A and B can be controlled independently

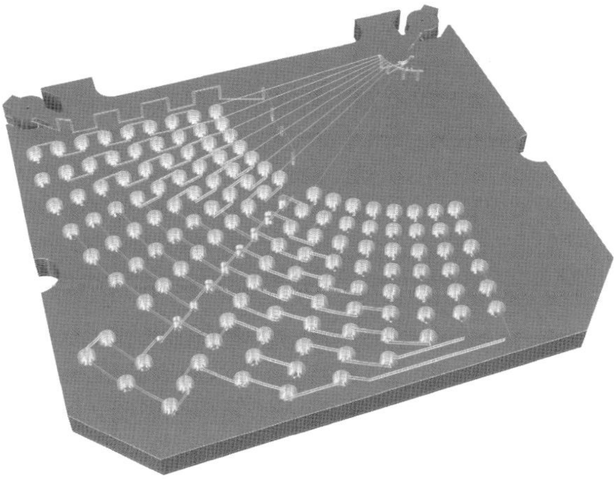

Figure 2.1.2 Design of the microfluidic optimized Jet Weaver mixer. Copyright by Agilent Technologies - Reproduced with Permission.

to classic HPLC systems, which typically have large-volume pulsation damper and mixer, for modern UHPLC systems, microfluidically optimized mixers are used, as exemplarily shown in Figure 2.1.2.

Thereby, it can be ensured that the mobile phase is mixed in the smallest possible volume so that modern UHPLC systems usually have a specified gradient delay volume between 50 and 400 μl (depending on the mixing system, autosampler configuration, and applied mixer). In comparison, conventional HPLC systems usually have a gradient delay volume in the range of 500 and 1500 μl. Because of the decreased gradient delay volume of UHPLC systems, the introduction of software-controlled algorithms was necessary to compensate for the different solvent properties such as change in volume, compressibility, and

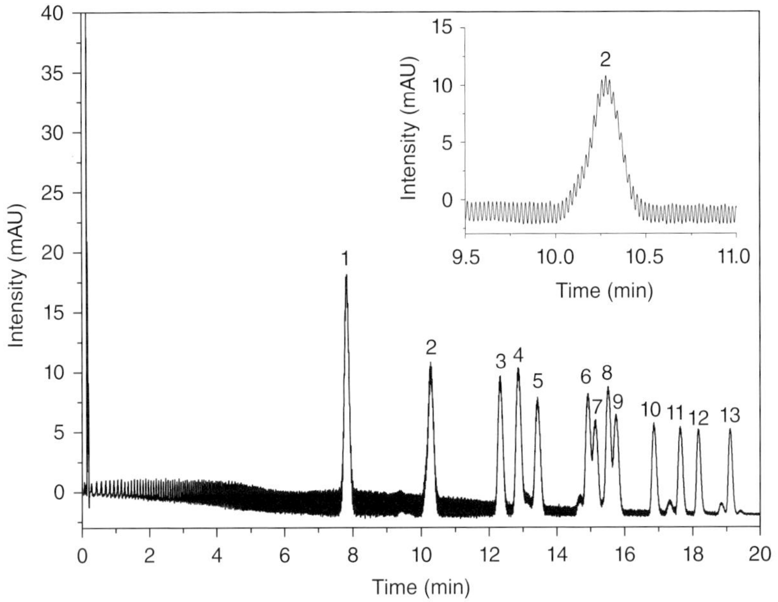

Figure 2.1.3 Separation of 13 aldehyde- and ketone 2,4-dinitrophenylhydrazine. Chromatographic parameter: mobile phase: (i) water and (ii) acetone; stationary phase: C18 reversed phase (50 mm × 2.1 mm, sub 2 μm); flow rate: 1 ml min^{-1}; gradient: 5–95% acetone in 30 min; and injection volume: 1 μl; UV detection: at 360 nm.

viscosity when mixing the mobile phase. If such adjustments are not made, pressure fluctuations (ripples) occur, which lead to a "noisy" baseline especially for spectroscopic detectors. As a consequence, the signal-to-noise ratio is negatively affected and an increased limit of detection (LOD) and limit of quantitation (LOQ) can be observed. An example for an inadequate compensation of the different solvent properties for a UHPLC system (High-Pressure mixing) is shown in Figure 2.1.3.

As clearly depicted, the resulting pressure fluctuations lead to a very noisy baseline using acetone as organic solvent. This phenomenon is increasingly pronounced in the first 10 min of the chromatogram. During this time frame, the organic content of the mobile phase comprises between 5% and 35%. By increasing the acetone content in the mobile phase to 95%, a continuous reduction of the noise to an acceptable level is observed (20–30 min).

At this point, it should be mentioned that the described pressure fluctuations could not be observed when using acetonitrile or methanol as organic modifier in the mobile phase.

Further investigations showed that the systematic pressure fluctuations observed by the UV detector can be correlated with the flow rate, the piston stroke as well as the volume and efficiency of the applied mixer. In general, the optimal mixing volume appears to depend upon the manufacturer and the mixer type as previously described by Dong in the context of the applicability of trifluoroacetic acid (TFA) as a mobile-phase additive [4, 5]. In order to reduce the noise, it would be possible to increase the volume between the pump and the column by increasing the volume of the mixer. However, this is associated with an increase of the gradient delay volume, which is disadvantageous with respect to fast UHPLC separations at low flow rates. However, there are some applications that require a larger volume for optimal mixing of the mobile phase. One example is the aforementioned use of TFA as an additive in the mobile phase. In the low wavelength range (≤220 nm), an inadequate mixing of the mobile phases leads to an increased noise with UV detection. This topic is extensively discussed in Chapter 4.

In summary, it should be noted that modern UHPLC pumps, regardless of whether it is a Low- or High-Pressure mixing system, should be able to deliver the flow rate with an accuracy of ±1% and a precision of ≤0.1% independent of the respective system pressure [3, 6, 7].

2.1.2.2 Gradient Delay Volume

The gradient delay volume, also known as *Dwell Volume*" or *Delay Volume*, describes the volume from the point of mixing of the mobile phase to the head of the column. It should not be confused with the extra-column volume (dispersion volume, see Chapter 3). Contributions to the gradient delay volume include the pump (Low-Pressure mixing systems), the mixer, the autosampler (including the loop), as well as all connection tubings between the individual assemblies and the column (see Figure 2.1.1). Due to this reason, the gradient delay volume is different for the various available UHPLC systems. The decisive advantage of modern UHPLC systems compared to conventional HPLC systems is the significant reduction of the gradient delay volume. This volume is responsible for the involuntary isocratic step at the beginning of each gradient method. Moreover, it is an important parameter during method transfer between different UHPLC and HPLC systems. In order to compensate the isocratic step at gradient elution, most software packages provide the possibility of a delayed sample injection to synchronize the injection of the sample and the actual start of the solvent gradient. Although the isocratic step at the beginning can be compensated, the analysis time is extended by the duration of the time delay. Therefore, a low gradient delay volume is desirable. Figure 2.1.4 illustrates the influence of gradient delay volume. In this double logarithmic plot, the time frame of the isocratic step is shown in dependence of the flow rate for six different gradient delay volumes.

At a flow rate of 0.3 ml min^{-1}, for instance, and a gradient delay volume of 300 µl, an isocratic step of 1 min results (Figure 2.1.4 (❶)),whereas a gradient delay volume of 500 µl leads to an isocratic step of 1.7 min (see Figure 2.1.4 (❷)).In contrast, when a traditional HPLC system with a gradient delay volume of 1 ml is used at a flow rate of 0.3 ml min^{-1}, the resulting isocratic step at the beginning of a gradient elution extends for more than 3 min (see Figure 2.1.4 (❸)).The isocratic step at the beginning of a separation influences particularly the early eluting analytes. The chromatographer is not able to influence the retention of an analyte eluting within this time by the solvent gradient. A large gradient delay volume is particularly critical for separations at low flow rates, which are usually used for the hyphenation with MS (see Chapter 5). Therefore, the user should know the

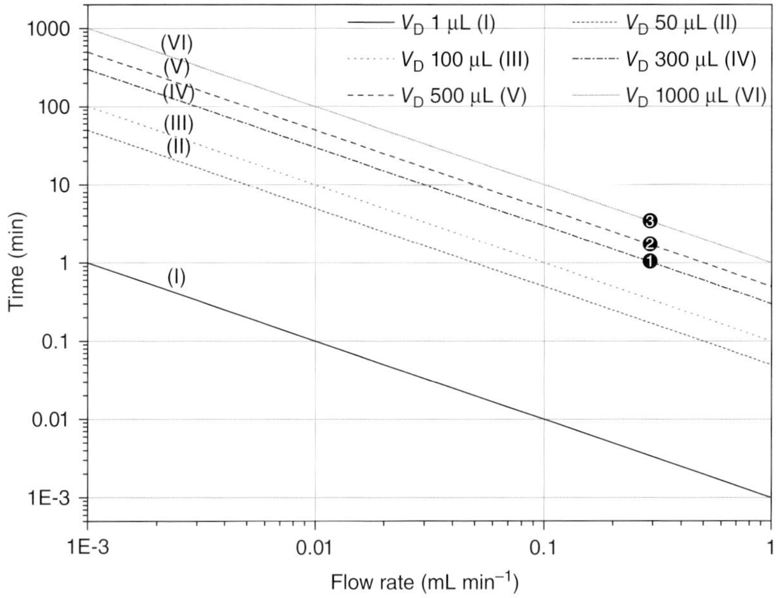

Figure 2.1.4 Representation of the gradient delay time depending on the flow rate for six different gradient delay volumes.

gradient delay volume of the LC system to get an idea when the real solvent gradient reaches the column. This is also important for equilibration of the column as well as for programming the gradient table. As a rule of thumb, the volume of the gradient should be at least as big as the sum of the column dead volume and the gradient delay volume.

Further practical details regarding the gradient delay volume are discussed in Section 9.1. For example, the determination of the gradient delay volume and the influence on real separations are shown.

In summary, it should be noted that modern UHPLC systems should have a low gradient delay volume. This is especially important for the applicability of columns with an inner diameter ≤2.1 mm and low flow rates of ≤500 µl min^{-1}, such as the hyphenation of UHPLC and MS (see Chapter 5). A specified gradient delay volume of ≤200 µl is desirable. For applications at high flow rates (≥500 µl min^{-1}), the influence of an increased gradient delay volume is less pronounced because the time difference between programmed and effective gradient is marginal and can often be neglected. For the sake of completeness, it should be mentioned that the gradient delay volume has no influence on isocratic separations.

2.1.3 Autosampler

Another challenge for the manufacturer of UHPLC systems is the development or revision of the autosampler in order to inject very small samples volumes (e.g., 100 nl) in a reproducible manner into the system at elevated pressures. Depending on the autosampler design, pressure drops of several 100 bars are possible during the injection. Therefore, the mechanical stability of the column has to be considered. In addition, it should also take into account that for solvent gradients methods, pressure differences of several hundred bars are possible during a chromatographic run.

In general, two main types of autosampler designs have been established at modern UHPLC systems: on the one hand, a fixed-loop and, on the other hand, a variable flow-through concept, both of which are discussed in more detail in the following sections.

2.1.3.1 Fixed-Loop Autosamplers

With fixed-loop autosamplers, the injection needle is not part of the high-pressure flow path. The needle is used to aspirate and dispense the sample out of the vial into the sample loop. For the actual injection, the sample loop is switched into the high-pressure flow path. As a consequence, only the sample loop and not the needle is rinsed by the mobile phase during the chromatographic separation. The filling and subsequent injection are schematically shown in Figure 2.1.5.

In order to fill the sample loop, two concepts are used, which are shown schematically in Figure 2.1.6: on the one hand, the so-called Pulled-Loop design and, on the other hand, the Pushed-Loop design. With a Pulled-Loop autosampler (Figure 2.1.6a), the needle is positioned in the desired sample vial. Afterward, by a syringe or peristaltic pump, the sample is drawn directly into the sample loop. The design of a Pulled-Loop autosampler is more simple compared to that of a Pushed-Loop autosampler (Figure 2.1.6b) because no low-pressure needle seat is needed. However, the consumption of sample solution is very high. For example, for an injection of 20 μl, sample volume up to 100 μl may be required. As clearly depicted in Figure 2.1.6a, the needle and all required connection tubings between the needle and the valve must be filled with the

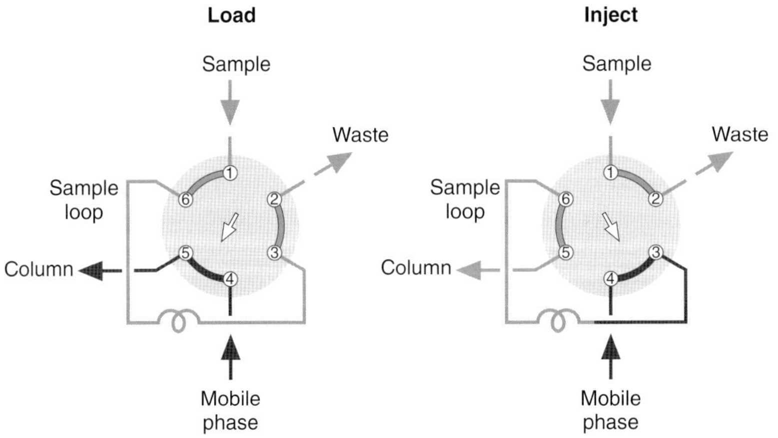

Figure 2.1.5 Flow path of the mobile phase of a fixed-loop autosampler (pushed-loop design) during the filling of the sample loop and injection. Reproduced with permission of VICI AG International.

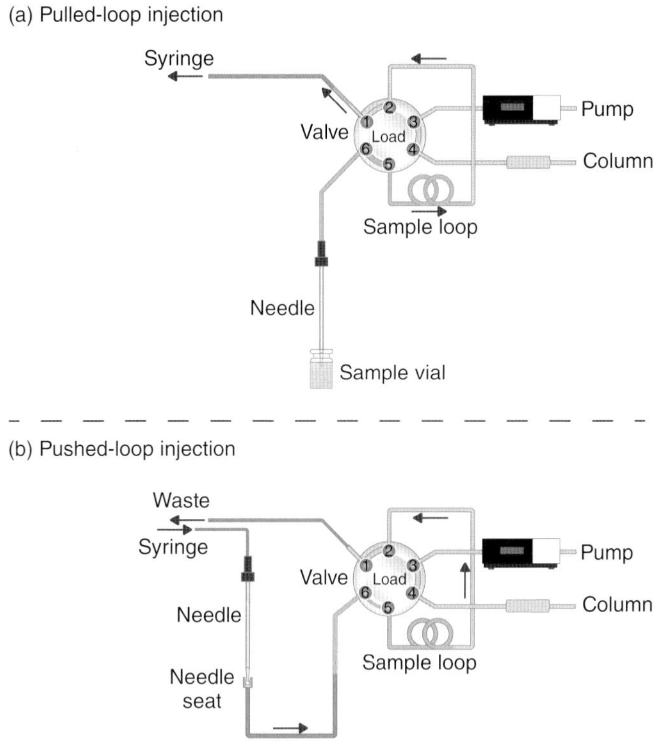

Figure 2.1.6 Schematic design of (a) pulled-loop design and (b) pushed-loop design during the filling of the sample loop. The arrows show the flow path of the sample. Reproduced with permission of Thermo Fisher Scientific.

sample before the sample loop is filled. This sample volume is usually lost in the standard injection mode.

In comparison, using a fixed-loop autosampler, which operates according to the Pushed-Loop design, the sample is first drawn into a syringe. Afterward, the syringe is positioned in a Low-Pressure needle seat, and the previously aspirated sample is pushed into the sample loop. Generally, the Pushed-Loop design requires a lower sample volume than the Pulled-Loop design. However, the Low-Pressure needle seat can be a source of carryover. Therefore, this assembly should be adequately flushed between injections. Nevertheless, most manufactures prefer the Pushed-Loop concept for their fixed-loop autosamplers.

Furthermore, with fixed-loop autosamplers, two different injection modes can be distinguished, depending on the volume, which is actually filled into the sample loop. Fixed-loop autosamplers are able to perform a full-loop injection as well as a partial loop injection. With full-loop injection, the volume of the sample loop defines the maximum possible injection volume. If the sample loop is only partially filled, the chromatographer should only use a specific aliquot of the available sample loop volume as injection volume. When transferring the sample plug into the sample loop, issues can occur due to the parabolic flow profile of the liquid in the loop. This is demonstrated in Figure 2.1.7. When a liquid flows through

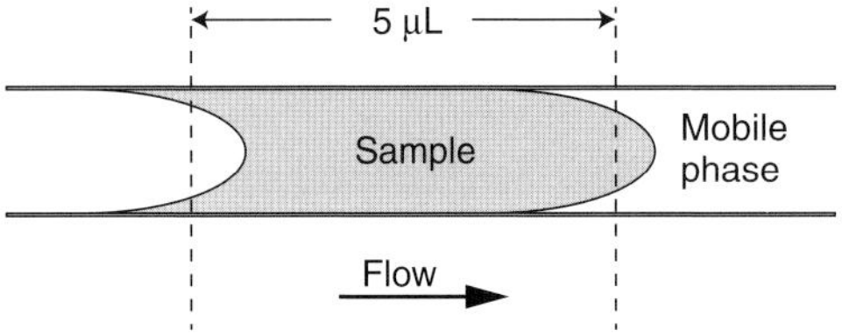

Figure 2.1.7 Simplified representation of the laminar flow profile of a sample plug into a sample loop. Reproduced with permission of CTC Analytics AG.

a capillary, the flow velocity of the liquid close to the capillary wall is about half of the flow velocity compared to the center of the capillary, due to friction. As an example, the filling of a 5 μl sample loop with a sample volume of 5 μl (highlighted by the dashed lines in Figure 2.1.7) is considered. Because of the parabolic flow profile, some components at the beginning of the sample plug already leave the sample loop toward the waste while other ingredients at the end of the sample plug have not reached the sample loop. In addition, the actual injection of the sample is applied in opposite flow direction (Backflush). Therefore, an additional mixture between the sample and the mobile phase takes place.

As a rule of thumb, if full-loop injection is used, the sample loop should be overfilled three to five times. With regard to the example of filling a 5 μl loop, a sample volume between 15 and 25 μl should be flushed through the sample loop before injection. This is similar with partial-loop injection. The injection volume should be selected in the range of 20–60% of the sample loop volume. In other words, for partial filling of a 5 μl sample loop, 1–3 μl should be injected. If higher injection volumes (≥200 μl) are required, such as in the "Large-Volume Injection" (LVI), which is discussed in Section 9.6, 20–80% of the available loop volume can be utilized. When a sample loop of 200 μl is used, a volume between 40 and 160 μl should be injected [8]. If the injection volume is not within this range, usually higher standard deviations of the peak area and height of the analytes can be observed, which lead to a decreased reproducibility of the method. Several manufacturers are aware of this problem and allow only an adjustment of the injection volume within the trusted range. In any case, it makes sense to consult the manufacturer as to whether the adjustment of the injection volume is oriented on the installed sample loop or not. If this is not the case, the user should consider the aforementioned comments when selecting the injection volume.

2.1.3.2 Flow-Through Autosamplers

In contrast to fixed-loop autosamplers, with flow-through autosamplers, the needle is part of the high-pressure flow path. In addition, the needle seat and the needle seat capillary (connection between the needle seat and the injection valve) are components of the high-pressure flow path. Therefore, the needle and the needle seat need to be pressure-stable. Further information about the applied materials is presented in more detail in Chapter 7. For this autosampler type, sample loops with different loop volumes can also be installed. Upon injection, the sample loop and the needle are switched into the flow path. Thereby, the needle and the sample loop are flushed with the mobile phase during the entire chromatographic run. The flow path of the mobile phase for a flow-through autosampler is schematically shown in Figure 2.1.8a for sample aspiration and in Figure 2.1.8b for the injection.

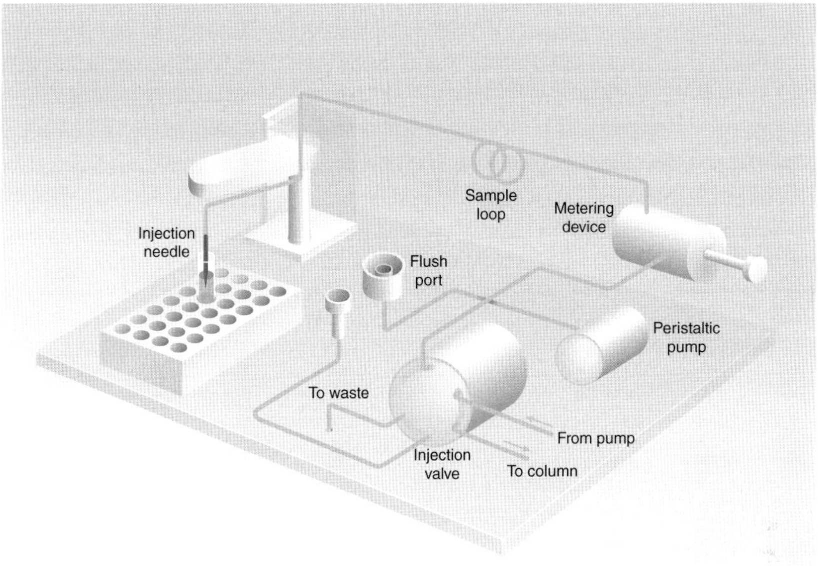

(a)

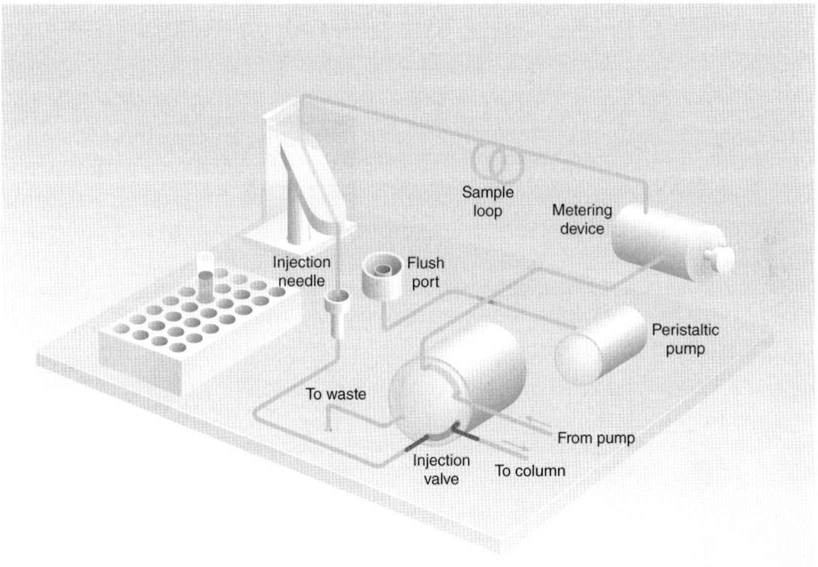

(b)

Figure 2.1.8 Flow path of the mobile phase for a flow-through autosampler during aspiration (a) and injection (b) of the sample. Copyright by Agilent Technologies - Reproduced with Permission.

2.1.3.3 Review of the Advantages and Disadvantages of Fixed-Loop and Flow-Through Autosamplers

The individual autosampler designs have both advantages and disadvantages. One benefit of fixed-loop autosamplers is the decreased contribution to peak dispersion when small sample loops (≤5 μl) are used. In addition, the resulting gradient delay volume is reduced compared to flow-through autosamplers.

Therefore, this design is suitable for the application of 2.1-mm-ID as well as 1-mm-ID columns. However, several wash solvents are required for fixed-loop autosamplers in order to avoid carryover, because the needle and other parts of the injection path are switched out of the flow direction of the mobile phase for the duration of the chromatographic separation. Furthermore, it should be mentioned that for such autosamplers, the use of sample loops with a higher volume (≥20 μl) lead to increased peak dispersion. In addition, the consumption of sample solution with fixed-loop autosamplers is higher compared to flow-through autosamplers as the connection tubings as well as the groove of the rotor of the injection valve must be filled with sample when loading the sample loop. This sample volume is lost. Using flow-through autosamplers, the entire aspirated volume is usually transferred to the column.

Another and often neglected point of modern UHPLC autosamplers is the cycle time between the injections. In particular, the working steps are important, which take place independently of the chromatographic separation [9]. This includes, for example, the positioning of the *x/y/z* arm of the autosampler, the sample vial, the filling of the sample into the sample loop, the injection of the sample as well as the time needed for the single-wash steps. While the sample throughput with conventional HPLC separations with run times of more than 20 min is not affected by a preparation time of 1–2 min, the influence on UHPLC separations with method run times less than 5 min might be an issue. For variable flow-through autosamplers, a preparation time of 1 min can lead to a reduction in sample throughput of 20% because the autosampler needle is usually switched into the flow path for the entire analysis time. In order to further minimize the cycle time of flow-through as well as fixed-loop autosamplers, manufacturers provide the possibility of an overlapped sample preparation. During one chromatographic separation, the injection of the next sample is prepared. Users are encouraged to consult the autosampler manufacturer for an overview of the different settings and options.

Moreover, a cycle time below 5 min requires a sufficient sample capacity of the autosampler. For instance, for an overnight sequence of 16 h and a cycle time of 5 min, the autosampler has to offer space for approximately 190 sample vials.

In summary, it should be noted that modern UHPLC autosamplers should be able to inject the smallest specified injection volume (often 100 nl) with an accuracy of 1–2% and a relative standard deviation below 2%. This also applies to the reproducibility of the injection precision. The relative standard deviation of the injection precision should be below 2% for the smallest specified sample volume. The aforementioned values should be understood as guidelines, which should be achieved by every UHPLC autosampler. The accuracy and precision of the injection volume and the associated precision of the peak area and height of the analytes depend on the respective injector type (fixed-loop, flow-through

Table 2.1.2 Comparison of the advantages and disadvantages of a fixed-loop and flow-through autosamplers with reference to [10, 11].

Properties	Fixed-loop autosamplers	Flow-through autosamplers
Influence of the gradient delay volume	Less pronounced using small sample loops	Significant influence
Contribution to band broadening	Less pronounced using small sample loops	Larger influence
Variable injection volume	Limited to the size of the loop, loop change frequently required	Usually supports larger range of injection volumes
Carryover	Vulnerable, requires several wash solvents	Less vulnerable
Sample consumption	Some sample is lost	Sample is usually injected completely
High-pressure needle seal	—	May be prone to leakages

design), the injection technique (full loop, partial loop) as well as on the pressure but also on the volume of the samples syringe and metering device. Finally, a comparison of the individual advantages and disadvantages of the fixed-loop and flow-through autosamplers is given in Table 2.1.2.

2.1.4 Column Oven

Regarding the tempering of the mobile and stationary phase, the modern UHPLC systems were further developed and revised compared to conventional HPLC systems. As a consequence, modern UHPLC systems offer the possibility to apply elevated temperatures (up to 150 °C). For classic HPLC systems, device-related maximum temperatures ≤100 °C are common. The reason why elevated temperatures are useful was shown by Teutenberg [12]. Some key factors are as follows: change in selectivity, improvements of the peak shape, reduction of retention time, reduced consumption of organic solvents, decreased system pressure, more efficient separations, and applicability of special hyphenation techniques (e.g., isotope-ratio MS).

Regardless of which temperature range is used, the column should always be tempered to ensure reproducible chromatography. As a rule of thumb, increasing the temperature by 1 °C leads to a reduction of the retention factors in approximately 2% for analytes in RP-HPLC. This is demonstrated in Figure 2.1.9. As depicted in Figure 2.1.9, the temperature was slightly changed in a range between 23.6 and 30 °C for the separation of different polycyclic aromatic hydrocarbons (PAHs). It should be noted that the separations were carried out at a system pressure ≤400 bar. Therefore, it can be concluded that no influence due to frictional heating occurs. It can be clearly seen that with increasing temperature, shorter retention times are observed. Additionally, a change in resolution between the

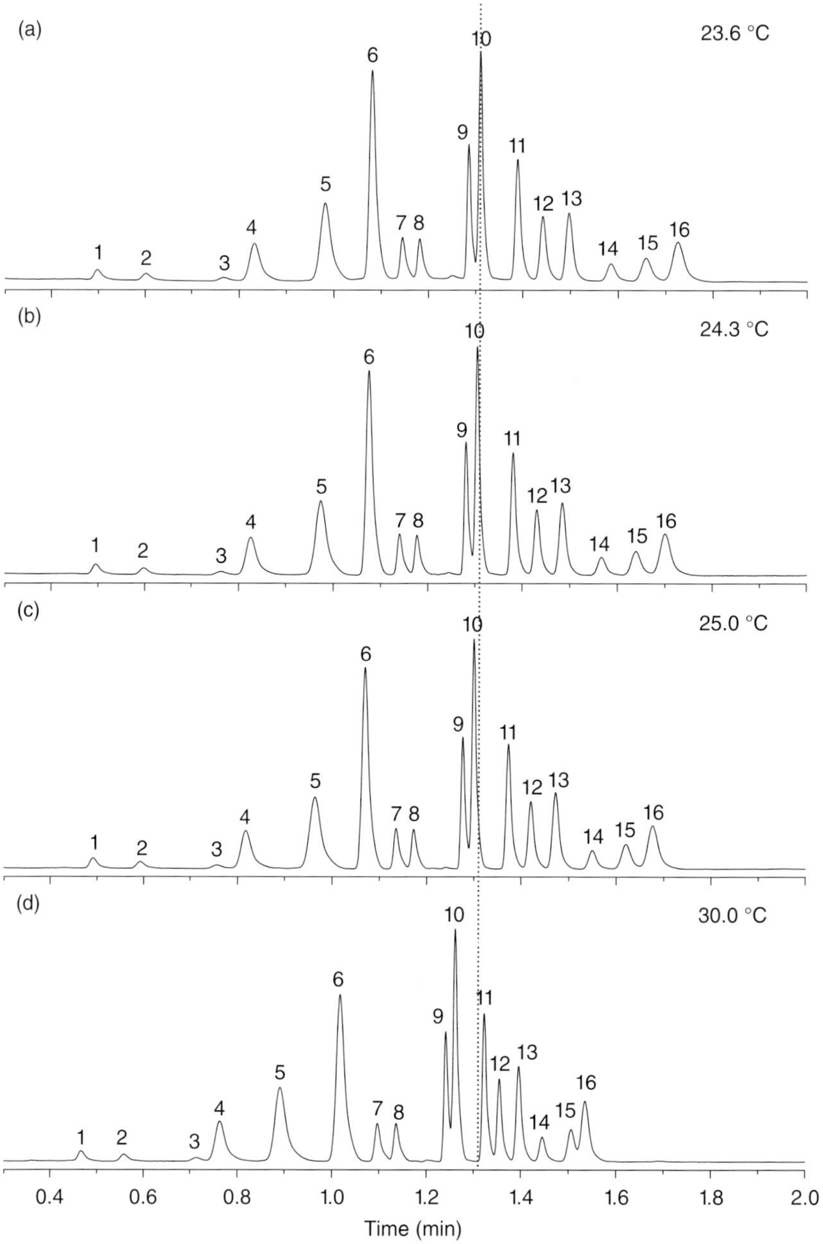

Figure 2.1.9 Separation of 16 PAHs at different temperatures. Chromatographic parameters: mobile phase: (A) water and (B) acetonitrile; stationary phase: YMC-PAH (50 mm × 2.1 mm, 3 μm); flow rate: 1.3 ml min^{-1}; gradient: 0–0.8 min, 58% B, 0.8–1.13 min, 58–100% B, 1.13–2.0 min, 100% B; injection volume: 0.5 μl; temperature: see figure; and UV detection: at 254 nm. Analytes: (1) naphthalene, (2) acenaphthylene, (3) acenaphthene, (4) fluorene, (5) phenanthrene, (6) anthracene, (7) fluoranthene, (8) pyrene, (9) benz[*a*]anthracene, (10) chrysene, (11) benzo[*b*]fluoranthene, (12) benzo[*k*]fluoranthene, (13) benzo[*a*]pyrene, (14) dibenz[*a*,*h*]anthracene, (15) benzo[*g*,*h*,*i*]perylene, and (16) indeno[1,2,3-*c*,*d*]pyrene.

peaks can be noted. As an example, the critical resolution between the peak pair 15/16 continuously decreases with increasing temperature from 1.51 at 23.6 °C to 0.99 at 30 °C. When working in a non-air-conditioned laboratory, such temperature changes, as shown in Figure 2.1.9, are quite common. In other words, a baseline separation, which was achieved in the morning at a temperature of 23.6 °C, cannot be reproduced under certain circumstances in the afternoon at a temperature of 30 °C in the laboratory. This is not only true for UHPLC but also for HPLC applications. Furthermore, a method transfer between different UHPLC/HPLC systems is only possible if the temperature of the separation is controlled and managed.

In addition, in UHPLC applications with a back pressure ≥400 bar, a noticeable heat generation can be observed because of the friction between the mobile and stationary phase. Because of the generated frictional heating, the resulting formation of radial and axial temperature gradients might lead to temperature differences of about 20 °C between the column inlet and outlet [13, 14]. This phenomenon is discussed in Chapter 2.2 in more detail. Due to this, a tempering of the mobile and stationary phase is necessary to ensure reproducible UHPLC separation at high pressures (≥400 bar).

Another often disregarded temperature-related topic is the preheating of the mobile phase before entering the column. Inadequate preheating often results in less efficient separations because the peak shape is negatively influenced [15]. This topic is discussed again in Section 9.4. In order to ensure a sufficient preheating of the mobile phase, active or passive tempering can be selected. For example, using passive preheating, the mobile phase is passed through a long capillary in the column oven. Thereby, the temperature of the mobile phase and that of the column oven become almost comparable. An active preheating requires a direct heating/cooling of the mobile phase within the column oven. The temperature of the oven should not be influenced as far as possible. One interesting option is provided in Shimadzu's column oven (Nexera UHPLC) because the active preheating can also be used to cool down the mobile phase before entering the column. This possibility enables the compensation of any eventual negative effects due to the friction-induced generation of heat. Regardless of how the mobile phase is tempered, the following applies: every additional volume prior to the column increases the gradient delay volume and can also contribute to an undesired band broadening, which is particularly noticeable in isocratic separations.

Another useful tool of modern column compartments is the integration of column switching valves, which are provided by every UHPLC manufacturer. However, these valves must be critically reviewed in terms of the increase in gradient delay volume and the extra-column volume because usually two additional valves are installed: one before the column and one behind the column. Advantages of this switching design include, for example, the possibility to choose between different columns during a sequence without the need to change the columns manually. Several configurations are available regarding the number of columns, which can be installed, and how these are controlled. Some vendors use two valves inside or outside the column oven (e.g., Waters, Thermo Scientific, KNAUER), whereas others use only one in the oven (e.g., Agilent). Here, the temperature stability of the rotor needs to be considered. Not every manufacturer adjusts the maximum allowed temperature of the rotor of the column switching valve to the maximum oven temperature. The maximum available temperature of the column oven is often significantly higher than the maximal specified temperature of the rotor in the switching valve. As a consequence, the erosion of the rotor is increased, and a replacement of the rotor after only a few 100 switching operations may be required. Furthermore, the increased abrasion of the rotor leads to blockages in the column and capillaries. This is particularly problematic for columns that are packed with sub 2 μm particles. For such phases, column inlet frits with pore diameter of 0.5 μm are typically used, which clog very quickly. In contrast, inlet frits of approximately 2 μm are used for columns packed with 3–5 μm particles. Therefore, conventional HPLC columns are less prone to clogging.

In summary, it should be noted that a modern UHPLC system should provide the possibility to temper the mobile and stationary phase (≥100 °C). Furthermore, the tempering, irrespective of whether passive or active, should not contribute to a disproportionate band broadening. For the preheating of the mobile phase, short capillaries with small inner diameter (e.g., ≤130 μm) should be used. This also applies to the usage of column switching valves and particularly for the connection capillary to the detector.

Regarding spectroscopic detectors that are very sensitive to small temperature changes, the following pieces of advice should be considered. When working at elevated temperatures (≥80 °C), it is useful to temper the mobile phase after leaving the column to a defined temperature similar to the temperature of the detector cell. Otherwise, fluctuations of the peak area and peak height of the analytes can be observed.

2.1.5 Detectors

In addition to the separation column, the detector is the second most important component of a modern UHPLC system and needs special attention. In the following section, the explanation of physicochemical basics and schematic system designs of UHPLC detectors is not discussed in detail. These have already been described in the scientific literature [6, 7, 9, 16–20]. Furthermore, the characteristics of the hyphenation to MS are not further explained. This topic is discussed in detail in Chapter 5.

In Table 2.1.3, a short overview of the most commonly used UHPLC detectors is given.

The detectors listed in Table 2.1.3 have both advantages and disadvantages. However, all devices have to satisfy the general requirements of UHPLC. When working with 2.1-mm-ID columns and sub 2 μm particles, peak widths of 1 s or less can be expected. Therefore, modern detectors have to provide a sufficient data acquisition rate as well as fast response. As a general rule, at least 15–20 data points per peak should be recorded in order to represent the peak shape with sufficient accuracy and to ensure proper peak integration [16, 17]. For a chromatographic peak with a width of 1 s, an acquisition rate of at least 15–20 Hz is required. Usually, an acquisition rate of 40 Hz is adequate. Higher data acquisition rates as well as very low time constants for the response of the detector lead to large data sets and increased noise. Therefore, the acquisition rate and the

Table 2.1.3 Overview of the common HPLC-/UHPLC detectors.

Detector	Selectivity	Detection range	Linear range	Gradient suitability	Advantages	Limitations
Refractive index	Universal	µg	10^3	No	Universal, wide range of applications, nondestructive, cheap, easy to use	Poor sensitivity, influenced by smallest temperature and flow fluctuations
Conductivity	Selective	ng–µg	10^4	Yes	Ideal for ion chromatography, detection of inorganic ions and organic acids, highly selective, low costs, nondestructive	Requires suppression of the conductivity of the mobile phase for optimal detection limit, requires special HPLC systems and columns, not all substances can be detected
Light scattering	Almost universal	pg–ng	10^3	Yes	Detection of many nonvolatile substances	Requires volatile mobile phase (also buffer), optimization sometimes difficult, problem at reproducibility of the methods
CAD	Universal	pg–ng	10^4	Yes	Highest sensitivity of all universal detectors, all nonvolatile and medium volatile substances can be detected, easy to use	Requires volatile mobile phase (also buffer),
UV/Vis	Selective	pg–ng	10^5	Yes	"Almost" universal in the lower UV wavelength range (e.g., 200 nm), qualitative and quantitative, DAD often allows determination of the peak purity/spectra homogeneity, spectra libraries, nondestructive, reliable, low costs, easy to use	Chromophore necessary, solvents must be UV/Vis transparent
Fluorescence	Highly selective	fg–pg	$10^3 – 10^4$	Yes	Highly selective and sensitive	Not all analytes have native fluorescence, often derivatization is required, loss of performance due to quenching
Electrochemical detector	Highly selective	fg–pg	10^5	Yes (limited)	Highly selective and sensitive	Mobile phase must be conductive, vulnerable to baseline noise, contamination of the electrodes, substances must be oxidizable or reducible, gradient elution can cause loss of sensitivity

time constant of the detector should be chosen with respect to the expected peak width.

Narrow peaks require not only a higher acquisition rate and low time constants for the response but also a smaller volume of the detector cell than conventional cells [16]. In this context, it should be noted that manufacturers sell their detectors for both the conventional HPLC and UHPLC. However, the cell volumes of many detectors are often too large for UHPLC applications using short columns with a smaller inner diameter (e.g., 50 mm × 2.1 mm). If, for example, a classic UV flow cell with 10 μl volume is used for the detection of separations with columns of ≤2.1 mm inner diameter and sub 2 μm particles, the additional extra-column volume leads to an unacceptable band broadening. Although a simple decrease of the cell volume reduces the extracolumn volume, the signal intensity is negatively influenced at the same time due to the smaller illuminated light path. This challenge for UV/Vis detectors was solved by Waters in 2004 with the introduction of the liquid-core optical waveguide technology (total reflection technology). This cell technology is shown in Figure 2.1.10.

Similarly to an optical fiber, the UV/Vis light is coupled to the liquid-core optical waveguide and is passed by means of total reflection on the inner Teflon AF wall through the detection capillary. As a result, less intensity loss occurs, which leads to a lower baseline noise. In addition, through total reflection, the light path is increased, and the volume of the detection capillary can be reduced to 500 nl at the same time. In this way, the extra-column volume is reduced, and simultaneously, the signal intensity is increased. Thereby, this cell technology is ideally suited for the detection of separations using columns with small inner diameter and sub 2 μm particles. The approach of total reflection in a liquid-core optical waveguide was taken up by other manufacturers. Agilent Technologies presented the so-called Max-Light flow cell, Thermo has introduced LightPipe flow cells, KNAUER has marketed LightGuide flow cells, and also Shimadzu has introduced a new diode array detector, which is based on the liquid-core optical waveguide technology.

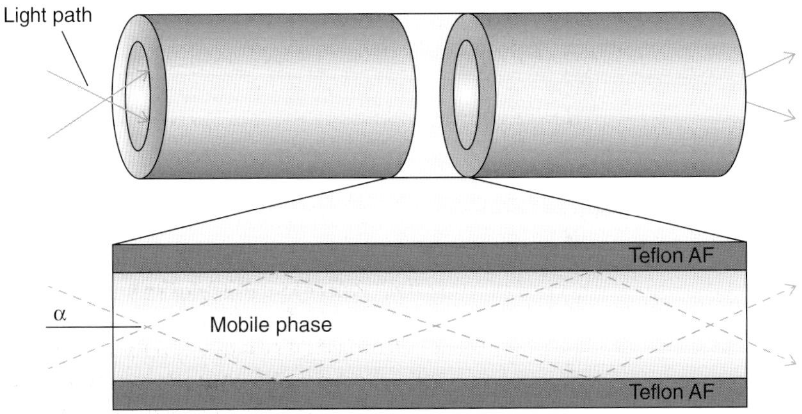

Figure 2.1.10 Schematic design of a UV cell based on the liquid-core optical waveguide technology. Reprinted with permission from Waters GmbH.

In summary, modern UHPLC detectors should be able to record chromatographic data with a low response time constants (≤50 ms) and a fast acquisition rate of at least 40 Hz. Furthermore, the geometric volume of the detector cell should not contribute to excessive peak broadening. In general, the smaller the volume of the detector cell, the lower the contribution to peak dispersion. This point is sometimes critical because often a compromise between a still acceptable band broadening and minimum required signal intensity has to be found. This is especially difficult when using columns with small inner diameter (≤2.1 mm). This topic will be taken up from a practical perspective in Section 9.7.

2.1.6 Capillaries and Fittings

Capillaries and fitting are used in HPLC/UHPLC to ensure the transport of the sample and the mobile phase through the chromatographic system. Unfortunately, they are often the reason for loss in separation efficiency as well as leakages.

Capillaries can be classified into low-pressure and high-pressure capillaries. Usually, plastic-based capillaries are used for the transport of the solvent to the pump or to connect the detector to the solvent waste. For the connections in the high-pressure part of a UHPLC system, capillaries with a significantly higher pressure stability are required. Plastic capillaries made of polyether ether ketone (PEEK) are suitable for pressures up to approximately 400 bar. These capillaries are flexible and can be easily cut and adapted to the individual requirements. PEEK capillaries can be used within this pressure range without the risk of bursting. The more problematic part is the plastic fittings in the high-pressure area of the HPLC system (autosampler, column head). At pressures ≥200 bar, the fittings can slip out of the seat, which usually leads to leakages.

For applications with pressures ≥200 bar, stainless steel capillaries are standard. Compared to PEEK capillaries, stainless steel capillaries are difficult to adjust and cut to a high quality by the user. It is recommended to purchase electrolytically precut and cleaned capillaries. Both PEEK and stainless steel capillaries are commercially available in a wide range with different outer and inner diameters [21, 22]. Further information about the different materials of the individual capillaries can be found in Chapter 7.

Compared to classic HPLC, the selection of suitable capillaries has a significantly higher importance in UHPLC. Usually, the influence of band broadening in the capillaries (autosampler to column, column to detector) can be neglected for HPLC applications using conventional columns (150 mm × 4.6 mm). In contrast, with UHPLC applications using 2.1-mm-ID and 50-mm-long columns, the band broadening in the connection capillaries has a great influence on separation efficiency [6, 23–25].

For the applicability of UHPLC systems in combination with short columns (≤50 mm) and small inner diameter (≤2.1 mm), it is recommended to use short capillaries with small inner diameter (≤130 μm) both for the connection between autosampler and column and between column and detector. Otherwise, an increased peak broadening can be observed.

Moreover, for applications with low flow rates ($\leq$0.5 ml min^{-1}), such as in the hyphenation to mass spectrometry with electrospray ionization (ESI-MS), the inner diameter as well as the length of the transfer capillary should be minimized [26]. Often with LC/MS, long distances (50 cm or more) have to be covered between the column outlet and the MS inlet. In order to minimize the peak dispersion in the connection capillaries, the usage of capillaries with an inner diameter of 100, 63, or 50 μm is advantageous [26]. However, it should be noted that the reduction of the inner diameter of the capillaries leads not only to a decreased band broadening but also to an increased system back pressure. A reduction of the inner diameter of a capillary by half results in a 16 times higher pressure. Furthermore, capillaries with small inner diameter are more vulnerable to blockages, for example, by pump wear, rotor abrasion of the autosampler, and components from the sample or solvent.

In order to connect the capillaries to the individual modules of the UHPLC system and the separation column, suitable fittings are required. These fittings usually consist of a nut and a ferrule. Similarly to the connecting capillaries, low- and high-pressure fittings can be differentiated made of plastic or stainless steel, respectively. Here, stainless-steel-based high-pressure fittings are discussed in more detail due to the higher risk of band dispersion when disregarded [27]. Dealing with PEEK capillaries and fittings is usually easier to accomplish. The principle of stainless-steel-based fittings is shown in Figure 2.1.11a.

The screw and the ferrule are first pushed over the stainless steel capillary. After this step, the capillary is guided until it reaches the end of the conical union. Finally, the connection is tightened using a wrench. When tightening, the ferrule deforms and nestles itself to the conical connection ensuring a tight fit of the capillary in the union. In this way, high-pressure resistant and dead-volume-free connections can be made. Depending on the manufacturer of the fittings, the ferrules and corresponding unions have different designs and vary additionally in the installation depth (pilot) of the capillary into the union, as depicted in Figure 2.1.11b. Therefore, fittings from different vendors should not be mixed. For example, screw and ferrules of manufacturer A should not be used in combination with a union of manufacturer B. Otherwise, dead volumes can occur, which lead to an undesired peak broadening (see Figure 2.1.11c). Particularly problematic are all capillaries and fittings, which are part of the sample band flow path, for example, between the autosampler and the column head and between the column outlet and the detector. These connections have to be loosened and retightened during column change. Again, there are manufacturer-dependent differences in column hardware such as the shape of the ferrule or the installation depth of the capillary (pilot) into the column. The user should know the individual differences of the manufacturer. Usually, stainless steel capillaries are used before the column, and as described earlier, the installation depth of the capillary into the column is determined by the initial tightening of the ferrule. During this step, a dead volume can occur at the column head if the installation depth of the capillary is different from the mounting depth of the column. This phenomenon is shown in Figure 2.1.11c. If the installation depth is too short and the capillary does not reach the bottom of the union, a mixing volume arises. Dead volume and leakages also occur when the mounting depth of the capillary is too long and

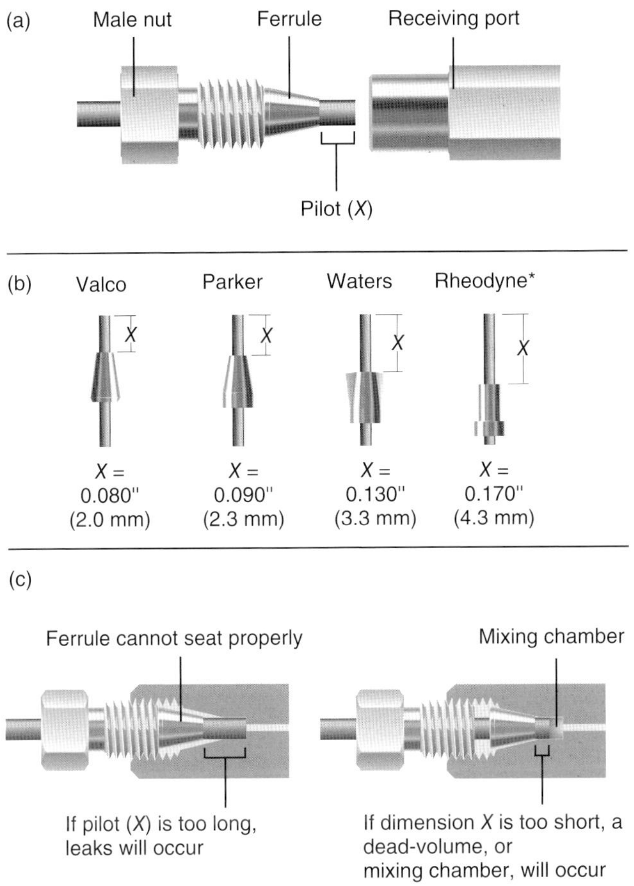

Figure 2.1.11 Overview and function of stainless-steel-based fittings. Shown is only a selection, which does not claim to be complete. * used for e.g. injection or column switching valves. Reproduced with permission of IDEX Heath & Science GmbH.

the ferrule does not reach the end of the union. In Figure 2.1.12, two resulting chromatograms with a correctly and poorly installed fitting at the column head are shown.

It can be seen that the poor fitting causes an increased tailing. As a result, all peaks elute as broader bands. This also becomes clear considering the peak capacity (P_c). A reduction of approximately 45% from 42 to 23 for the peak capacity can be observed. The previously described characteristics also apply to the connection between the column outlet and the detector. A dead volume behind the column also leads to peak dispersion. Peak broadening always occurs in isocratic separations regardless of whether dead volumes exist at the head or outlet of the column. In gradient separations, however, this is not the case. The analytes are ideally enriched at the head of the column assuming sufficient retention and elute only at a certain organic content in the mobile phase from the separation column. The resulting compression of the peak bands may compensate for a previous peak dispersion caused by a dead volume at the column head.

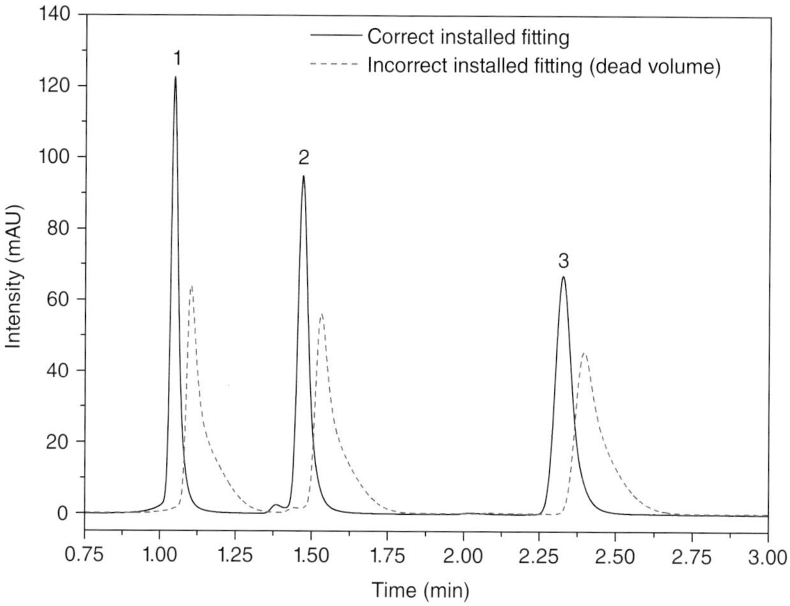

Figure 2.1.12 Comparison of two isocratic separations at a correct (—) and an incorrect (---) installation of the fitting at the column head. Chromatographic parameters: mobile phase: 50/50 (v/v) acetone/water; stationary phase: Agilent Zorbax SB C18 (50 mm × 2.1 mm, 1.8 μm); flow rate: 0.5 ml min^{-1}; injection volume: 0.2 μl; temperature: 30 °C; and UV detection: at 360 nm. Analytes: (1) formaldehyde-2,4-dinitrophenylhydrazone, (2) acetaldehyde-2,4-dinitrophenylhydrazone, and (3) acrolein-2,4-dinitrophenylhydrazone.

In contrast, a dead volume between the column outlet and the detector cannot be compensated using a solvent gradient, and this is particularly problematic for columns with an inner diameter of ≤2.1 mm compared to 4.6-mm-ID columns. The issue described earlier has been recognized by some UHPLC manufacturers and suppliers. In response, they have developed universal and reusable fittings. Some examples are Viper [28], OPTILOK [29], and VHP [21]. For further information, see Section 12.1.

Acknowledgment

The authors would like to thank Dr Kathryn Monks for the fruitful discussions and her help during the preparation of this chapter.

References

1 Foley, J.P., Crow, J.A., Thomas, B.A., and Zamora, M. (1989) *J. Chromatogr.*, **478**, 287.
2 Fekete, S., Kohler, I., Rudaz, S., and Guillarme, D. (2014) *J. Pharm. Biomed. Anal.*, **87**, 105.

3 Snyder, L.R. and Dolan, J.W. (2007) *High-Performance Gradient Elution: The Practical Application of the Linear-Solvent-Strength Model*, John Wiley & Sons, Inc., Hoboken, NJ.
4 Dong, M.W. (2013) *LCGC North Am.*, **31**, 868.
5 Noland, P., Hopper, T.N., Dong, M.W., Guo, Y., Maloney, T.D., and Xu, R.N. (2010) in *Chromatography – A Science of Discovery* (eds R.L. Wixom and C.W. Gehrke), John Wiley & Sons, Inc., Hoboken, NJ, p. 344.
6 Snyder, L.R., Kirkland, J.J., and Dolan, J.W. (2010) *Introduction to Modern Liquid Chromatography*, John Wiley & Sons, Inc., Hoboken, NJ.
7 Shackman, J.G. (2013) in *Liquid Chromatography – Fundamentals and Instrumentation* (eds S. Fanali, P.R. Haddad, C.F. Poole, P.J. Schoenmakers, and D. Lloyd), Elsevier, Amsterdam, p. 306.
8 Meyer, V. (2013) *Pitfalls and errors of HPLC in pictures*, Wiley-VCH, Weinheim; John Wiley, Chichester.
9 Swartz, M. (2009) *LCGC North Am.*, **27**, 1052.
10 Dolan, J.W. (2001) *LCGC North Am.*, **19**, 386.
11 Dolan, J.W. (2001) *LCGC North Am.*, **19**, 478.
12 Teutenberg, T. (2010) *High-Temperature Liquid Chromatography: A User's Guide for Method Development*, RSC Publishing, Cambridge.
13 Novakova, L., Veuthey, J.L., and Guillarme, D. (2011) *J. Chromatogr. A*, **1218**, 7971.
14 Gritti, F. and Guiochon, G. (2008) *Anal. Chem.*, **80**, 5009.
15 Wolcott, R.G., Dolan, J.W., Snyder, L.R., Bakalyar, S.R., Arnold, M.A., and Nichols, J.A. (2000) *J. Chromatogr. A*, **869**, 211.
16 Swartz, M. (2010) *J. Liq. Chromatogr. Related Technol.*, **33**, 1130.
17 Swartz, M. (2010) *LCGC North Am.*, **28**, 880.
18 Morgan, N.Y. and Smith, P.D. (2011) in *Handbook of HPLC* (eds D. Corradini and T.M. Phillips), CRC Press/Taylor & Francis, p. 231.
19 Poole, C.F. (2003) *The Essence of Chromatography*, Elsevier, Amsterdam, Boston, MA.
20 Meyer, V.R. (2010) *Practical High-Performance Liquid Chromatography*, John Wiley & Sons, Inc., Hoboken, NJ.
21 IDEX Health & Science http://www.idex-hs.com (accessed 04 February 2014).
22 VICI AG International http://www.vici.com/ (accessed 04 February 2014).
23 Wu, N.J. and Bradley, A.C. (2012) *J. Chromatogr. A*, **1261**, 113.
24 Wu, N.J., Bradley, A.C., Welch, C.J., and Zhang, L. (2012) *J. Sep. Sci.*, **35**, 2018.
25 Fekete, S. and Fekete, J. (2011) *J. Chromatogr. A*, **1218**, 5286.
26 Spaggiari, D., Fekete, S., Eugster, P.J., Veuthey, J.L., Geiser, L., Rudaz, S., and Guillarme, D. (2013) *J. Chromatogr. A*, **1310**, 45.
27 Stankovich, J.J., Gritti, F., Stevenson, P.G., and Guiochon, G. (2013) *J. Sep. Sci.*, **36**, 2709.
28 Thermo Fisher Scientific Inc. http://www.dionex.com/en-us/products/accessories/reagents-accessories/viper-fingertight/lp-81335.html (accessed 04 February 2014).
29 Optimize Technologies http://www.optimizetech.com (accessed 04 February 2014).

2.2 The Thermostate of Columns – A Minor Matter

Michael Heidorn and Frank Steiner

It should already be accepted by any experienced HPLC user that a column thermostat needs to be applied with an HPLC system to obtain reproducible results from run to run. However, thermal modes of column thermostats, hence the way how a column is thermostatted, can tremendously differ from each other. This chapter introduces the different thermal modes and technical realizations of column thermostats. The thermal mode of a column thermostat has a major impact on the actual separation temperature within a column. However, a regulated column thermostat temperature is generally identified as the actual column temperature, more precisely the mobile- and stationary-phase temperature inside that column, which controls the separation.

Therefore, column temperature equilibration is a central part of the LC method preconditions: an analysis does not start as long as the column thermostat temperature did not stay within a defined range for a certain time period. However, it is a reality that this time period is typically set on a generic default value, without much thought about it, let alone being tailored to the application specifics; nevertheless, the equilibration time needed for proper column thermostatting strongly depends on the thermal mode of the column thermostat. Unfortunately, when a thermal equilibration time is set too short, the column does not adopt column thermostat temperature comprehensively prior to the first injection, and retention time shifts may result (for the first runs). In fact, the more elegant approach for thermal column equilibration is to wait until a constant system back pressure is obtained. At constant system back pressure, a constant separation temperature within the column and a steady-state of all other LC conditions can be assumed.

The actual separation temperature inside that column still remains an assumption: since mobile phase flows continuously, the column separation temperature also depends on the temperature of the mobile phase entering the column. If mobile-phase and column thermostat temperatures do not match, the resulting separation temperature within the column deviates from the column thermostat temperature set by the user. Due to friction between mobile phase and column packing, frictional heat can occur in a column. Then, the resulting separation temperature within a column is higher than the appointed thermostat temperature. Finally, the column thermostat type determines if and to which extent temperature gradients occur inside a column. Temperature gradients, in turn, lead to an uneven heat pattern. Separation efficiency, retention times, selectivity, and thereby, resolution (cf. Purnell equation) can be affected.

All kinetic and thermodynamic retention mechanisms of the HPLC theory are temperature dependent. A change in separation temperature affects retention time (cf. van 't Hoff equation), selectivity, and separation efficiency (cf. van Deemter equation). The extent of such changes also depends on the physicochemical properties of the substances to be separated. Even a small temperature change can lead to a shift in the ionization behavior, for example, if the pK_a of this substance and that of the mobile phase are very similar. The result could be

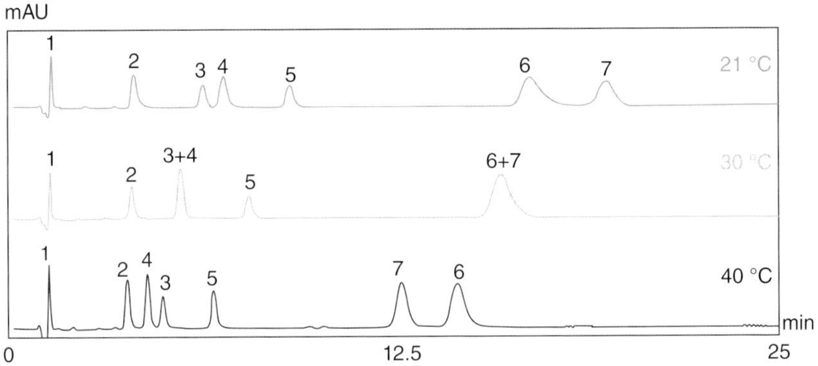

Figure 2.2.1 The same HPLC separation at different temperatures.

an abrupt or atypical change in retention time of this substance. As a result, the selectivity of an affected peak pair might change as well. In general, retention times decrease with increasing temperature though that decrease may vary strongly for substances of a sample mixture. In such cases, selectivities change or the elution order reverses. Quite contrarily, the separation efficiency often benefits from increased temperatures due to improved mass transfer of the analytes toward the stationary phase (because LC analyses are most often carried out at flow rates higher than the van Deemter optimum). This is also shown in Figure 2.2.1. Same substance peaks are narrower at a higher temperature, and thereby, better separation efficiencies are achieved. All substances are baseline separated at 21 °C, some substances coelute at 30 °C, and a reversed elution order takes place at 40 °C (see substance peaks 1–7). Accurate temperature control is very important for this application: the same substances, which coelute at 30 °C, would still be separated at 25 or 35 °C.

A precise column temperature control depends on the column thermostat choice. Temperature accuracy shall not matter in this context but the thermal operation principle of the column thermostat does. Column thermostats are offered in a diverse range of variants – from rudimentary heating pockets to high-tech modules with heating and cooling functions. Between the heating unit of the thermostat and the column wall, a medium such as air or metal is used for heat transfer. The greater the thermal conductivity of the heat transfer medium, the more efficient the heat conduction will be and the more heat is transferred per time. Hence, a differentiation regarding thermal modes of column thermostats can become important. Section 2.2.1 deals with distinctions on thermal coupling between appointed column thermostat temperature and corresponding column wall temperature.

If mobile phase flows into a column, a temperature exchange between column wall and column center occurs as soon as the temperatures of the mobile phase and column wall are different. In a column thermostat that provides high heat transfer efficiency, the column wall is effectively maintained at the temperature of the surrounding column thermostat. Heat exchange from the column thermostat heater to column wall is continuous and fast. This leads to radial temperature gradients within the column. However, if the heat transfer efficiency of the column thermostat is negligibly low and the flow rate of the mobile phase is high, the column packing adopts nearly the temperature of the mobile phase after a certain period of time. These topics are covered in Section 2.2.2.

Starting at column back pressures of approximately 400 bar, measurable frictional heat is produced within a column. With a column thermostat that provides fast heat transfer, a certain part of this frictional heat is dissipated through the column wall. This continuous heat flow from column interior to column wall forms radial temperature gradients within the column: in the column wall regions, the column thermostat temperature dominates, whereas a higher temperature remains in the column center. The nondissipated portion of frictional heat remains in the column and is transported by mobile phase into flow direction toward the column end. The overall result is an uneven temperature distribution inside the column: an axial temperature gradient is formed in addition to the radial temperature gradients. If the heat transfer efficiency of the column thermostat is negligibly low, only little frictional heat is dissipated through the column wall. Then, a pronounced axial temperature gradient is formed, and radial temperature gradients are absent or formed in the area of the column wall only. Section 2.2.3 deals with this subject in more detail.

Radial temperature gradients, whether due to a temperature difference between the column wall and the influent mobile phase or due to dissipated frictional heat via the column wall, usually lead to a loss in separation efficiency. An axial temperature gradient changes the average temperature within a column and can lead to retention time shifts. This, for example, could further lead to coelution of a critical peak pair, which would be adequately resolved at a lower separation temperature (see Figure 2.2.1).

In a method transfer between two HPLC systems providing different column thermostat types, separation performance differences and shifts in the retention times can additionally occur when mobile-phase preheating is not considered. A method transfer can fail although exactly the same column and nominally the same thermostat temperature were used, and all other method parameters were maintained as well. In a method speed-up from an HPLC column to a UHPLC column or in a method transfer between UHPLC systems having different column thermostat types, different analytical results are even more likely. In Section 2.2.4, both possibilities are presented. Successful method transfer with regard to temperature control is discussed. Furthermore, some hints are proposed to tackle unwanted thermal effects in method development.

2.2.1 Thermal Modes of Column Thermostats

Nowadays, the most commonly used HPLC column thermostats are air baths, followed by metal contact thermostats. Liquid baths are only partially suitable for HPLC and are, therefore, rarely used. A metal contact thermostat features a direct thermal contact to the column wall. Two metal elements, commonly made of aluminum, are used. Both metal elements provide cavities, which, when put

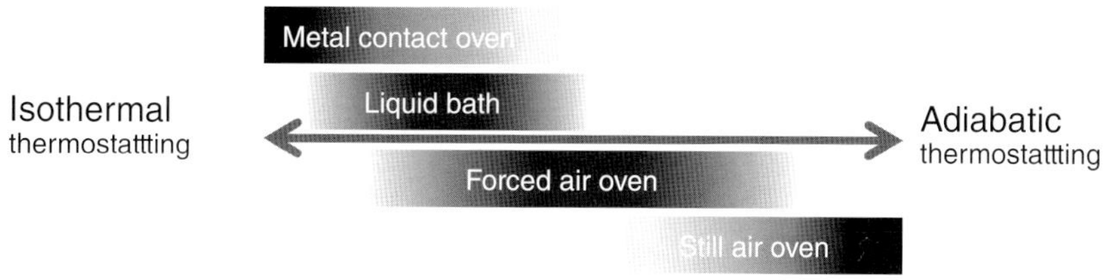

Figure 2.2.2 Classification of column thermostat types in thermal modes.

on each other, result in a slot having the diameter and length of a certain column, and hence, the metal elements shall perfectly enclose the corresponding column. However, such an encapsulation is not given if the diameter of the slot does not exactly match the diameter of the column and shape of the column end fittings. Then, a comprehensive contact between the metal furnace and the entire column wall is not attainable. To compensate for this deficiency, metal fabric is widely used, especially in heating pockets. It can entirely adapt to the outer shape of the column, ensuring a maximum of surface contact area.

Liquid baths have the great disadvantage that a column has to be immersed directly into the liquid. A possible leak of the column cannot be detected without high effort. Additionally, it may take a long period of time until a liquid bath offers a constant temperature after it had been changed. A stirrer can remedy this issue to a certain extent. Air baths are available in two main variants. With a still-air thermostat, stagnant air is heated in a chamber. A forced-air thermostat enforces air circulation in the column chamber by the use of a fan.

Due to different heat transfer efficiencies, column thermostat types can be classified into the so-called thermal modes; that is, the way a column wall is thermostatted (see Figure 2.2.2). The more effective a column wall is maintained against internal and external influences at a predetermined temperature, the more isothermal the mode of operation. Caused by the excellent heat conductivity of metals, a metal contact thermostat forces a column wall quickly and effectively to a predetermined temperature. A liquid bath shows a similar heat transfer efficiency due to the good heat conduction of liquids and the high heat capacity of most heating liquids; furthermore, the heat transfer efficiency can also be improved by stirring the liquid. Accordingly, metal contact thermostats and liquid baths operate a column wall isothermally.

Air offers very poor heat conduction and is, therefore, often used as an insulator. In a forced-air thermostat, an improved heat transfer is achieved by directed airflow paths and forced convection through ventilation. The degree of isothermal operation mainly depends on the extent of air circulation. However, a maximum achievable isothermal operation is given by a certain airflow rate. Any further increase in the airflow rate does not further improve the isothermal operation mode. Lowering the airflow rate, on the other hand, decreases the forced convection and, hence, the isothermal character of the operation mode. When the airflow rate is set to zero, adiabatic operation is achieved. This empowers a forced-air thermostat to provide a wide range of thermal modes. However, a heat transfer to the column wall via metal is, in general, up to five times greater compared to the heat transfer via forced air (see Figure 2.2.2). Hence, a forced-air thermostat cannot provide the same degree of isothermal operation as a metal contact thermostat.

If a column does not exchange any heat with its environment, it is operated under adiabatic conditions. A still-air thermostat offers the lowest heat transfer efficiency of all column thermostat types. Heat transfer capacities of the thermal conductivity of air and the natural convection process are negligibly low. The thermal radiation of the heating element is in general very low and thus plays only an insignificant role for the heat transfer capability of virtually any thermostat apart from the still-air concept; with the latter ones, the thermal radiation of the heat spreader itself gains relevance. However, even after long periods of time, a column does not always thoroughly adopt the preset column thermostat temperature. Moreover, the column temperature is prone to change from processes within the column. The column wall temperature is not effectively controlled by a still-air thermostat. The column wall is therefore operated under adiabatic conditions.

Finally, none of all the different column thermostat types can provide a physically perfect isothermal or adiabatic column wall operation. Both thermal conditions are mixed up with each other to some extent. Those effects can be drastically different, even for two different variants of the same column thermostat type. Forced-air thermostats are the best example as they provide the biggest range of thermal modes between adiabatic and isothermal operation conditions (see Figure 2.2.2).

Column compartments differ in their heat transfer efficiencies:

- In an ideal-isothermal operation, the column wall rapidly and effectively adopts the prescribed column thermostat temperature, since it is directly coupled to the heating element of the thermostat. Tentative changes to the column wall temperature from thermal processes within the column are suppressed.
- In an ideal-adiabatic operation, the column wall does not exchange heat with the column thermostat; that is, any changes to the column wall temperature from processes within the column are not suppressed. Nevertheless, the initial temperature of the column wall might be determined by the column thermostat temperature or ambient temperature.
- No column thermostat type operates a column wall under ideal-isothermal or ideal-adiabatic conditions.
- Metal contact thermostats and liquid baths operate a column wall quasi-isothermally.
- The airflow rate of a forced-air thermostat determines how isothermally a column wall is operated. In comparison to a forced-air thermostat, however, the isothermal characteristic of a metal contact thermostat is up to five times higher.
- Still-air thermostats operate a column wall quasiadiabatically.
- Column thermostats of the same type or design may differ in their thermal operation mode.

2.2.2 Temperature Differences between Column and Mobile Phase

Some column thermostats have a passive mobile phase or eluent preheater. Such a preheater consists of a capillary with a certain length. This capillary is either embedded in the heating unit or in a special heat exchanger plate. Such heat exchanger plates are available in different volumes and are mounted directly on the heater surface. The effective heating power of a passive eluent preheater is determined by the flow rate of the mobile phase and the heat exchange surface of the embedded capillary. The higher the volume flow rate, the higher the required heat exchanger surface of the embedded capillary.

For sufficient eluent preheating at higher mobile-phase flow rates, long capillaries would be needed, which lead to adverse extracolumn volumes. In addition, a passive preheater volume should always be adjusted to a modified flow rate. A single passive preheater with a lower extracolumn volume is, therefore, often used for all flow rates. As a consequence, the mobile phase is not heated up to the intended temperature at higher flow rates. Active eluent preheaters provide their own heating system. To reach and maintain the target temperature, they adapt their required heat output to the mobile-phase flow rate. The length of the heated capillary is constant, and its extracolumn volume is kept to a minimum. The need for mobile-phase preheating is more and more highlighted in the literature and application notes, especially with HPLC separations at increased separation temperatures.

Figure 2.2.3 shows the same HPLC separation at 70 °C, with the chromatograms only differing by the use of an eluent preheater. The wall of the column is heated up to 70 °C by forced air in both cases. When no preheater is used, mobile phase at approximately ambient temperature flows into the column. Then, the column head is cooled down harshly. The entire separation temperature within the column decreases, which leads to retention time shifts. As a consequence, selectivities of sample compounds can significantly change (see Figure 2.2.1). Furthermore, peak broadening or even peak distortion occurs, especially with later eluting peaks (see Figure 2.2.3). Such a peak distortion

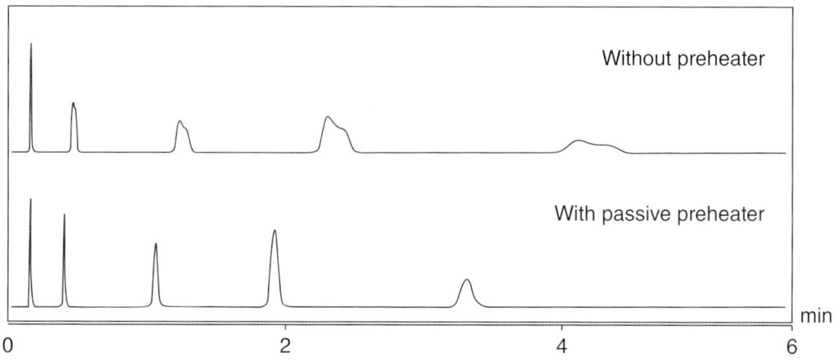

Figure 2.2.3 The same HPLC separation in a forced-air thermostat at 70 °C with passive preheater and without preheater.

resembles a double peak. When an unknown sample is analyzed, this "double peak" could be misinterpreted as two different substance peaks, although it is only one substance.

The mobile-phase temperature should not deviate by more than 6 °C from the column thermostat temperature. Hence, if a laboratory is air-conditioned at the very common room temperature of 22 °C, thermal imbalances inside a column may occur as soon as the column thermostat temperature exceeds 28 °C. However, column thermostat temperatures between 30 and 40 °C are used traditionally for the majority of applications. As a result of those very common conditions, a certain temperature distribution can be formed within the column. The pattern and degree of this temperature distribution depend on the column thermostat type and if eluent preheating is used or not. Figure 2.2.4 shows purely hypothetical examples of an ideal-isothermal (left) and ideal-adiabatic (right) column thermostat with eluent preheating (top) or without (bottom). However, there is no column thermostat available that provides ideal-isothermal or ideal-adiabatic operation. Each column thermostat provides a hybrid of the two modes (see Figure 2.2.2).

When column thermostat and incoming mobile phase do not have the same temperature, thermal differences between adiabatic and isothermal column thermostats become relevant (Figure 2.2.4c,d). With an isothermal column thermostat, the column wall continuously stays at column thermostat temperature. As soon as a mobile phase of a different temperature enters the column, a heat flow starts between the column wall and the column interior. The magnitude of this radial heat flow decreases from the column inlet toward the column outlet (see Figure 2.2.4c). The volumetric flow moves the mobile phase that was heated up at the column wall into the flow direction. The continuing process of thermal conduction from the column wall region further into the column center is, thereby, axially moved from the column inlet to the outlet.

As a result, radial temperature gradients are formed from the column wall to the column center. These are distinct for each column cross section in axial z position, so from column inlet to column outlet. A thermal imbalance arises inside the column (Figure 2.2.4c). The value of a radial temperature gradient due to a temperature difference between an isothermal column wall and incoming mobile phase can be calculated using the following formula:

$$\Delta T_{\text{radial, thermal imbalance}} = \frac{u \cdot \rho \cdot c_{\text{p}} \cdot \Delta T_z \cdot r^2}{4 \cdot \lambda \cdot \varepsilon} \tag{2.2.1}$$

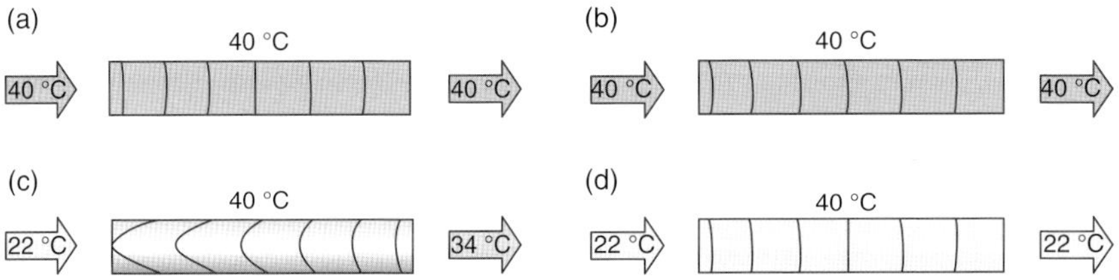

Figure 2.2.4 Temperature profiles in a column, resulting from the mobile-phase temperature and either ideal-isothermal (left) or ideal-adiabatic (right) mode.

In that formula, u stands for the linear flow rate, ρ for the density of the mobile phase, c_p for the specific heat capacity of the mobile phase, ΔT_z for a maximum temperature difference between the mobile phase and the column wall within the respective column cross section in axial z-position, r for the column radius, λ for the thermal conductivity of the mobile phase, and ε for the porosity of the column packing.

Radial temperature gradients are shown in Figure 2.2.4 by black characteristics. If such a characteristic is straight, the same temperature exists in this column cross section. If such a characteristic is of parabolic shape as shown in Figure 2.2.4c, the column center is at a lower temperature compared to the column wall within a part of the included column cross sections. This is caused when the incoming mobile phase has a lower temperature compared to the column. Moreover, these characteristics show the linear velocity distribution in a defined column cross section. The linear velocity depends on the viscosity of the mobile phase while the viscosity, in turn, is highly dependent on the temperature. In Figure 2.2.4c, radial temperature gradients lead to parabolic curves with same linear velocity. Due to temperature differences, a different flow velocity is formed in the column center of this examined column cross section than in the column wall region. As a result, the linear velocity distribution in this column cross section is similar to an inverse-laminar flow profile. Contrary to Figure 2.2.4c, a laminar flow profile is caused when the incoming mobile phase has a higher temperature compared to the column. Such a temperature-dependent flow rate distribution is superimposed by the general laminar flow profile, which is always induced when mobile phase is pumped through a packed column. In each case, peak broadening results from radial temperature gradients, which is perceived as a peak broadening in a detector.

The larger the extent of isothermal column operation, the more pronounced the radial temperature gradients, which are formed when incoming mobile phase and column thermostat temperature do not match (cf. Figure 2.2.4c as an extreme example). At the same time, the mean temperature within the column changes continuously. Heat from the column wall moves into the column center if a cooler mobile phase enters a warmer, isothermally operated column. Less common, heat moves from the column center to the column wall if a warmer mobile phase enters a cooler, isothermally operated column. In both cases, a resulting radial temperature gradient is continuously transported by the mobile phase in the flow direction. Hence, an inhomogeneous, axial temperature gradient from the column inlet to the column outlet is formed as well. The magnitude of those radial temperature gradients and the average axial temperature gradient depend on the column diameter, flow rate and heat conducting properties of the mobile phase (cf. Formula 2.2.1).

With ideal-adiabatic column operation, the entire column takes the temperature of the mobile phase because the column wall is not maintained at the column thermostat temperature (Figure 2.2.4d). In reality, this case does not happen, since a column is not isolated in a column thermostat but is always thermostatted to some extent. A still-air thermostat approximates the ideal-adiabatic condition very well. However, a heat flux into the column is not completely suppressed since a stainless steel column acts as a kind of heat exchanger. Hence, a column wall is operated in an adiabatic manner but never in an ideal-adiabatic manner. A temperature difference between the mobile phase and column wall, however, is always much smaller than it would be with an isothermal column thermostat. The column wall is in a delayed and indirect heat contact with the heater of the adiabatic column thermostat. The mobile phase flows continuously into a column at its own temperature and thus dominates the heat introduction to the column. In general, only weakly pronounced radial temperature gradients are formed in the column wall regions. When the mobile phase and column have the same temperature, no temperature gradients are formed inside a column (Figure 2.2.4a,b) with both adiabatic and isothermal column thermostat – as long as no frictional heating occurs.

The impact of temperature differences between the mobile phase and the column on a UHPLC separation result is illustrated in Figures 2.2.5 and 2.2.6. If a forced-air thermostat is used and the mobile phase is not preheated to the column temperature, separation efficiency gets lost. Due to this, the critical peak pair merges into a double peak (see Figure 2.2.5). When a passive preheater is

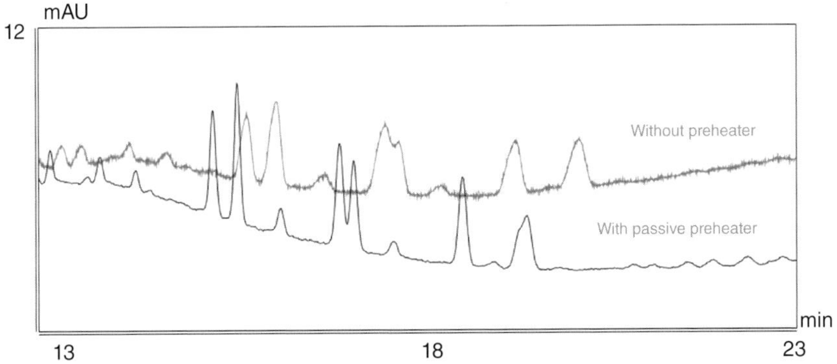

Figure 2.2.5 The same UHPLC separation with passive preheater or without preheater in a forced-air thermostat at 40 °C.

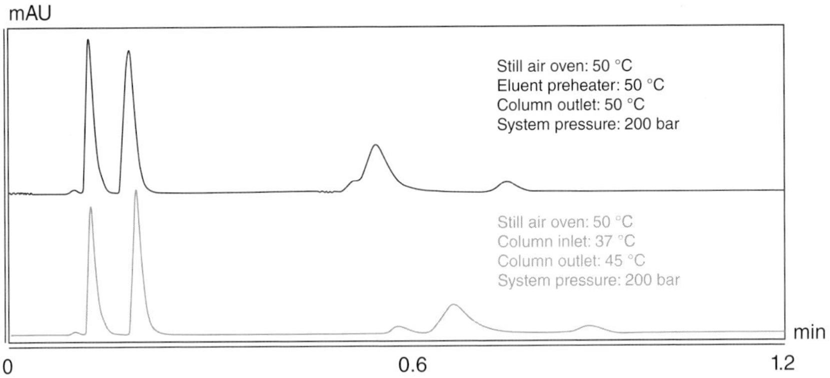

Figure 2.2.6 The same isocratic UHPLC separation in a still-air thermostat with active preheater and without preheater.

used instead, all peaks become narrower. Thereby, a better separation performance and a much higher resolution are achieved. Then, even the critical peak pair is resolved into two peaks, which can be reliably integrated (see Figure 2.2.5). If a still-air thermostat is used instead, changes to retention time occur if the mobile phase is not preheated to the column temperature (see Figure 2.2.6). In the shown example, a critical peak pair is not resolved when the mobile phase is preheated to the column temperature. Without preheating the mobile phase, however, the critical peak pair is baseline separated. This effect is based on the temperature-dependent selectivity (see Figure 2.2.1). Due to differences in the average temperature inside each of both columns, the observed selectivities differ notably although the column thermostat is operated at nominal 50 °C in both cases.

In Figure 2.2.6, the temperature of the mobile phase was measured as it passed the preheater, which did not actively heat. The measured 37 °C was solely caused by a passive mobile-phase preheating effect that occurred inside the capillary housed within the column thermostat to move from the autosampler to the column inlet. With an assumed ambient temperature of 20 °C, the mobile-phase temperature increased by 17 °C just by passive preheating. This can be explained by the low flow rate of 300 μl min^{-1} that was used.

The temperature of the mobile phase that flowed from the column had been measured in each example. The column effluent and column thermostat temperate were equal, whether the passive preheater was used or not in the forced-air thermostat (Figure 2.2.5). While the effluent had the same temperature as the still-air thermostat when the active preheater was used, the effluent became 5 °C colder (45 °C) when the preheater was not actively used (Figure 2.2.6). In a rough estimation, the average separation temperature inside the column can be calculated. It was approximately 40.5 °C, which equals the average between 37 °C (inlet temperature) and 45 °C (outlet temperature). This can explain the selectivity differences between the two separations shown in Figure 2.2.6, which just resulted from turning the active preheater either on or off – something nobody would assume to make such a difference.

Temperature differences between mobile phase and column wall lead to a thermal imbalance inside a column:

- In an adiabatically operated column, the column interior adopts the mobile-phase temperature.
- In an isothermally operated column, pronounced radial temperature gradients are formed.
- In reality, a stainless steel column always acts to some extent as a heat exchanger, which results in radial temperature gradients at least within the column wall region, regardless of the thermal operation mode of the column thermostat.
- However, the radial temperature gradients in quasi-adiabatically operated columns are negligible when compared to those of quasi-isothermally operated columns.
- The higher the linear flow rate, the smaller the radial temperature gradient.
- The smaller the column diameter, the smaller the radial temperature gradient.

2.2.3 Frictional Heat – Just a Phenomenon in UHPLC?

When a column is operated at pressures of 1000 bar, the column outlet becomes perceptibly hotter than the column inlet. This noticeable difference results from frictional heat, which is generated due to the high column back pressure. If a mobile phase shall flow through a column, the flow resistance of the column packing must be overcome. The smaller the particle size of the column packing, the higher that resistance will be. The pump builds up more and more pressure, until the column resistance is equalized. Then, the mobile phase is able to flow at a desired rate through the column packing. The hydraulic power, which needs to be applied by the pump, corresponds to the product of volumetric flow rate and pump pressure. The general system back pressure without column must also be overcome by the pump pressure, but it leaves no frictional heat in the column. Such a column-free system back pressure accounts for up to 10% of the entire pump pressure and should, therefore, be considered for the estimation of frictional heat values.

The hydraulic power, which is needed to overcome the column back pressure, is proportional to the reciprocal value of the particle diameter raised to the square. During the pressurization of the mobile phase, its volume is compressed. This compression work converts into heat, which dissipates to the environment before the mobile phase has reached the column inlet. However, the compressed mobile phase is forced to flow through the column, which leads to a certain frictional force between the mobile phase and the column packing. The hydraulic power, which is needed to overcome the column back pressure, is consumed by these frictional forces and, thereby, converted into frictional heat.

As soon as the mobile phase flows through a column, its volume re-expands, and natural decompression cooling occurs. Approximately 30% of the simultaneously generated frictional heat is directly compensated by this emerging decompression cooling effect. The remaining frictional heat is distributed in all directions within the column by thermal conduction. At the same time, frictional heat is transported by the mobile-phase flow toward the column outlet, which creates axial temperature gradient. Depending on how the frictional heat distribution is influenced by the thermal mode of the column thermostat, radial temperature gradients may result in addition.

The amount of total frictional heat ΔT_{axial} that is generated depends on the specific heat capacity C_{p} of the mobile phase and the column back pressure ΔP. ΔT_{axial} can be estimated with Formula 2.2.2. It shall be noted once again that a displayed pump pressure is the sum of the actual column back pressure and the column-free system back pressure. However, since Formula 2.2.2 should be used for quick estimations of the maximum frictional heat amount, the overall pump pressure value can be used:

$$\Delta T_{\text{axial, frictional heat}} = \frac{\Delta P}{C_{\text{p}}} \tag{2.2.2}$$

A value that is calculated according to Formula 2.2.2 corresponds to frictional heat that is theoretically generated along an entire column. For such a calculation,

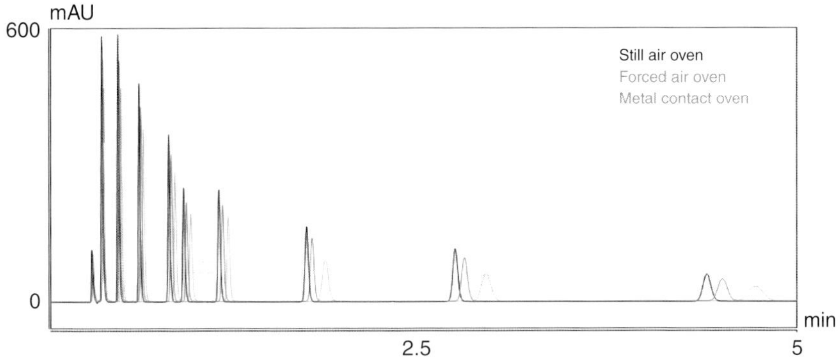

Figure 2.2.7 The same isocratic UHPLC separation operated in different column thermostats at 1000 bar.

only column back pressure and specific heat capacity of the mobile phase are decisive. The resulting value is up to 30% higher than an experimentally determined value. The main reason for this discrepancy is consumption of frictional heat by decompression cooling, which is not taken into account in Formula 2.2.2. When assuming a linear pressure drop across the entire column, however, this formula allows an easy estimation of frictional heat for each axial column cross section. For example, half of the calculated value represents the frictional heat that is produced up to the point at half the column length. Frictional heat is transported in flow direction by the volumetric flow. Since frictional heat is continuously produced from one axial cross section to the next one, it adds up constantly until it reaches the column outlet, as long as it is not dissipated through the column wall (Figure 2.2.7).

Under quasiadiabatic conditions, a column wall is not maintained at thermostat temperature. Frictional heat is not dissipated through the column wall, resulting in an axial temperature gradient. A value that is calculated by Formula 2.2.2 corresponds to the entire temperature increase from the column inlet to the column outlet and, hence, the axial temperature gradient. If this value is halved, not only the axial temperature increase up to the point at half the column in length is determined but also the average separation temperature within an adiabatic operated column.

Under isothermal conditions, frictional heat is dissipated through the column wall to the environment. Thereby, a radial heat flow from the column center to the column wall is established. Due to a continuous flow of mobile phase from the column inlet to the column outlet, frictional heat that was generated in the column center does not have enough time to completely move to the column wall of the same column cross section where the frictional heat was generated and a radial temperature gradient was formed. The higher the flow rate, the more pronounced the radial temperature. Residual frictional heat continuously adds up when it is transported from the column inlet to the column outlet by the mobile-phase flow. This results in several axial temperature gradients, which correlate to the radial temperature gradients of each column cross section. An axial temperature gradient in the column center is much more pronounced in comparison to one in

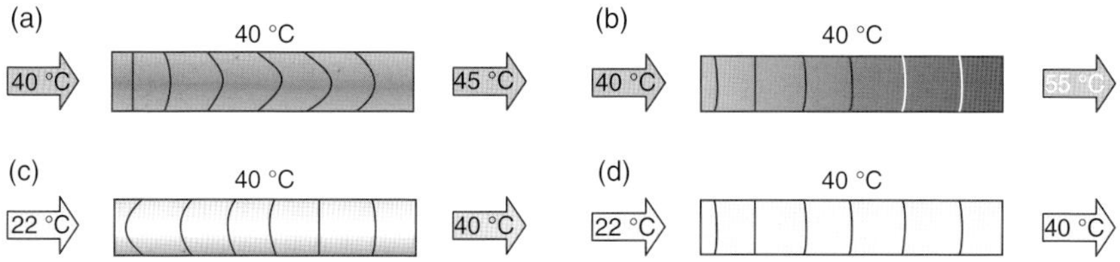

Figure 2.2.8 Temperature profiles, which occur inside a column that is placed into an ideal-isothermal (left) or ideal-adiabatic (right) column thermostat, further depending on the incoming mobile-phase temperature and the emerging frictional heat due to flow resistance.

the column wall region. The column center becomes, consequently, warmer than the column wall. A radial temperature gradient can be estimated according to Formula 2.2.3:

$$\Delta T_{\text{radial, frictional heat}} = \frac{u \cdot \Delta P_z \cdot r^2}{Z \cdot 4 \cdot \lambda \cdot \varepsilon} \quad (2.2.3)$$

where u is the linear flow rate, ΔP_z the column back pressure P at the axial column position Z, r^2 the square of the column radius r, λ the thermal conductivity, and ε the column permeability.

Figure 2.2.8 presents hypothetical temperature distributions inside a column that is placed into an ideal-isothermal (left) or ideal-adiabatic (right) column thermostat and with or without mobile-phase preheating. In an isothermally operated column, a large portion of the frictional heat is dissipated through the column wall, whereby radial temperature gradients arise. In the column center, temperature increases because a part of the frictional heat is not released via the column wall but is transported further by the volumetric flow of the mobile phase (Figure 2.2.8a). In Figure 2.2.8c, the mobile phase of lower temperature compared to the column wall flows into the column. The mobile phase is heated up at the column wall only. As a consequence, the mobile phase in the column center has a lower temperature than at the column wall. A thermal imbalance is formed (see Section 2.2.2), which counteracts the frictional heating effect. Although frictional heat is quickly dissipated at an isothermally operated column wall, it remains in the respective column center. This leads to a temperature increase in the column center (as in Figure 2.2.8a). If cooler mobile phase enters the column, this mobile phase is heated up by frictional heat in the column center and, much faster, in the column wall region by the isothermally operated column wall.

The radial temperatures are shown in Figure 2.2.8 by black, partly white characteristics. If such a characteristic is straight, the same temperature exists in the entire column cross section. If the mobile-phase temperature in the column center is higher than that at the column wall, the arc of the parabolic curve shows into the column outlet direction. If the parabolic-shaped characteristic points to the column inlet direction, the mobile-phase temperature in the column center is lower than that at the column wall. In Figure 2.2.8c, the entering mobile phase is cooler than the column wall, and a thermal imbalance is formed inside the column. The amplitude of the parabolic arc represents the strength of the temperature difference. Thus, a radial temperature gradient leads to a parabolic curve that represents a region of the same temperature. This parabolic curve also reflects the temperature-dependent mobile-phase viscosity and, hence, the viscosity-dependent laminar flow profile. This is superimposed by the general flow profile of the sheer pumping of mobile phase through the column (cf. Hagen–Poiseuille equation). Band broadening results, which is perceived as a peak broadening in the detector.

The more isothermal the column thermostat, the more pronounced radial temperature gradients result. Frictional heat that is not radially dissipated forms several axial temperature gradients. In an adiabatically operated column, a strong axial temperature is formed because no frictional heat is removed via the column wall; this, in return, eliminates radial temperature gradients, but frictional heat accumulates until the column outlet. If a mobile phase at a temperature lower than the thermostat temperature enters a column, an adiabatically operated column adopts the mobile-phase temperature. With an isothermal operated column, heat flows from the column wall to the column center, which forms again a radial temperature gradient. In the presence of frictional heat, an axial temperature gradient is additionally formed in an adiabatic column. With an isothermally operated column, frictional heating leads to radial temperature gradients and a small axial temperature gradient.

With conventional HPLC, operating pressures above 400 bar are very rare as the design of most HPLC hardware does not allow for higher pressures. Frictional heat arises, if at all, in a fairly negligible amount only. Nevertheless, it may make a difference whether the separation was carried out in a still-air or forced-air thermostat (Figure 2.2.9). Although the mobile phase was preheated to column thermostat temperature in both cases, significant differences between the chromatograms are observable. Dark gray parts mark sections of the chromatogram where the use of a forced-air thermostat yielded a better resolution, while light gray shades highlight those sections where better resolution was achieved by using a still-air thermostat. The examples in Figure 2.2.9 show that frictional heat is also produced at 400 bar. In direct comparison, retention times of distinctive peaks were shorter with the still-air thermostat, which indicates a higher effective separation temperature. Since in both cases the column thermostat and

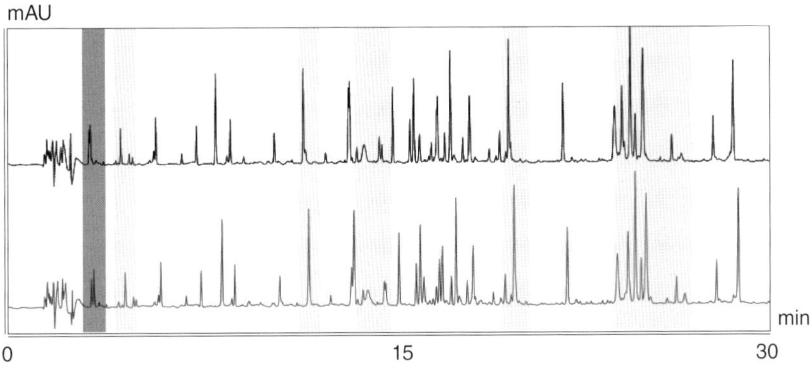

Figure 2.2.9 The same HPLC separation in a still-air (on top) and forced-air thermostat (bottom).

preheater temperatures were set to 80 °C, these retention time differences indicate the presence of frictional heat. With a forced-air thermostat, frictional heat was dissipated. With the still-air thermostat, in turn, it stayed inside the column. Nowadays, pressures of up to 1500 bar generate substantial frictional heat, and the interaction between frictional heat and column thermostat type gains importance.

Back in 1980, a simplified calculation model predicted already that radial or axial temperature gradients can arise inside a column at 400 bar column back pressure dependent on the thermal operation mode of the column thermostat. It was calculated that at a column back pressure of 1000 bar, an axial temperature increase of 34 °C is calculated with 100% acetonitrile as mobile phase, 30 °C with methanol, and 21 °C with water. Nowadays, more complex computational models, which include expanded equations , for example, for the heat transport, are used to calculate temperature gradients inside a column. Moreover, temperature and pressure dependencies of the mobile-phase viscosity, density, and heat capacity are taken into account. These parameters are recalculated recursively as they dynamically change from one column cross section to the next due to frictional heating. While the real heat flow situation inside a column can be simulated well with such a complex calculation model, the raw prediction of the temperature profile inside a column by the fairly simplified approximations of Formula 2.2.2 is typically sufficient for everyday lab work.

In UHPLC, commonly used columns have an internal diameter of 2.1 mm and are 50–100 mm long. The typical column packings are made from particles with a diameter of 1.7 or below up to 2.2 μm. The amount of frictional heat that results from a certain mobile-phase composition being pressed at 1000 bar through such a typical UHPLC column back pressure was calculated. Based on this, radial and axial temperature gradients were calculated for either a still-air thermostat or in a water bath. For the still-air thermostat, a radial temperature gradient of less than 3 °C and only in the region near the column wall and an axial temperature gradient of 20 °C were calculated. With the simple Formula 2.2.2, a frictional heat amount of 28 °C is calculated for the same conditions. This is, nevertheless, a good match between both results if a consumption of 30% by the decompression cooling is taken into account. For the same method conditions, a radial temperature gradient of 6 °C and, exclusively in the column center, an axial temperature gradient of 5 °C were calculated when the use of a water bath was simulated. Finally, all the calculated axial temperature increases were experimentally confirmed. For back pressures from 40 to 1000 bar, good correlations were found between the calculated prediction and the corresponding experimental result.

With expanded equations for the adsorption rate of a solute in porous and aqueous media as well as for the heat transfer, mass transfer, and flow rate, the band broadening or the concentration profile of a substance when flowing through a packed column was calculated. The temperature and pressure dependencies of the viscosity, density, heat capacity, and flow rate of the mobile phase were taken into account and calculated in a progressive manner. This approach allows for a very good prediction of a concentration profile inside a column. For isothermally operated columns, next to the ordinary band broadening, which is always caused by a laminar flow profile of the mobile phase, additional band broadening was calculated as a result of radial temperature gradients. This band broadening is detected as peak broadening, which results in a loss of separation efficiency.

For adiabatically operated columns, no additional peak broadening was determined. However, a strong axial temperature is formed by frictional heat. As a result, the mobile-phase viscosity decreases, and, in turn, the flow rate continuously increases from the column inlet to column outlet. Thermodynamic and kinetic retention processes vary in the axial direction of a column. However, same separation conditions are established in each column cross section because no radial temperature gradients are present within the column. Due to the axial temperature gradient, both retention factors and local mobile-phase linear velocities dynamically change for each column cross section.

Those calculations were experimentally verified. In a chromatogram, the peaks of an LC analysis were symmetric while using a still-air thermostat. However, the same LC analysis resulted in broadened, asymmetric peaks when a water bath was used instead, which also led to a separation efficiency loss. Generally, the isothermal and adiabatic modes have different effects on chromatographic results. In an isothermally operated column, the temperature inside the column increases very slightly because frictional heat is removed via the column wall. The smaller the column diameter and the lower the flow rate of the mobile phase, the more effectively the frictional heat is dissipated and, consequently, the smaller the radial temperature gradient within the column. The set column thermostat temperature and the actual average separation temperature within the column correlate closely to each other. As a result, retention factors remain constant.

In Figure 2.2.10, temperature profiles (black curves), flow rates (dotted gray curves), and the concentration profiles (in the form of peaks of certain width and height) are shown for different thermal conditions. If the temperature distribution inside the column is homogeneous, each column cross section offers the same flow rate. The concentration profile widens solely on the basis of general chromatography theory (cf. van Deemter equation). Should the column temperature increase homogeneously in the axial direction, the flow rate also increases homogeneously from one column cross section to the next. The ordinary widening of the concentration profile decreases. The retention time is shorter in comparison to the one observed with a homogeneous separation temperature because the average separation temperature and average flow rate are higher with the axial temperature gradient. Retention factors are no longer constant along the column. As a rule of thumb, an increase in the column back pressure from 70 to 750 bar leads to frictional heating and a decrease in retention factors of 12–15%.

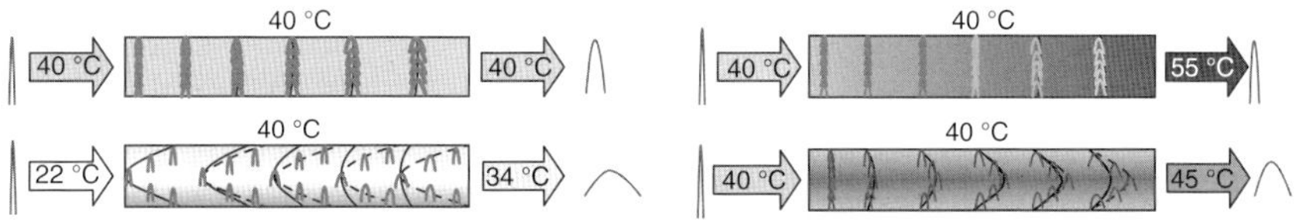

Figure 2.2.10 Concentration profiles within a still-air (on top) and forced-air thermostat (bottom).

If a mobile phase of lower temperature flows into a column of higher temperature, radial temperature gradients result. While the mobile phase is warmed up quickly at the column wall, the mobile phase in the column center remains cool. This results in a faster flow rate at the column wall, whereas the flow rate stays constant in the column center. Different flow rates in the column cross section cause band broadening. A detected substance peak is therefore much wider and flatter, and its retention time is longer than it would be at a homogeneous separation temperature and flow rate. Somehow, the same effects apply inversely when frictional heat is dissipated from a column. The flow rate in the column center increases and the flow rate at the column wall remains constant. Then, a detected substance peak is also wider and flatter, but its retention time is shorter than it would be at a homogeneous separation temperature and flow rate.

Starting at 400 bar column back pressure, frictional heat is generated in a measurable amount:

- If placed into an adiabatic column thermostat, a strong axial temperature and no radial temperature gradient are formed inside the column.
- If placed into an isothermal column thermostat, strong radial temperature gradients and a weak axial temperature are formed inside the column.
- The higher the linear flow rate, the greater the radial temperature gradient and the axial temperature gradient within the center of a column, which is placed into an isothermal column thermostat.
- The smaller the column diameter, the smaller the radial temperature gradient and the axial temperature gradient in the center of a column, which is placed into an isothermal column thermostat.
- Radial temperature gradients lead to a loss in separation efficiency.
- With radial temperature gradients, frictional heat is dissipated from the column, and the separation temperature rises slightly inside the column. Retention factors and selectivities are well scalable between HPLC and UHPLC conditions.
- Axial temperature gradients do not lead to any appreciable loss in separation efficiency.
- With an axial temperature gradient, the separation temperature inside a column increases from the column inlet to the column outlet. Retention factors and selectivities are not scalable between HPLC and UHPLC conditions.

2.2.4 Thermostatic Control in Method Transfer, Method Speed-Up, and Method Development

Even many skilled HPLC users are not entirely aware of temperature-related problems that may occur in a method transfer from one HPLC system to another. If the chromatograms are not very similar, a temperature-related problem is at least not the first reason being thought of. For example, a different extracolumn volume is quickly blamed when the separation efficiency differs too much between both HPLC systems. However, the extracolumn volume of both HPLC systems must be enormously different if significant changes in separation efficiency are observed with a typical HPLC column (4.6 mm × 100 mm, 5 μm).

Differences in gradient delay volume are quickly brought into play when retention times change systematically. However, this does not explain cases where retention times change in varying degrees to each other. If an HPLC method is transferred from one to another equipment provider, the thermal operation modes of the respective column thermostats can be completely different. This causes different results, especially when operating at separation temperatures above 30 °C.

Between the mobile phase and column thermostat temperature, a maximum difference of 6 °C should not be exceeded. This prevents separation efficiency losses due to radial temperature gradients. If a method is transferred between different HPLC systems, an effective eluent preheating should always be ensured. Most suitable is an active preheater. A passive preheater can also be used if a sufficient heat exchanger volume is suited to the applied flow rate. However, a resulting extracolumn volume of the passive preheater may also adversely affect the chromatographic result. Before using a passive preheater at high flow rates, it should be tested whether the additional extracolumn volume or radial temperature gradients have a bigger negative effect on the chromatographic separation. At system pressures below 400 bar, no appreciable frictional heat occurs. Then, using a metal contact or forced-air thermostat for temperature control is advantageous. Those thermostat types heat the column quickly to the desired temperature, and passive mobile phase preheating is much more effective than with a still-air thermostat.

In theory, a speed-up from HPLC to UHPLC conditions is easy and supported by computer programs or chromatography software. Based on the column geometry, the injection volume, flow rate, and, if necessary, solvent gradient are automatically calculated. The general aim of a method speed-up to UHPLC conditions is that same retention factors, selectivities, and similar separation efficiencies are obtained in a much shorter time compared to those previously reached with the prolonged HPLC method. A general approach is the use of smaller particles of an identical column chemistry. In practice, however, changes with respect to retention and selectivity factors are often observed. Furthermore, the separation efficiency is not always as high as predicted by theory.

With a speed-up of an HPLC method to UHPLC conditions, not using a mobile phase preheater alters the results unexpectedly. As an example, if a flow rate of 1 ml min^{-1} was used with a 4.6 mm × 150 mm, 5 μm HPLC column, a flow rate of 0.4 ml min^{-1} is used after migrating to a 2.1 mm × 100 mm, 1.9 μm UHPLC column. As a consequence of these different flow rates, the passive mobile-phase preheating is differently effective as long as the capillary is used for it. The mobile phase is heated more strongly in the UHPLC method and probably achieves the column thermostat temperature. In the HPLC method, the flow rate is much higher, and thereby, the mobile phase cannot adopt the column thermostat temperature most likely. Thus, while radial temperature gradients occur in the HPLC column as a result of a thermal imbalance between mobile phase and column temperature, such radial temperature gradients are at least strongly minimized within the UHPLC column. Hence, when switching to the UHPLC method, often a higher-than-expected separation efficiency results, but retention factors and selectivities tend to change.

However, these effects mainly depend on the column thermostat type. If the column thermostat of the HPLC system is a forced-air thermostat, passive mobile-phase preheating is more efficient than that in a still-air thermostat. How frictional heat is dissipated also depends on the respective operating mode of the column thermostat. If radial temperature gradients result, a loss in separation performance is caused. Various technical designs of column compartments are available to minimize frictional heating effects.

If a shorter analysis time and constant retention factors and selectivities are the desired goals of a method speed-up, then the use of a forced-air thermostat with active mobile-phase preheating is recommended. The diameter of the UHPLC column should be chosen to be as small as possible with a consideration of the extracolumn volume of the UHPLC system at the same time. When the highest possible separation efficiency is the goal of a method speed-up, and any change to the retention times and selectivities can be tolerated, then the use of a still-air thermostat with active mobile-phase preheating is recommended. However, both approaches limit the users in the sense that they have to choose either highest possible separation efficiency or constant retention factors.

There are also approaches that promise both maximum separation efficiencies and consistent retention factors. For example, several short UHPLC columns can be used with intermediate heat exchangers in series instead of a UHPLC column in full length. These can be connected to the desired length of the column. Possible additional dead volumes through capillary connections should, however, be kept in mind. This approach affects both isothermal and adiabatic UHPLC columns, since frictional heat is "gently" withdrawn and can form neither strong radial nor powerful axial temperature gradients. This approach is very simple, since only column hardware is changed. Frictional heating effects do not occur, and the choice of the column thermostat is not important anymore. For routine analysis laboratories, however, this column segment approach is less feasible.

In a method speed-up, the column back pressure of the targeted UHPLC method can be calculated, and by the use of Formula 2.2.2, the amount of frictional heat can be estimated. This calculated value corresponds to the axial temperature gradient in an adiabatic column or to the temperature value at column outlet. By halving this value, the average temperature increase is calculated. The average temperature inside the column determines the retention and selectivity factors, not the axial temperature gradient as such. In order to maintain the mean temperature between the HPLC and UHPLC methods, the average temperature increase by the frictional heating can be subtracted from the column thermostat temperature. This results in both maximum separation efficiencies and consistent retention factors with respect to the HPLC method. Starting at 100 bar and 30 °C, the same retention factors at 300 bar and 26 °C, 600 bar and 22 °C, or 1000 bar and 14 °C were obtained. This approach works well in still-air and forced-air thermostats, although higher separation efficiencies are achieved in a still-air thermostat. In a forced-air thermostat, radial temperature gradients occur, which have the same effect as mentioned before.

However, in a method using a mobile-phase gradient, the resulting frictional heat does not stay constant as soon as the mobile-phase composition changes. With a column back pressure of 1000 bar, an axial temperature increase of 34 °C is generated when 100% acetonitrile is used, 30 °C with 100% methanol, and 21 °C with 100% water. Hence, the frictional heat can fluctuate by up to 10 °C. Temperature fluctuations affect both the precision of retention time measurements and the measured separation efficiency. Temperature fluctuations lead to baseline noise because they occur during the analysis, and the temperature in the detector cell fluctuates as well. With steep solvent gradients, the equilibration time might be too short for the UHPLC column to reach a constant temperature in a steady state. It is, therefore, advisable to extend the equilibration of the UHPLC method to the point of time when a constant system back pressure has been achieved.

Literature

Excerpts of the following literature formed the basis of considerations and remarks in this chapter and are recommended to deepen information about the chapter subject.

1 Broeckhoven, K., Billen, J., Verstraeten, M., Choikhet, K., Dittmann, M., Rozing, G., and Desmet, G. (2010) *J. Chromatogr. A*, **1217**, 2022–2031.

2 Colon, L.A., Cintron, J.M., Anspach, J.A., Fermier, A.M., and Swinney, K.A. (2004) *Analyst*, **129**, 503–504.

3 Debrus, B., Rozet, E., Hubert, P., Veuthy, J.-L., Rudaz, S., and Guillarme, D. (2012) Method transfer between conventional HPLC and UHPLC, in *UHPLC in Life Sciences*, RSC Chromatography Monographs, vol. **16**, Royal Society of Chemistry.

4 Fallas, M.M., Hadley, M.R., and McCalley, D.V. (2009) *J. Chromatogr. A*, **1216**, 3961–3969.

5 Gritti, F. and Guiochon, G. (2007) *J. Chromatogr. A*, **1166**, 47–60.

6 Gritti, F. and Guiochon, G. (2008) *Anal. Chem.*, **80**, 6488–6499.

7 Gritti, F. and Guiochon, G. (2010) *Chem. Eng. Sci.*, **65**, 6310–6319.

8 Gritti, F. and Guiochon, G. (2013a) *J. Chromatogr. A*, **1289**, 1–12.

9 Gritti, F. and Guiochon, G. (2013b) *J. Chromatogr. A*, **1291**, 104–113.

10 Kaczmarski, K., Gritti, F., Kostka, J., and Guiochon, G. (2009a) *J. Chromatogr. A*, **1216**, 6560–6574.

11 Kaczmarski, K., Gritti, F., Kostka, J., and Guiochon, G. (2009b) *J. Chromatogr. A*, **1216**, 6575–6586.

12 Kostka, J., Gritti, F., Guiochon, G., and Kaczmarski, K. (2010) *J. Chromatogr. A*, **1217**, 4704–4712.

13 Neue, U.D., Kele, M., Bunner, B., Kromidas, A., Dourdeville, T., Mazzeo, J.R., Grumbach, E.S., Serpa, S., Wheat, T.E., Hong, P., and Gilar, M. (2009) Ultra-performance liquid chromatography technology and applications, in *Advances in Chromatography*, vol. **48**, CRC Press.

14 Nováková, L., Veuthey, J.-L., and Guillarme, D. (2011) *J. Chromatogr. A*, **1218**, 7971–7981.

15 Poppe, H., Kraak, J.C., Huber, J.F.K., and van den Berg, J.H.M. (1981) *Chromatographia*, **14**, 515–523.

16 Stankovich, J.J., Gritti, F., Stevenson, P.G., Beaver, L.A., and Guiochon, G. (2014) *J. Chromatogr. A*, **1325**, 99–108.
17 Wolcott, R.G., Dolan, J.W., Snyder, L.R., Bakalyar, S.R., Arnold, M.A., and Nichols, J.A. (2000) *J. Chromatogr. A*, **869**, 211–230.
18 Wu, N., Welch, C.J., Natishan, T.K., Gao, H., Chandrasekaran, T., and Zhang, L. (2013) Practical aspects of ultrahigh performance liquid chromatography, in *Ultra-High Performance Liquid Chromatography and Its Applications*, John Wiley & Sons, Inc.

3

The Issue of External Band Broadening in HPLC/UHPLC Devices

Monika Dittmann

3.1 Introduction

The separation power of modern (U)HPLC systems has dramatically increased over the past 10 years, enabling fast separation of complex mixtures. This development has been driven by two main factors:

1) Improvements in column technology delivering higher column plate numbers in shorter time
2) Next-generation HPLC instrumentation meeting enhanced requirements for the operation of high-efficiency columns.

The new column technologies comprise totally porous sub 2 μm diameter particles and particles that are superficially porous with 1–5 μm in diameter. Columns of 50 mm length packed with such particles can easily generate 10.000–15.000 plates in a few minutes [1–3].

HPLC instruments for the operation of such columns at high efficiency need to be capable of delivering solvents at ultrahigh pressure (>1000 bar) while conserving the column efficiency by minimization of extracolumn band broadening. The term *UHPLC* is nowadays generally used for the combination of columns and instruments capable of delivering high performance at high operating pressure.

Along with the development in UHPLC column technology, there is a trend to reduce the inner diameter of (U)HPLC columns from 4.6 to 2.1 mm or even 1 mm. The driving factors for the decrease in the internal column diameter can be summarized as follows:

Minimized impact of frictional heating: It has been shown that frictional heating at high operating pressures leads to axial and radial temperature gradients resulting in heterogeneous distribution of eluent velocity [4–6]. In the case of nonadiabatic column operation, this can seriously affect column efficiency. The resulting loss in column performance can be minimized by decreasing the column inner diameter.

Reduced sample dilution: When only limited amount of sample is available, a gain in sensitivity is obtained by decreasing the column diameter (for concentration

The HPLC Expert II: Find and Optimize the Benefits of your HPLC/UHPLC, First Edition.
Edited by Stavros Kromidas.

sensitive detectors, e.g., UV) since the smaller column volume leads to less dilution of sample components.

Improved mass spectrometric detection: In coupling UHPLC with MS detection, the maximum practical flow rate is often limited by a loss of sensitivity at higher flow rates observed for ion sources such as electrospray ionization. The flow rate dependency of the MS inlet depends on the design of the ionization sources [7–11].

Reduced solvent consumption: Since small particle columns are typically operated at higher linear velocities compared to standard HPLC columns, solvent consumption can become an issue when using large-bore (4.6 mm) columns.

The increase in column efficiency with smaller particles along with the decrease in column dimensions results in an overall decrease of peak volume in chromatographic separations. If the peak volumes become very small, the separation efficiency can become seriously compromised if the instrument contributions to zone dispersion before and after separation on the column (extracolumn band broadening) are not controlled carefully. Consequently, a clear understanding of the factors that affect extracolumn band broadening and its measurement are mandatory to allow uncompromised operation of UHPLC columns and systems.

When the contributions to dispersion are independent and uncorrelated, the total dispersion of a chromatographic system is given by

$$\sigma^2_{\mathrm{v,total}} = \sigma^2_{\mathrm{v,column}} + \sigma^2_{\mathrm{v,instrument}} \tag{3.1}$$

In Equation 3.1, the zone dispersion is described by the variance of the broadened zone in volume units.

With the increasing use of small-bore columns for UHPLC, the impact of extracolumn band broadening is gaining importance in particular in conjunction with column performance testing in isocratic mode [12–20].

In the following sections, we will discuss how the apparent peak width of a column will be affected by instrument contributions to dispersion.

3.2 Theoretical Background

3.2.1 Efficiency and Resolution of Modern UHPLC Columns

The resolution R_s between two peaks in a chromatogram is given by

$$R_s = \frac{V_{\mathrm{r},2} - V_{\mathrm{r},1}}{2 \cdot (\sigma_{\mathrm{v},1} + \sigma_{\mathrm{v},2})} \tag{3.2}$$

where $V_{\mathrm{r},i}$ is the retention volumes of a peak and $\sigma_{\mathrm{v},i}$ is the standard deviation in volume units. The resolution equation could also be written in time units, but for the investigation of extracolumn contributions to peak broadening, it is most convenient to use volumetric units.

With

$$\sigma_{\mathrm{v},1} \cong \sigma_{\mathrm{v},2} \text{ and } \sigma_{\mathrm{v}} = \frac{V_{\mathrm{r}}}{\sqrt{N}}$$

Equation 3.2 can be transformed into the well-known Purnell equation [21, 22]:

$$R_s = \frac{\sqrt{N}}{4} \cdot \left[\frac{\alpha - 1}{\alpha}\right] \cdot \left[\frac{k_2}{k_2 + 1}\right] \tag{3.3}$$

N	=	plate number
k	=	retention factor
α	=	selectivity factor k_1/k_2.

The difference in retention volumes between two peaks is a thermodynamic property and depends on the stationary phase and eluent properties as well as temperature. In this chapter, we will only discuss the peak widths generated by modern (U)HPLC columns and how they are impacted by system dispersion.

As σ is related to the plate number N of a column, we will first take a look at the basics of column efficiency for columns packed with spherical particles.

Modern analytical (U)HPLC columns are packed with spherical particles ranging in diameter between 1.3 and 5 μm. The vast majority of commercially available column packings consist of fully porous particles, but also partially or superficially porous particles have been introduced to the market in the past few years. The column dimensions range from 1 to 4.5 mm in diameter and 30 to 250 mm in length.

The band broadening induced by the chromatographic process is essentially an interplay of three mainly random processes: (i) the molecular diffusion that takes place in the fluid phase used as the eluent and along the adsorbent surface on which molecules are adsorbed; (ii) the mass transfer kinetics that govern the speed at which equilibrium between the mobile and the stationary phase is achieved; and (iii) the dispersion that originates from irregular velocity distribution in the space between the particles [23–25].

For the purpose of this book, we will use the van Deemter equation as most experimental data can be fitted to this equation.

$$H = A \cdot d_p + \frac{B \cdot D_m}{u} + C \cdot \frac{d_p^2}{D_m} u \tag{3.4}$$

As advocated by Giddings [25], it is useful to write Equation 3.4 in its reduced form

$$h = A \cdot \nu + \frac{B}{\nu} + C \cdot \nu \tag{3.5}$$

with

$$\nu = \frac{u \cdot d_p}{D_m} \text{ and } h = H/d_p$$

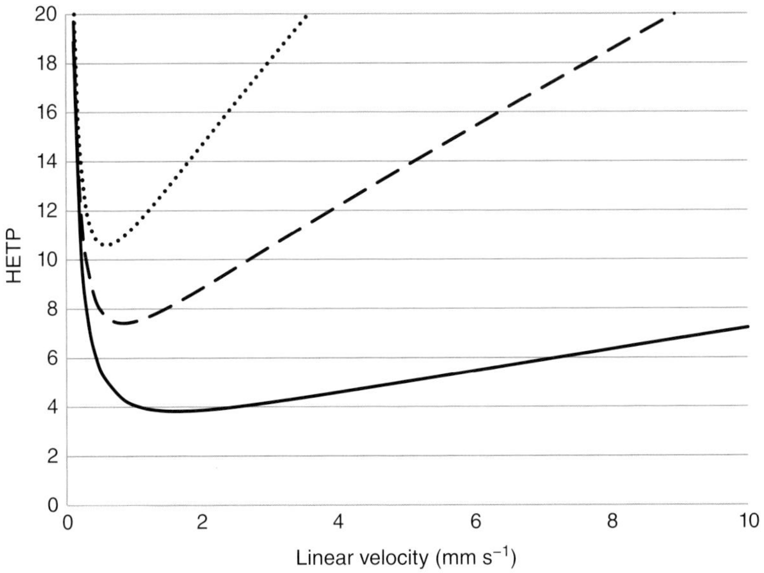

Figure 3.1 HETP values vs. linear velocity for particles of 1.8, 3.5, and 5 μm in diameter.

u	=	linear solvent velocity
d_p	=	particle diameter
D_m	=	diffusion coefficient of analyte
H	=	height equivalent to a theoretical plate
A, B, C	=	empirical parameters obtained by fitting to experimental data
h	=	reduced plate height (H/d_p)
ν	=	reduced velocity.

The parameters A, B, and C are empirical and are usually determined by fitting Equation 3.5 to experimental data. Typical values for the van Deemter parameters are $A = 1$, $B = 2$, and $C = 0.1$.

Figure 3.1 shows the typical HETP curves for fully porous particles of 1.8, 3.5, and 5 μm diameter. The resulting plate numbers for a 100 mm column around the optimum velocity are 25 000, 14 000, and 9000, respectively.

3.2.2 Estimation of Column Peak Volumes

In order to understand the requirements for modern (U)HPLC systems, we need to estimate the peak volumes eluting from a column depending on the particle size and column dimensions.

In the absence of any external contributions, the peak variance generated by a column for isocratic separations is given by

$$\sigma^2_{\mathrm{v,column}} = \frac{V_0^2}{N} \cdot (1 + k)^2 \tag{3.6}$$

with V_0 = column dead volume.

The column dead volume V_0 is related to the column dimensions (radius r and length L) and total porosity ε_T

$$V_0 = r^2 \cdot \pi \cdot L \cdot \varepsilon_T \tag{3.7}$$

In isocratic elution, the retention factor k of a peak is given by

$$k = \frac{V_r - V_0}{V_0} \tag{3.8}$$

For gradient separations, the peak volume is governed by the retention factor at the point of elution k_{elution}

$$\sigma_{\text{v,column}} = \frac{V_0}{\sqrt{N}} \cdot (1 + k_{elution}) \tag{3.9}$$

k_{elution} depends on the gradient slope b [26]:

$$k_{\text{elution}} = \frac{k_{\text{start}}}{b \cdot \left(k_{\text{start}} - (V_d/V_0)\right) + 1} \tag{3.10}$$

with b given by

$$b = S \cdot \Delta\phi \cdot \frac{V_0}{V_g} \tag{3.11}$$

V_g	=	gradient volume
V_d	=	dwell volume
Φ	=	range of eluent composition in volume fractions
S	=	slope for linear solvent strength relationship ($\ln k = \ln k_0 + S\phi$)
k_0	=	retention factor at $\phi = 0$.

For $k_{\text{start}} > 1$ and $V_d = 0$, Equation 3.10 reduces to

$$k_{\text{elution}} = \frac{1}{b} \tag{3.12}$$

In this case, the expression for the peak volume σ of a column in gradient separations can be written as

$$\sigma_{\text{v,column}} = \frac{V_0}{\sqrt{N}} \cdot \left(1 + \frac{1}{\Delta\varphi \cdot S} \cdot \frac{V_g}{V_0}\right) \tag{3.13}$$

It is important to keep in mind that in isocratic separations, the retention factor increases with increasing retention time, while the k_{elution} values in gradient separation depend mainly on the slope b of the gradient and thus on the ratio of V_0/V_g, the range of elution composition, and the S value. Figure 3.2 shows an isocratic separation and a gradient separation both completed in about 5 min. While the k values in isocratic mode range from 0 to 17, the k_{elution} values of all components (except t_0 marker) are ~4 based on the gradient conditions.

Table 3.1 gives the typical values for peak volumes for different column dimensions at k, $k_{\text{elution}} = 5$. As we can see from Table 3.1, the peak volumes produced for modern UHPLC column packed with 1.8 μm with dimensions of 2.1 mm × 50 mm are on the order of a few microliters.

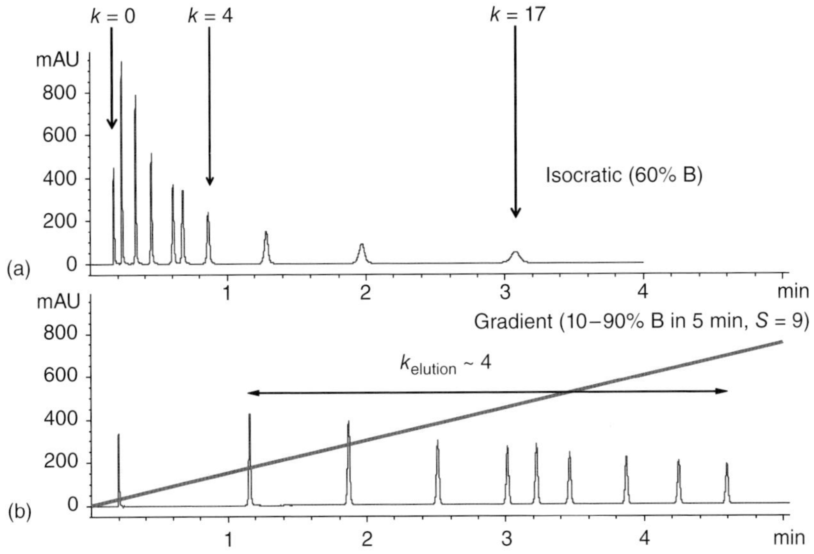

Figure 3.2 Retention factors k for isocratic (a) and k_{elution} for gradient separations (b). While in isocratic separations, each analyte has a different k value and thus a different peak width, and the k_{elution} values and thus peak width in gradient separations are similar for all analytes (they depend on the gradient slope b).

Table 3.1 Typical values for peak standard deviation in isocratic and gradient separation separations.

Column diameter (mm)	Column length (mm)	Particle diameter (μm)	Plates	V_0 (μl)	k or k_{elution}	Peak standard deviation (σ_v) (μl)
4.6	100	5	10 000	800	5	48
3	100	3.5	15 000	350	5	17
2.1	100	1.8	25 000	170	5	6
1	100	1.8	25 000	40	5	2

3.3 Extracolumn Dispersion in (U)HPLC Systems

3.3.1 Sources of External Band Broadening in HPLC/UHPLC Systems

The instrument contributions to the system can be written as the sum of the individual contributions from the system components between sample introduction and detection.

$$\sigma^2_{\text{v,instrument}} = \sigma^2_{\text{v,injector}} + \sigma^2_{\text{v,pre-column tubing}} + \sigma^2_{\text{v,post-column tubing}} + \sigma^2_{\text{v,detector}} + \tau^2_{\text{detector}} \cdot F^2 \tag{3.14}$$

In the following sections, we will investigate how the individual system components contribute to the total system dispersion.

3.3.1.1 Injection Systems

In the ideal case, when the injected sample is considered a perfectly rectangular plug, the variance of the injection is given by

$$\sigma^2_{\mathrm{v,injection}} = \frac{V^2_{\mathrm{inj}}}{12} \qquad (3.15)$$

Depending on the characteristics of the injector, the experimental variance can differ significantly from the value predicted by Equation 3.15. In particular, for small injection volumes, the additional contributions from the injector components such as valves, needle seats, and seat capillaries can become dominant and significantly contribute to overall dispersion.

In most modern (U)HPLC systems, two type of injectors can be found, namely fixed-loop (FL) designs and flow-through needle (FTN) designs (see Figure 3.3).

In the fixed-loop design, a loop with a specified volume attached to a two-position valve is filled with sample through a metering device/syringe, and the content of the loop is injected into the column. Some instruments offer the possibility to inject only a portion of the loop. In terms of dispersion, the FL design has the advantage that the sample has to pass fewer components and can have lower dispersion if very small sample loops and small injections volumes are used.

In the FTN design, the sample is drawn into a needle and, after injection, has to pass the needle seat, the seat capillary, and the valve. The FTN design offers greater flexibility in terms of sample volume, washing procedures, sample preparation, and cleanup and usually has excellent carryover specifications. Because the sample passes the needle seat, the seat capillary, and the injection valve, the contributions to dispersion are typically larger than those of an FL injector. With careful design of the needle seat and seat capillary, it is, however, possible to achieve a dispersion performance similar to that of an FL-type injector.

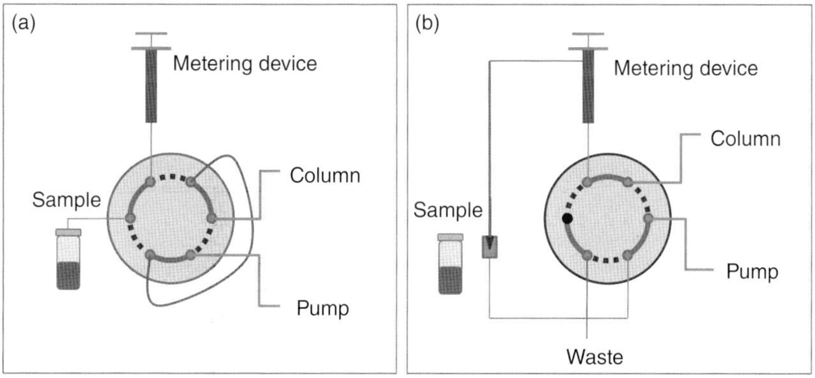

Figure 3.3 Schematic representation of a fixed-loop (FL) injector (a) and a flow-through needle injector (b).

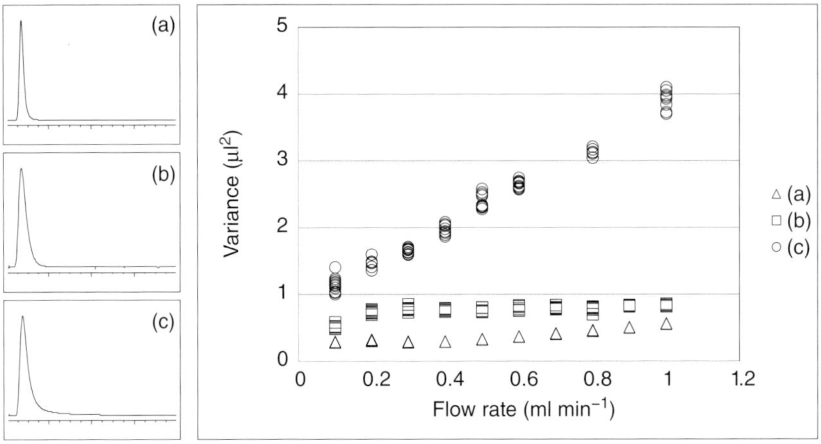

Figure 3.4 System variances (from 5 sigma width) for different injector configurations: fixed-loop (a), optimized flow-through needle (b), and standard flow-through needle (c). The injectors were connected to a 250 nl flow cell by a 0.050 mm × 340 mm stainless-steel capillary. The injection volume was 0.15 μl.

Figure 3.4 compares the experimental variances for a prototype nanovalve with an internal loop of 150 nl, a conventional FL injector with partial-loop injection on a 2 μl loop, two FTN injectors with optimized seat/seat capillary design, and an FTN injector with conventional seat/seat capillary design. In the optimized FTN designs, the internal volume of the needle seat was 0.13 μl and a seat capillary id of 75 μm compared to the seat volume of 0.36 μl and the seat capillary id of 120 μm in the conventional design.

To minimize the remaining system variance in these experiments, the injector was connected to a prototype 250 nl detection cell by a 0.05 mm × 340 mm capillary, and the injection volume was 0.15 μl. The variance contribution from the capillary connection and detection cell was estimated to be <0.15 μl^2 under the experimental conditions.

As the injected volume becomes larger, it dominates the total contributions from the injection. Figure 3.5 shows experimental variance data for different injection volumes (0.1–20 μl) compared to the theoretical value (from Equation 3.15) for three different flow rates. The data were generated using an optimized FTN injector connected to a 250 nl flow cell by a 0.050 mm × 340 mm stainless-steel capillary. The contribution from the capillary connection and detection cell was estimated to be <0.15 μl^2 under the experimental conditions.

3.3.1.2 Tubing

The variance contributed from open tubes under laminar flow conditions (parabolic flow profile) is described by the Golay equation

$$\sigma^2_{\mathrm{V,cap}} = \frac{V^2_{\mathrm{cap}}}{L} \cdot \left(\frac{2 \cdot D_{\mathrm{m}}}{u} + \frac{r^2}{24 \cdot D_{\mathrm{m}}} \cdot u \right) \tag{3.16}$$

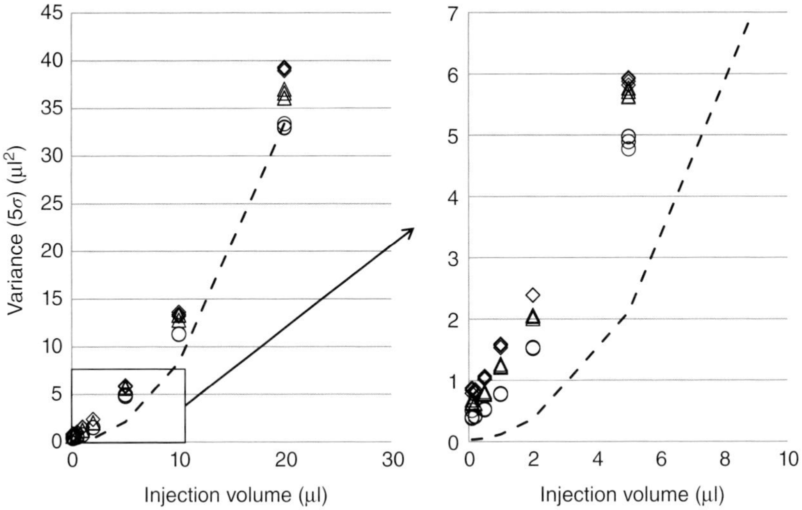

Figure 3.5 System variances (from 5 sigma width) for different injection volumes at flow rates of 0.1 (○), 0.5 (Δ), and 1 ml min^{-1} (◇). The dotted line corresponds to the theoretical values calculated from Equation 3.15.

L	=	length of tubing
r	=	radius of tubing.

As Atwood and Golay have shown in [27], Equation 3.16 is only valid for relatively long tubes, as in short tubes, the parabolic flow profile is not fully established. Several authors have shown that Equation 3.16 typically overestimates the dispersion at flow rate >0.1–0.2 ml min^{-1} [19, 28, 29].

Figure 3.6 compares experimental variances for tubes of different diameters with the theoretical values derived from the Golay equation. A prototype nanovalve with an internal loop of 150 and a 250 nl detection cell were used for this comparison, and the combined variance of the valve and the detection cell was about 0.2 μl^2. As can be seen in the figure, experimental variances are in all cases lower than predicted by the Golay equation. In contrast to the prediction, there is only a small dependence on flow rate, and also, the length dependence of the capillary variance is much smaller than that predicted.

The reduction in dispersion from capillary connection is, however, associated with an increase in pressure drop. The back pressure of a capillary of given length and diameter is given by the Hagen–Poiseuille equation.

$$\Delta P = \frac{8 \cdot \eta \cdot L}{r^4 \pi} \cdot F \tag{3.17}$$

η	=	viscosity of solvent.

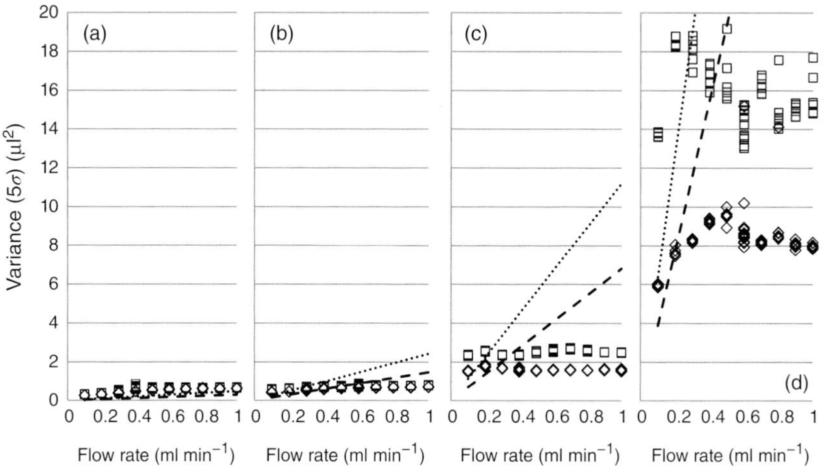

Figure 3.6 System variances (from 5 sigma width) for configurations with stainless-steel capillaries of different diameters of 0.055 mm (a), 0.070 mm (b), 0.080 mm (c), 0.095 mm (d), and 0.110 mm (e) with lengths of 250, 500, and 1000 mm. The dashed lines represent the theoretical variance derived from the Golay equation.

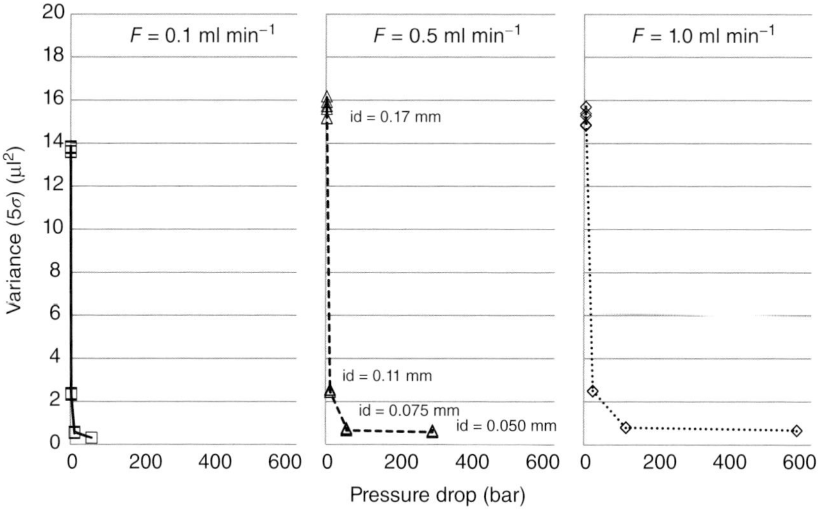

Figure 3.7 Plot of system variance versus pressure drop for capillaries with different nominal diameters and lengths of 250, 500, and 1000 mm at flow rates of 0.1, 0.5, and 1 ml min^{-1} (viscosity = 1 cp).

Figure 3.7 shows experimental variances for capillaries with nominal diameters of 0.055, 0.070, 0.08, 0.095, and 0.110 mm plotted versus the resulting back pressure for different flow rates. It becomes obvious that in the flow rate range between 0.3 and 1 ml min^{-1}, the small decrease in variance between 0.070 and 0.055 mm ID capillaries is counteracted by an increase in back pressure of almost a factor of 3. From this picture, it can be concluded that the optimum capillary id for the flow rate range between 0.3 and 1 ml min^{-1} is 70–110 μm. A decrease of the capillary id only results in higher back pressure with no significant improvement in dispersion. For flow rates of 0.1 ml or lower (operation of 1 mm columns), the use of 0.05–0.06 mm ID capillaries might be useful.

3.3.1.3 Fittings and Connections

As all components of a chromatographic system are minimized to reduce extra-column band broadening, great care has to be taken to ensure proper connection of capillaries to injectors, columns, and detection cells. As columns from different manufacturers are based on different column hardware, it is easy to create void volumes when attaching a fitting to columns from different manufacturers due to the different geometries of the receiving ports [18] (see Figure 3.8). As these connections can still be leak-tight, the resulting void volume is not always obvious to the user. Until recently, the standard fittings type for high-pressure applications have been Swagelok-type fittings that are not removable once tightened. This type of fitting should not be used interchangeably between different types of columns, injectors, or cells as there is always the danger of introducing a leak (capillary sticking out too far) or a void (capillary too short).

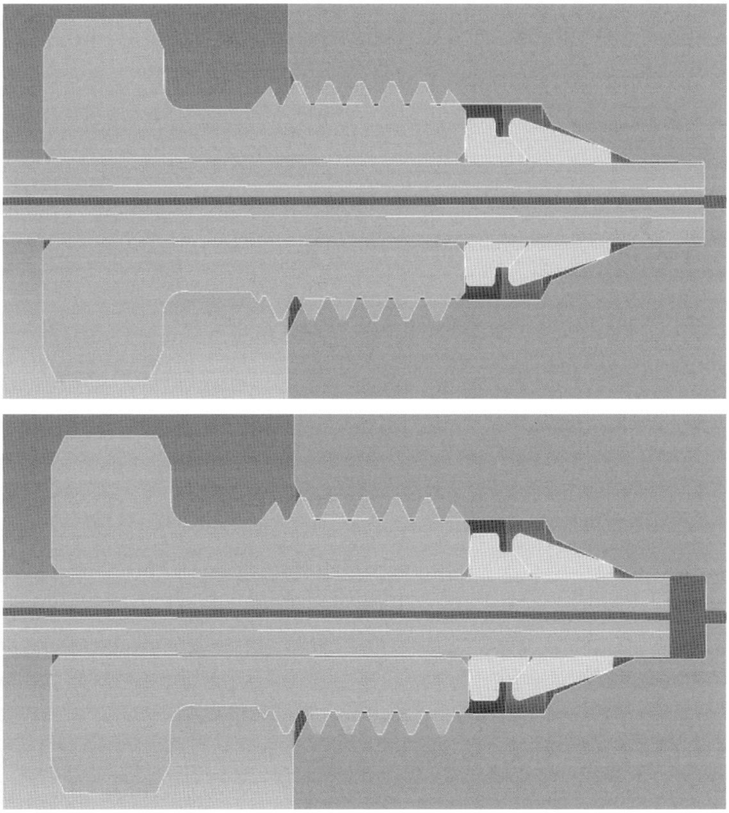

(a) (b)

Figure 3.8 Example of a void volume created by nonmatching column capillary connections.

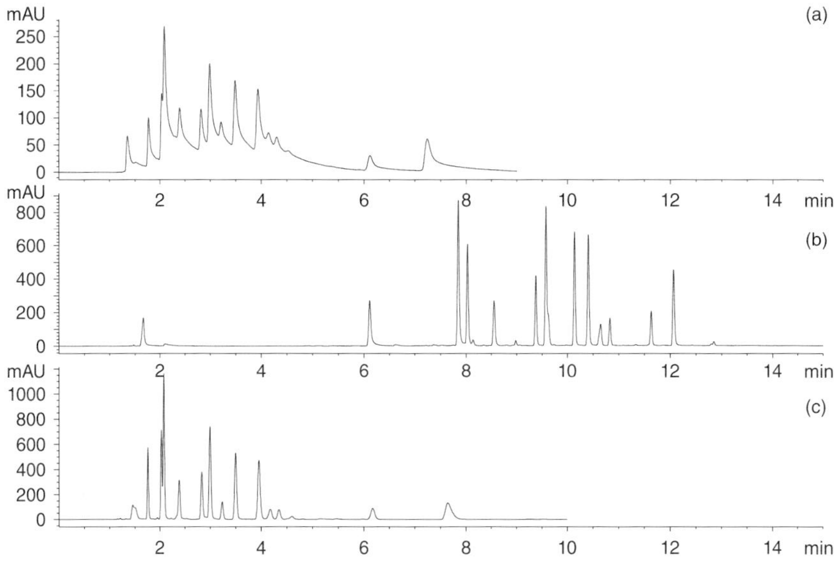

Figure 3.9 Impact of improper capillary connection at the injection valve on separations on a 1 mm column. (a) Isocratic separation with *bad connection*, (b) gradient separation with *bad connection*, and (c) isocratic separation with *proper connection*.

In the past years, removable fittings have been developed by many manufacturers, which allow proper positioning of the capillary inside the receiving port every time a connection is made [30]. However, also with these types of fittings, great care has to be taken to avoid any possible void volumes.

An example of how a badly connected capillary can destroy column performance is shown in Figure 3.9 for a 1 mm × 150 mm 1.7 μm column. In this case, the capillary between the injector and the column had a void volume at the injector valve port. The isocratic separation (Figure 3.9a) showed very broad, badly distorted peaks, while the gradient separation (Figure 3.9b) was almost unaffected due to focusing of the sample at the column entrance. After replacing the bad capillary connection, the column showed good performance in isocratic mode as well (Figure 3.9c)

3.3.1.4 Heat Exchangers

Many (U)HPLC systems are equipped with column ovens and frequently use a heat exchanger (HE) to preheat the solvent before it enters the column. These heat exchangers are either integrated into the column oven or consist of a separate part that provides heat transfer and is connected to the inlet of the column. In some instruments, the heat exchanging part is integrated into the capillary connecting the column to the autosampler. Depending on the type of column to be operated and the flow rates applied, the volume of a heat exchanger can range from about 6 μl (for flow rates >1 ml min^{-1}) down to 0.6 μl. It is not straightforward to estimate the dispersion of a heat exchanger as this depends not only on the volume but also on the geometry of the individual part. In order to determine

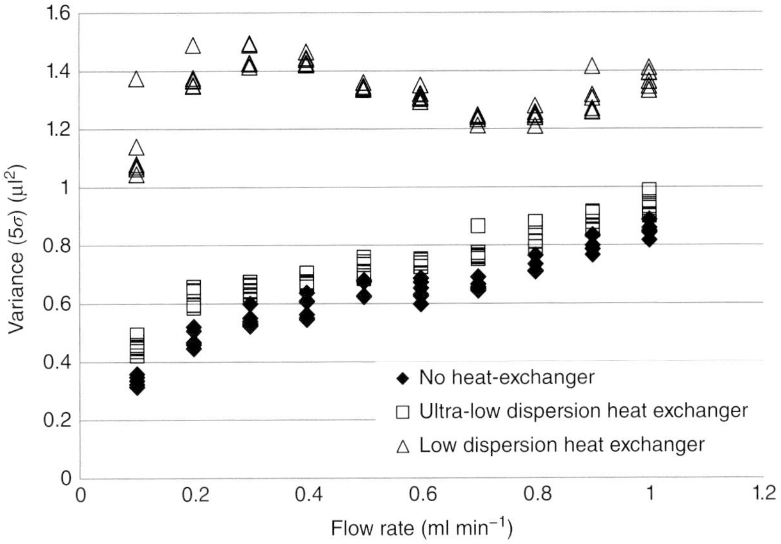

Figure 3.10 System variances (from 5 sigma width) for different heat exchangers.

the contribution of the heat exchanger to the total dispersion, it is best to measure the system variance with and without the HE.

Figure 3.10 compares the variance of a system (nanovalve + 250 nl cell) without HE, with a 0.6 μl HE and a 1.6 μl HE. The heat exchangers were connected to the injector and the cell by 0.055 mm × 250 mm capillaries. The additional variance from the 0.6 μl HE is about 0.1 μl^2 while the 1.6 μl HE contributes between 0.3 and 0.4 μl^2 depending on the flow rate.

3.3.1.5 Detection

The contributions to extracolumn band broadening from the detector can either result from volumetric contributions (volume of detection cell, volume of internal capillaries, etc.) or from time-related effects such as acquisition rate and software data filters.

3.3.1.5.1 *Optical Detectors* The theoretical variance value (perfectly rectangular plug) given for a certain detection volume is given by

$$\sigma^2_{\mathrm{v,detection}} = \frac{V^2_{\mathrm{detection}}}{12} \tag{3.18}$$

Equation 3.18, however, largely underestimates the real variance generated by a detection cell for small detection volumes due to internal cell geometry, internal connection capillaries, and so on.

Thus, the nominal cell volume can only be a rough guideline, and the true cell dispersion has to be determined experimentally.

We have measured the variance (determined from peak width at 4.4% of peak height for cells with different nominal cell volumes (Figure 3.11)). Two injection configurations were used: (i) an Agilent 1290 Infinity Multisampler

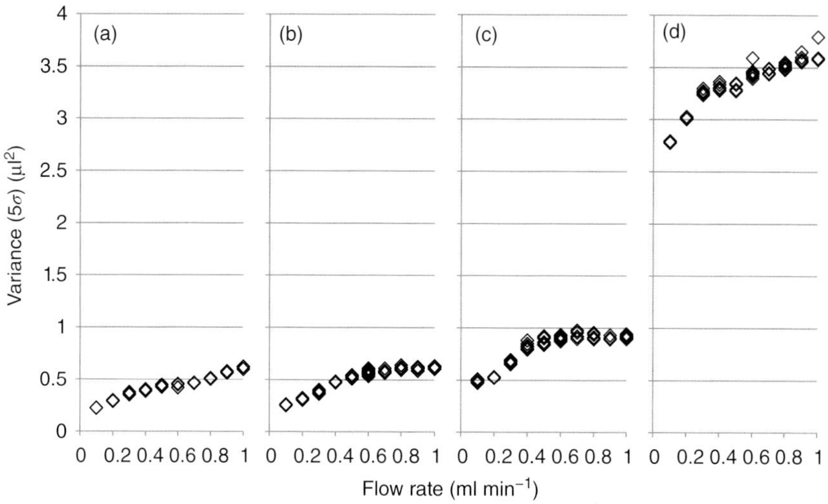

Figure 3.11 System variances (from 5 sigma width) for different detection cells. (a) 80 nl cell, (b) 250 nl cell, (c) 800 nl and (d) 2400 nl cell connected to a nanovalve with internal groove (square) and to a low-dispersion FTN injector (1290 Infinity II Multisampler) (circle).

with a low-dispersion (0.075 μm) seat capillary and (ii) a prototype nanovalve with an internal loop of 150 nl. In both cases, the detector was connected by a 0.050 mm × 250 mm capillary. The variance contribution from the nanovalve and the capillary was about 0.15 μl^2.

A reduction in cell volume is usually associated with an increase in detection noise (reduced light throughput with decreasing volume) and thus a decrease in sensitivity [31]. The detection cell should therefore be carefully selected based on the dispersion and the sensitivity requirements.

Besides the volumetric contributions, time-based factors such as acquisition rate and peak filters can contribute to peak broadening [32]. Figure 3.12 shows the changes in peak shape and baseline noise for a peak of 0.005 min width at half height as a function of the acquisition rate and filter constant. With decreasing data rate, the peak does not only become wider but also changes shape as the steep front is broadened more than the back end of the peak.

3.3.1.5.2 *Mass Spectrometers*

There have been few reports about the contributions to system dispersion coming from mass spectrometers (MS) [11, 33–35]. In [33], different MS systems with electrospray-ionization (ESI) have been compared, and the difference in dispersion has been mainly attributed to the differences in pre-MS system configuration (connection capillaries), while the mass spectrometer did not seem to have a large impact.

We have compared the variance of a highly optimized HPLC system with a UV detector (postcolumn variance = 0.2 μl) to that of an ESI-MS system in gradient mode and found no significant difference in peak width for the two detection modes for peaks with a width >0.05 min. Figure 3.13 shows peak widths obtained with UV and ESI-MS detection for columns of different diameters (all Zorbax

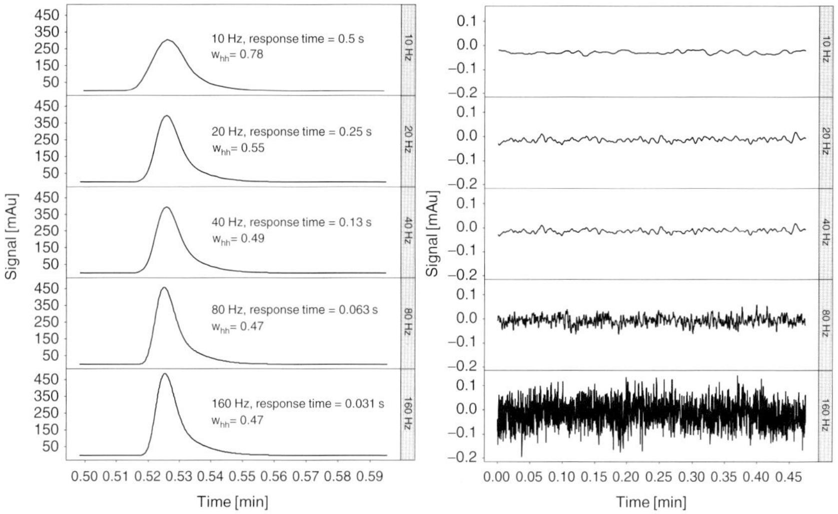

Figure 3.12 Impact of data acquisition rate and peak filter on peak width and detection noise.

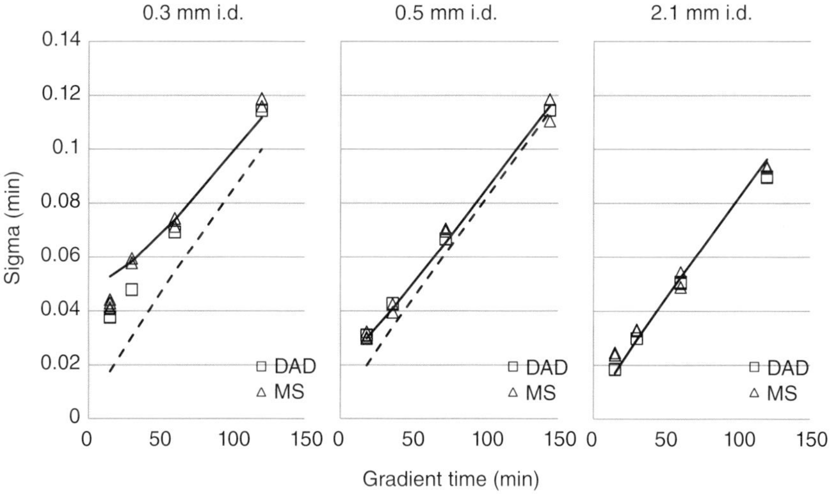

Figure 3.13 Peak widths (50% of height) on 150 mm Zorbax C18, 3.5 μm columns measured on a system with DAD-UV detection (rectangle) and a QQQ-MS with ESI–Jet Stream Interface (triangle). The lines represent the estimated peak widths (Equation 3.13) without external contributions (— — —) and for an external variance of 0.2 μl^2 (————).

C18, 3.5 μm, 150 mm). For peaks with a smaller width, a significant impact of acquisition rate and peak width filters was observed [36].

3.3.2 Determination of External Band Broadening

3.3.2.1 Analysis of Extracolumn Volume without Column (Short Circuit Method)

The most commonly used procedure to measure extracolumn dispersion of a system is to replace the column by a zero-dead volume connector and inject a small volume of a tracer solution into the system [30]. The system variance can then be calculated from the width of the resulting peak. Most chromatographic data systems provide values for the peak width at different positions of the peak (at half height, at 4.4% of the peak height (5 sigma width) and the tangent width (see Figure 3.14)). These peak widths can be converted to variance values by

$$\sigma_v^2 = \sigma_t^2 \cdot F^2 = \left(\frac{w_{hh}}{2.355}\right)^2 \cdot F^2 = \left(\frac{w_{5s}}{5}\right)^2 \cdot F^2 = \left(\frac{w_{tan}}{4}\right)^2 \cdot F^2 \tag{3.19}$$

Equation 3.20 is only valid for peaks that can be described by a Gauss function. If the peak shape deviates considerably from a Gaussian shape (tailing or fronting), the variances calculated from the different peak width measurements are not identical. In particular, the variance determined from the width at half height can differ significantly from that determined from the 5 sigma width.

The most accurate way to describe the variance of a peak is through statistical moments [37]. As the variance values determined from this method are very sensitive to the integrations settings, it is difficult to use in practice and will not be further discussed here [38].

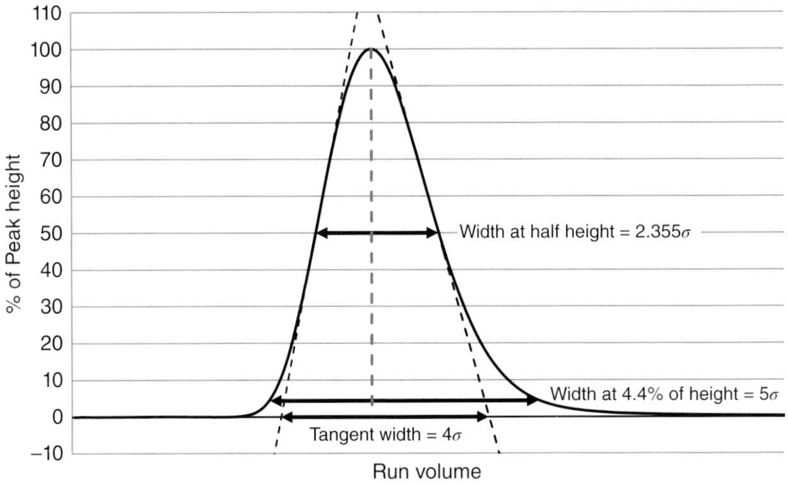

Figure 3.14 Relation between peak widths (at different heights) and standard deviation of a Gaussian peak.

As system dispersion is influenced by diffusion of the analyte used, it is important to use the same experimental conditions (analyte, eluent composition, injection volume, etc.) when comparing variances of different system configurations. It is also important to select the analyte concentration low enough to not exceed the linear range of the detector and use appropriate settings for data acquisition rate and peak width filters in the chromatography software.

3.3.2.2 Analysis of Extracolumn Volume Including a Column

An alternative method to determine the system variance with a column installed has been proposed by Lauer and Rozing [39].They injected a mixture of components that have similar diffusion coefficients under the conditions used and plotted the measured variance versus t_r. The slope of this plot equals $1/N_{\text{column}}$ and the intercept equals $\sigma^2_{\text{external}}$. This method assumes that the HETP value (or plate number) is independent of k.

Figure 3.15 compares the two methods for the determination of system variance for two different system configurations,

1) UHPLC system A, equipped with FTN injector, 0.110 mm × 340 mm capillary between FTN and heat exchanger, a 1.6 μl heat exchanger, a

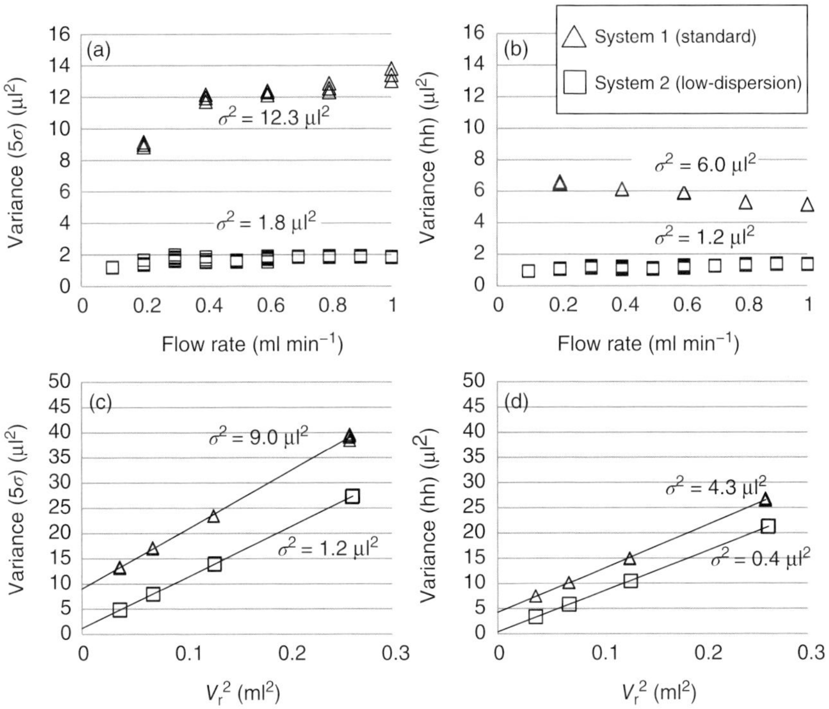

Figure 3.15 Determination of system variance with and without column. (a) Variance determined without column (5 sigma width), (b) variance determined without column (half-height width), (c) variance determined with column (5 sigma width), and (d) variance determined with column (half-height width). Configuration of system 1 and 2 and conditions are described in the text.

0.110 mm × 220 mm capillary between the column and the detector, and a 2.4 μl detection cell.

2) UHPLC system B, equipped with optimized FTN injector, 0.075 mm × 340 mm capillary between FTN and heat exchanger, a 1.0 μl heat exchanger, a 0.075 mm × 220 mm capillary between the column and the detector, and a 0.8 μl detection cell.

Figure 3.15a,b show the system variance determined without column from the 5 sigma width (a) and the width at half height (b) for different flow rates. In Figure 3.15c,d, the variances (from 5 sigma width (c) and width at half height (d)) of different components (acetophenone, propiophenone, butyrophenone, and valerophenone) separated in isocratic mode on a 2.1 mm × 50 mm column (Zorbax Eclipse Plus C18 1.8 μm, 2.1 mm × 50 mm, 0.6 ml min^{-1}, 60% ACN/40% H_2O, 30 °C) are plotted versus the square of the retention volume V_r. The intercept of the line fitted to the data points is equal to the variance of the system. According to Figure 3.15, the two methods do not give exactly the same result, which is probably due to the fact that the plate number of the column is not completely independent of the retention factor. However, both methods can be employed to compare systems with different configurations.

3.4 Impact of External Contributions in Different Application Areas

3.4.1 Impact on Isocratic Separations

In isocratic separations, where each compound in a chromatogram has a different retention factor k, the variance of an early eluting analyte can differ dramatically from that of a late eluting one. This means that compounds with low retention factors will be much more affected by extracolumn contributions than compounds with high k.

While the true plate number of a column is related to the variance produced by the column through Equation 3.6, the plate number observed with a specific HPLC system is given by

$$N_{\text{observed}} = \frac{V_0^2}{\sigma_{\text{v,total}}^2} \cdot (1 + k)^2 \tag{3.20}$$

where $\sigma_{\text{v,total}}^2$ is given by Equation 3.1.

Figure 3.16 shows an estimate of apparent plate number (values are based on HETP data for 1.8 μm fully porous particles measured on a 4.6 mm column) as a function of k for different column diameters (L = 50 mm) and different values of external variance. For compounds eluting at low retention factors (<4), a significant loss of apparent plate numbers is observed even for larger bore columns. While there is only a moderate loss in plate number for 3.0 and 4.6 mm columns at k values below 4, it becomes obvious that even for very small values of external

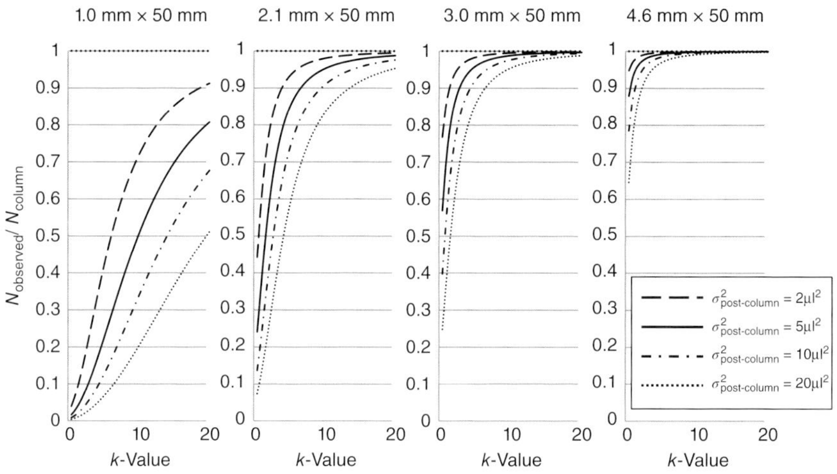

Figure 3.16 Estimated ratio of observed versus intrinsic plate number for different system variances in isocratic mode (50 mm columns with different diameters, particle size = 1.8 μm).

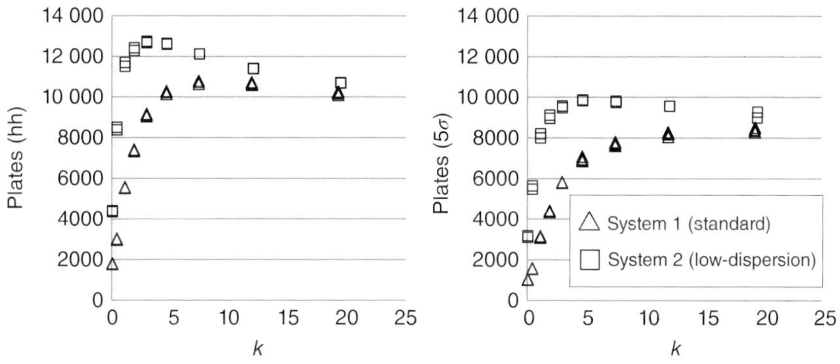

Figure 3.17 Plate numbers determined in isocratic mode on a 2.1 mm × 50 mm column based on 5 sigma width (a) and half-height width (b) with different system configurations. Systems 1 and 2 are described in the text.

variance, the apparent column efficiency in 2.1 and 1 mm columns is significantly impacted.

Figure 3.17 shows experimental plate numbers measured on a 2.1 mm × 50 mm, 1.8 μm columns with two different system configurations: UHPLC system 1 ($\sigma^2 \sim 2.2\,\mu l^2$) and UHPLC system 2 ($\sigma^2 \sim 10\,\mu l^2$) as in Figure 3.15. Plate numbers were determined from width at 5 sigma heights and width at half height (Zorbax Eclipse Plus C18 1.8 μm, 2.1 mm × 50 mm, 0.6 ml min^{-1}, 60% ACN/40% H_2O, 30 °C).

At low k values, we observe a significant loss in the observed plate number, and even with the low dispersion configuration, it is not possible to observe the true plate number of a small-dimension column. Figure 3.18 shows chromatograms obtained with the two system configurations in isocratic mode at 0.6 ml min^{-1}.

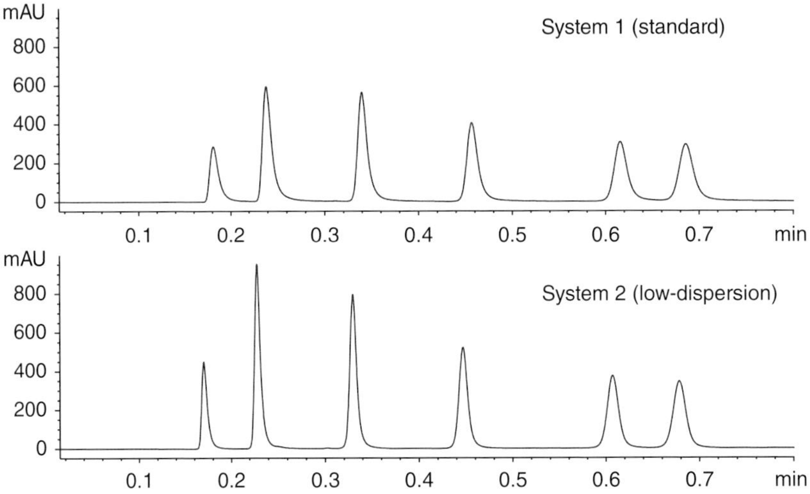

Figure 3.18 Comparison of an isocratic separation with system configurations 1 and 2 (as described in text). Conditions: Zorbax Eclipse Plus C18 1.8 μm, 2.1 mm × 50 mm, 0.6 ml min^{-1}, 60% ACN/40% H_2O, 30 °C.

3.4.2 Impact on Gradient Separations

In gradient separations, we are dealing with a different situation as long as the analytes elute in the gradient

1) In most gradient applications, the sample is focused at the head of the column; thus, the extracolumn contributions from the point of injection up to the column head are eliminated. This is particularly important for large-volume injections.
2) The variance of a peak in elution mode is determined by the retention factor at the point of elution (k_{elution}), see Equation 3.10. As k_{elution} depends on the gradient slope (see Equations 3.11–3.13), it is similar for all components in a sample provided that they have similar S values. For small molecules, S is around 10, and for peptides, S can be 40 or larger.

An analyte with an S value of 10 has a k_{elution} value of ~5 for $t_g = 5$ and k_{elution} value of ~30 for $t_g = 30$ (50 mm column operated at 6 mm s^{-1} and a gradient from 10 to 90% B).

A frequently used measure of column performance in gradient separations is peak capacity PC given by Neue [40, 41]

$$\text{PC}_{\text{column}} = 1 + \frac{V_g}{4 \cdot \sum_{i=1}^{n} \sigma_{\text{v,column}}} \tag{3.21}$$

and

$$\text{PC}_{\text{observed}} = 1 + \frac{V_g}{4 \cdot \sum_{i=1}^{n} \sigma_{\text{v,total}}} \tag{3.22}$$

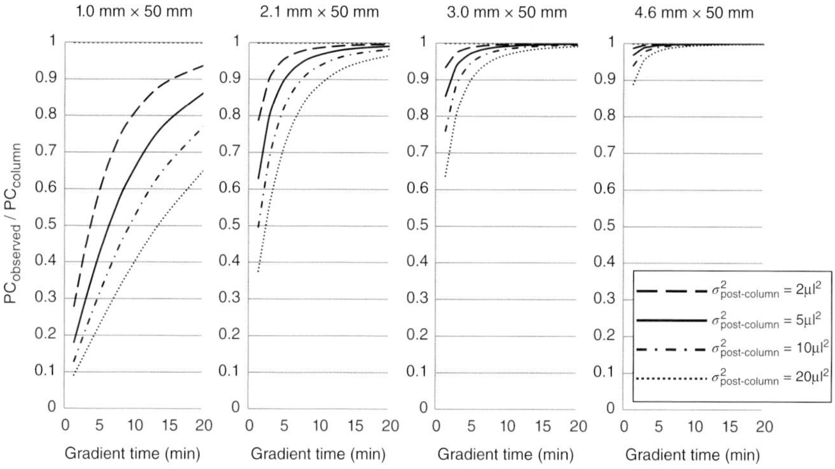

Figure 3.19 Estimated peak capacity ratio ($PC_{observed}/PC_{column}$) in gradient mode for columns 50 mm columns of different lengths with respect to different system variances.

Figure 3.19 shows the estimated peak capacity ratios ($PC_{observed}/PC_{column}$) for columns of different diameters ($L = 50$ mm) in gradient mode versus gradient time operated at a linear flow velocity of 6 mm s^{-1}. The lines in Figure 3.19 correspond to different values of extracolumn variance (2, 5, 10, and 20 µl^2). It has to be noted here that these calculations refer to the extracolumn variance that affects the separation, in gradient mode when the sample is focused at the head of the column, that is, the postcolumn variance.

While 4.6 mm × 50 mm columns are hardly affected by postcolumn dispersion even for very short gradient times, the situation changes, as the column diameter is reduced. For gradient times shorter than 5 min, the postcolumn variance should not be larger than 2 µl^2 to retain at least 90% of the original peak capacity.

With 1 mm columns, the gradient time should be larger than 20 min (at a flow rate of 0.14 ml min^{-1}) at a postcolumn variance of ~2 µl^2 in order to not sacrifice too much resolution.

Figure 3.20 shows the experimental peak capacities for gradients with different gradient times measured on 2.1 mm × 50 mm, 1.8 µm columns (Zorbax Eclipse Plus C18 1.8 µm, 2.1 mm × 50 mm, 0.6 ml min^{-1}, 10–90% ACN in 2, 5, 10, and 20 min, 30 °C).

Two different system configurations were used: UHPLC system 1 ($\sigma^2 \sim 2.2$ µl^2) and UHPLC system 2 ($\sigma^2 \sim 10$ µl^2) as in Figure 3.15. Peak capacities were determined according to Equation 3.22 from width at 5 sigma heights and width at half height. For relatively shallow gradients (large t_g), the difference observed is very small (<3%) while for steep gradients ($t_g = 2$ min), a difference peak capacities of about 10% is observed.

Figure 3.21 shows gradient separations obtained with the two system configurations at $F = 0.6$ ml min^{-1} and a gradient time of 2 min.

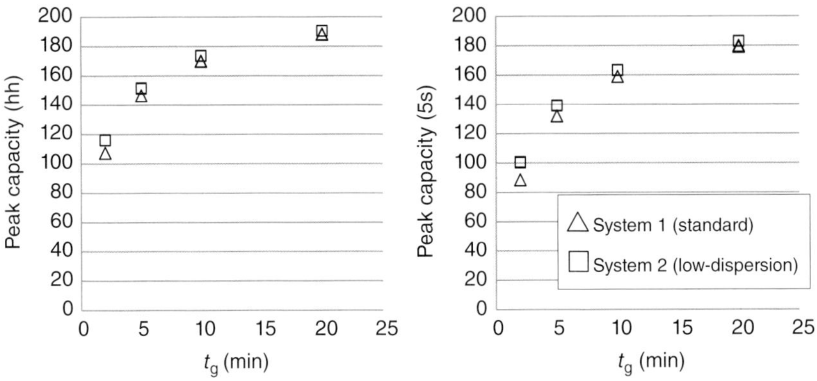

Figure 3.20 Peak capacities determined in gradient mode on a 2.1 mm × 50 mm column based on 5 sigma width (a) and half-height width (b) with different system configurations. Systems 1 and 2 are described in the text.

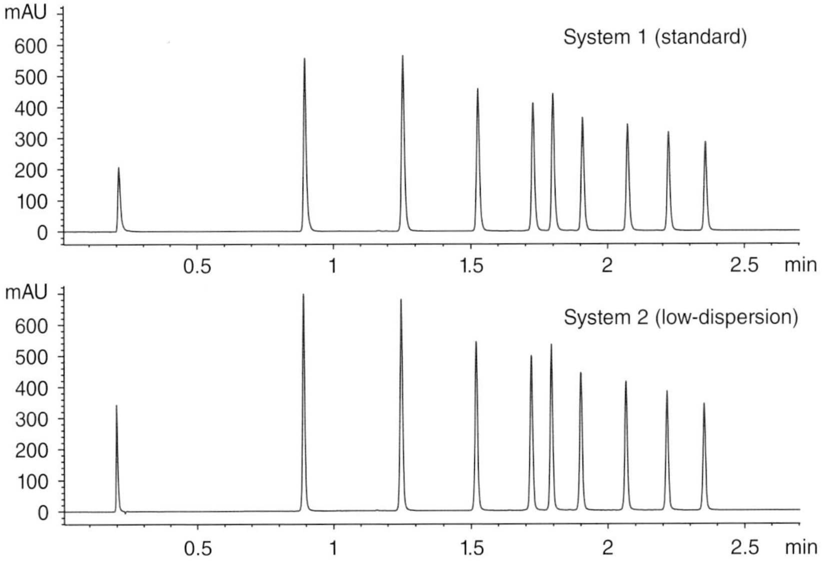

Figure 3.21 Comparison of a gradient separation with system configurations 1 and 2 (as described in text, Zorbax Eclipse Plus C18 1.8 μm, 2.1 mm × 50 mm, 0.6 ml min^{-1}, 10–90% ACN in 120 s, 30 °C).

3.5 Optimization of HPLC/UHPLC Systems

The system variance of a (U)HPLC system is not necessarily a property of the system itself but of the system configuration in terms of injector parts, connection tubing, and detection device. Many instrument manufacturers offer a choice of injection loops, seat capillaries, connection capillaries, heat exchangers, and detection cells to allow customers to modify their (U)HPLC system in order to reduce system dispersion. Several authors have investigated the impact of different system configurations on the overall column performance and possibilities to reduce system dispersion [12–16, 20, 42, 43].

The configuration of an HPLC system has to be selected carefully depending on the applications to be performed. The choice of injection system, connection capillaries, and detection cells has to be adapted to the requirements in flexibility, sensitivity, and flow rate range as the cost of minimizing extra-column volume is higher pressure drop and increased detector noise.

In the following section, we will discuss some application examples using 2.1 or 1 mm columns.

3.5.1 Testing of Column Performance

The performance of columns is usually tested in isocratic mode with an emphasis on low k components, as those yield the highest plate numbers. The requirements for injector flexibility, carryover, and detection sensitivity are relatively low. The pressure drop generated by the connection capillaries can be tolerated.

For 2.1 mm × 50 mm columns, with a plate number of 15 000 (packed with 1.8 µm particles or smaller), the column peak variances is ∼5 µl^2 for a compound with a k value of 2. To measure the true plate number of this column, a reduction in extracolumn variance to a value of ≤1 µl^2 is required as can be seen from Figure 3.16. This can be achieved by using a fixed-loop injector with a loop volume (injection volume) of ≤1 µl or a highly optimized FTN injector. The diameter of connection capillaries should be ∼75 µm, where a reduction in id only increases the back pressure but does not yield a significant improvement in dispersion. The detection cell should have a volume between 250 and 800 nl. It has to be noted that the nominal volume of the cell is not as important as the actual dispersion the cell produces.

For 1.0 mm × 50 mm plate number of 15 000, the column peak variance is ∼1.5 µl, so even a reduction of system variance down to <1 µl^2 will still yield a 40% loss in the observed plates so that it becomes extremely difficult to measure the true performance of such a column at low k values. With a system variance of 1 µl^2, 90% of the true plate height is observed at k' of 15.

3.5.2 Other Isocratic Separations

If isocratic separations are to be performed with certain requirements for flexibility, carryover, and detection sensitivity, the system components cannot always be chosen freely. In separations where resolution of a peak pair is relevant and not absolute maximum plate number, one has to keep in mind that the resolution is related to the square root of plate number (Equation 3.3) so that a loss of 25% on apparent plate is reflected in a about 13.5% loss in resolution. If the separation conditions cannot be selected such that the critical peaks elute at retention factors >5, a reduction in system variance to a value of about 2 µl^2 is advisable to avoid loss in resolution. This can be achieved by using an optimized FTN injector, 0.075 mm connection capillaries, and an 800 nl flow cell (see Figure 3.15). The use of a smaller flow cell gives only a marginal improvement (see Figure 3.15) at the price of increased noise.

3.5.3 High-Resolution Gradient Separations

In most gradient separations, the critical peaks elute in the gradient and are focused at the head of the column. Thus, only the system contributions after the column (postcolumn connection capillary, detection cell) contribute to dispersion. In contrast to isocratic separations, the analytes in a gradient chromatogram show similar variances as the variance is determined by the retention factor at the point of elution k_{elution}.

In applications that aim at achieving high peak capacities, ratios of V_0/V_g of the order of 0.01 or lower are employed, resulting in values for k'_{elution} of 10 or larger ($S = 10$, $\Delta\phi = 0.8$) and column variances of >60 μl^2.

$$V_g = \text{gradient volume}(t_g \times F)$$

Under these conditions, the difference in peak capacity is minor, even for 2.1 mm × 50 mm columns, when using different system configurations (see Figure 3.20). For this type of application, the type of injector is of minor importance (sometimes even large volume injections are performed to improve sensitivity), and it might be useful to use larger connection capillaries (0.110 mm ID) and detection flow cells of 2–3 μl volume to enhance system robustness and detection sensitivity.

3.5.4 Fast Gradient Separations

For fast gradient separations (high throughput, 2D-LC), the ratio of V_g/V_0 is between 0.01 and 0.05 with k_{elution} values of between 2 and 10. For these applications, the volume of connection capillaries and detection cells should be reduced to 0.075 mm ID and 800 nl cell volume.

Very fast gradient separations (as required in 2D-LC) are usually conducted at high temperatures and high flow rates (up to 3 ml min^{-1} for a 2.1 mm × 50 mm column). Because the dead times are low, short gradient times can be employed while still maintaining a relatively low gradient slope b.

Figure 3.22 shows examples of gradient runs on a 2.1 mm × 50 mm column at 1.1 ml min^{-1} and a gradient time of 30 s. The k_{elution} value under these conditions was ~2, and the resulting column variance was ~12 μl^2. The separation was performed with three different postcolumn system configurations

1) 0.075 mm × 220 mm connection capillary + 250 nl cell
2) 0.075 mm × 220 mm connection capillary + 800 nl cell
3) 0.110 mm × 220 mm connection capillary + 2400 nl cell.

The difference in peak capacity is very small for the two low-dispersion configurations while the configuration with higher dispersion yields a loss of ~15% in peak capacity.

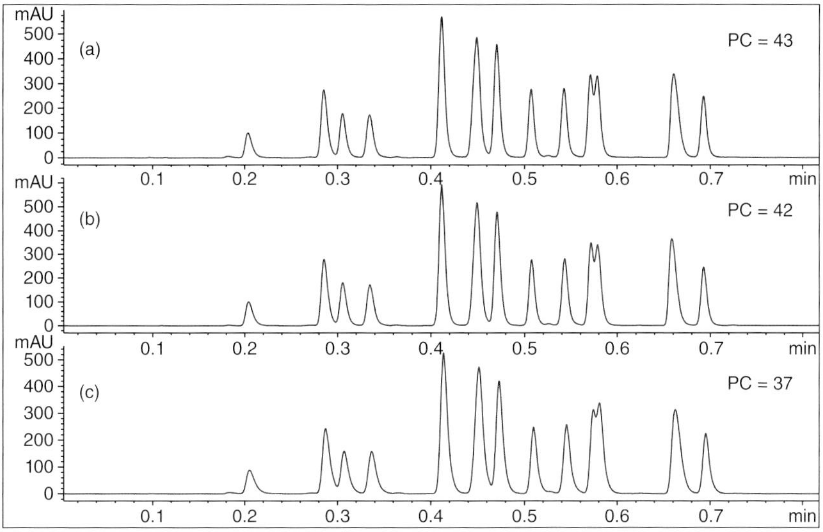

Figure 3.22 Comparison of a very fast gradient separation with different postcolumn configurations: (a) 250 nl cell + 0.075 mm × 220 mm capillary ($\sigma^2 = 0.5\ \mu l^2$), (b) 800 nl cell + 0.075 mm × 220 mm capillary ($\sigma^2 = 1\ \mu l^2$), and (c) 2400 nl cell + 0.110 mm × 220 mm capillary ($\sigma^2 = 4\ \mu l^2$). Conditions: PoroShell Eclipse Plus 2.7 μm, 2.1 mm × 50 mm, 1.1 ml min^{-1}, 30–60% ACN in 30 s, 60 °C.

3.6 Conclusions

Over the past years, a lot of effort has been put into improving the extracolumn dispersion of modern (U)HPLC systems to keep up with the developments of column efficiency and to enable the operation of short small-bore columns. As we have seen earlier, components of a system (injector parts, connection capillaries, heat exchanger, and detection cells) have to be chosen carefully based on the requirement of the separation. It is not always advisable to select the lowest volume configuration as this might compromise sensitivity, flexibility, and pressure range.

List of symbols

b	gradient slope
D_m	diffusion coefficient
Dp	particle size
F	volumetric flow rate
h	reduced plate height
H	height equivalent to a theoretical plate
k	retention factor
$k_{elution}$	retention factor at the point of elution
k_{start}	retention factor under gradient starting conditions

N	plate number
S	slope of linear solvent strength relationship
u	linear velocity
V_0	column dead volume
V_d	gradient delay volume
V_g	gradient volume
V_r	retention volume

Greek symbols

ε_T	total porosity
ϕ	volume fraction of eluent
η	viscosity
σ_v	standard deviation in volume units
σ_t	standard deviation in time units
τ	detector time constant
ν	reduced velocity

References

1 Nguyen, D.T.-T., Guillarme, D., Rudaz, S., and Veuthey, J.-L. (2006) *J. Sep. Sci.*, **29**, 1836–1848.
2 Fekete, S., Oláh, E., and Fekete, J. (2012) *J. Chromatogr. A*, **1228**, 57–71.
3 Gritti, F. and Guiochon, G. (2012) *J. Chromatogr. A*, **1228**, 2–19.
4 Desmet, G. (2006) *J. Chromatogr. A*, **1116**, 89–96.
5 Gritti, F. and Guiochon, G. (2008) *Anal. Chem.*, **80**, 5009–5020.
6 Gritti, F., Martin, M., and Guiochon, G. (2009) *Anal. Chem.*, **81**, 3365–3384.
7 Abian, J., Oosterkamp, A.J., and Gelpi, E. (1999) *J. Mass Spectrom.*, **34**, 244–254.
8 Hopfgartner, G., Wachs, T., Bean, K., and Henion, J. (1993) *Anal. Chem.*, **65**, 439–446.
9 Hopfgartner, G., Bean, K., Henion, J., and Henry, R. (1993) *J. Chromatogr. A*, **647**, 51–61.
10 Legido-Quigley, C., Smith, N.W., and Mallet, D. (2002) *J. Chromatogr. A*, **976**, 11–18.
11 Rodriguez-Aller, M., Gurny, R., Veuthey, J.-L., and Guillarme, D. (2013) *J. Chromatogr. A*, **1292**, 2–18.
12 Alexander, A.J., Waeghe, T.J., Himes, K.W., Tomasella, F.P., and Hooker, T.F. (2011) *J. Chromatogr. A*, **1218**, 5456–5469.
13 Fekete, S., Kohler, I., Rudaz, S., and Guillarme, D. (2014) *J. Pharm. Biomed. Anal.*, **87**, 105–119.
14 Gritti, F. and Guiochon, G. (2011) *J. Chromatogr. A*, **1218**, 4632–4648.
15 Gritti, F. and Guiochon, G. (2010) *J. Chromatogr. A*, **1217**, 7677–7689.
16 Gritti, F., Sanchez, C.A., Farkas, T., and Guiochon, G. (2010) *J. Chromatogr. A*, **1217**, 3000–3012.

17 Sanchez, A.C., Anspach, J.A., and Farkas, T. (2012) *J. Chromatogr. A*, **1228**, 338–348.
18 Stankovich, J.J., Gritti, F., Stevenson, P.G., and Guiochon, G. (2013) *J. Sep. Sci.*, **36**, 2709–2717.
19 Heinisch, S., Desmet, G., Clicq, D., and Rocca, J.-L. (2008) *J. Chromatogr. A*, **1203**, 124–136.
20 Wu, N., Bradley, A.C., Welch, C.J., and Zhang, L. (2012) *J. Sep. Sci.*, **35**, 2018–2025.
21 Purnell, J.H. (1960) *J. Chem. Soc.*, 1268–1274.
22 Purnell, J.H. (1959) *Nature*, **184**, 2009.
23 van Deemter, J.J., Zuiderweg, F.J., and Klinkenberg, A. (1956) *Chem. Eng. Sci.*, **5**, 271–289.
24 Gritti, F. and Guiochon, G. (2013) *J. Chromatogr. A*, **1302**, 1–13.
25 Giddings, J.C. (1966) *Dynamics of Chromatography*, Marcel Dekker Inc., New York, pp. 119–224.
26 Schellinger, A.P. and Carr, P.W. (2005) *J. Chromatogr. A*, **1077**, 110–119.
27 Atwood, J.G. and Golay, M.J.E. (1981) *J. Chromatogr. A*, **218**, 97–122.
28 Fountain, K.J., Neue, U.D., Grumbach, E.S., and Diehl, D.M. (2009) *J. Chromatogr. A*, **1216**, 5979–5988.
29 Verstraeten, M., Liekens, A., and Desmet, G. (2012) *J. Sep. Sci.*, **35**, 519–529.
30 Majors, R.E. (2014) *LC GC North Am.*, **32**, 840–853.
31 Kraiczek, K.G., Rozing, G.P., and Zengerle, R. (2013) *Anal. Chem.*, **85**, 4829–4835.
32 Gritti, F., Felinger, A., and Guiochon, G. (2006) *J. Chromatogr. A*, **1136**, 57–72.
33 Spaggiari, D., Fekete, S., Eugster, P.J., Veuthey, J.-L., Geiser, L., Rudaz, S., and Guillarme, D. (2013) *J. Chromatogr. A*, **1310**, 45–55.
34 O'Mahony, J., Clarke, L., Whelan, M., O'Kennedy, R., Lehotay, S.J., and Danaher, M. (2013) *J. Chromatogr. A*, **1292**, 83–95.
35 Eugster, P.J., Biass, D., Guillarme, D., Favreau, P., Stöcklin, R., and Wolfender, J.-L. (2012) *J. Chromatogr. A*, **1259**, 187–199.
36 Buckenmaier, S., Miller, C.A., Van De Goor, T., and Dittmann, M.M. (2015) *J. Chromatogr. A*, **1377**, 64–74.
37 Grushka, E., Myers, M.N., Schettler, P.D., and Giddings, J.C. (1969) *Anal. Chem.*, **41**, 889–892.
38 Stevenson, P.G., Gao, H., Gritti, F., and Guiochon, G. (2013) *J. Sep. Sci.*, **36**, 279–287.
39 Lauer, H.H. and Rozing, G.P. (1981) *Chromatographia*, **14**, 641–647.
40 Neue, U.D. (2008) *J. Chromatogr. A*, **1184**, 107–130.
41 Neue, U.D. (2005) *J. Chromatogr. A*, **1079**, 153–161.
42 Gritti, F. and Guiochon, G. (2012) *J. Chromatogr. A*, **1238**, 77–90.
43 Sanchez, A.C., Friedlander, G., Fekete, S., Anspach, J.A., Guillarme, D., Chitty, M., and Farkas, T. (2013) *J. Chromatogr. A*, **1311**, 90–97.

4

The Gradient; Requirements, Optimal Use, Hints, and Pitfalls

Frank Steiner

4.1 Instrumental Influences in Gradient Elution – An Overview

Many HPLC users consider the instrumental details of gradient generation as the most influential contributor on retention time and retention time precision as well as instrument-to-instrument reproducibility. This assessment is certainly fair, although other instrumental aspects such as column thermostatting may often be equally important. Deviations in gradient formation mainly derive from a fundamentally different technical implementation of gradient generation, namely high-pressure mixing as opposed to low-pressure mixing gradient generation. It is important that these two techniques physically deviate in much more aspects than the primary mixing point of the eluent components alone. More differences may result from the technical design of the individual mixers as well as the place in the fluidics where the mixer is localized. It must be noted that the true mixing device even of a low-pressure mixing gradient pump is normally located behind the pumping device, thus in the high-pressure part. Section 4.1 explains the principle technical implementation of both variants, its main characteristics, and in particular, the related impact on the chromatographic results. Section 4.2 is a deep dive into the technical details and chromatographic consequences, and it describes ways to characterize the specific performance of gradient instruments. It also gives some orientation to make the appropriate decision on what instrument is suited for the given application needs and how gradient methods can be successfully transferred between different instruments.

4.1.1 The Gradient Delay Effect and the Gradient Dwell Volume of a System

The fundamental technical principles of gradient formation are already described in Chapter 2.1 of this book, but several aspects will be repeated and discussed in depth in order to provide the appropriate context. A given extent of delay between the programmed mobile composition over time and the arrival of this composition at the column inlet is inherent to any HPLC gradient instrument, regardless

The HPLC Expert II: Find and Optimize the Benefits of your HPLC/UHPLC, First Edition.
Edited by Stavros Kromidas.

of its technical concept of gradient generation. This delay in time corresponds to the quotient of the volume between the physical mixing point of the different eluent components downstream to the column inlet as the counter and the applied flow rate as the denominator. The volume to be overcome is referred to as gradient dwell volume or gradient delay volume and abbreviated as GDV. The GDV for the entire instrument can be as low as 100 µl for modern UHPLC instruments but can also be several milliliters with conventional low-pressure mixing gradient HPLC instruments. Contributors to the total system GDV are the gradient pump with its inherent design and mixer volumes as well as the autosampler or injection valve fluidics and the connection tubing between the pump and the injector and between the injector and the column. The common assumption is that the main GDV contributor is the gradient pump and its mixer, but this does not always apply for modern instruments. When highly optimized high-pressure mixing gradient pumps are combined with integral-loop (also called split-loop or flow-through needle) autosamplers that provide wider injection volume flexibility, the autosampler's GDV contribution may significantly exceed the pump contribution. This is of relevance, because it is possible to switch the autosampler during the run back to the load positions after its sample loop is flushed out. While this operational principle can substantially reduce the gradient delay effect, it also has some implications and will be discussed further in Section 4.2.8.

In any case, it is crucial for the HPLC user to know the accurate value of the effective GDV of its instrument in operation. This will positively contribute to a better understanding of the obtained retention times in comparison to other instruments or to isocratic methods carried out on the same instrument. It is also an important prerequisite for successful transfer of gradient methods between different instruments, also in the context of the application of software to assist such a method transfer. While virtually all instrument manufacturers specify the GDV of their products in some way, care must be taken if these specified values are to be used for calculations and predictions or if instruments from different vendors are to be compared for performance criteria. The reasons for deviations can be manifold, but the most prominent ones are as follows:

- Insufficient or unclear differentiation between pump module GDV and total system GDV
- Changes to the exact configuration of mixers or connection capillaries
- Pressure dependence due to compressibility of liquids and pulse dampers
- The specific method applied for GDV determination (most important).

It does in no way express mistrust toward a manufacturer's specified GDV but is a good laboratory practice to determine the effective system GDV with a clearly defined method for each individual instrument. This accounts for an objective way to compare different instruments and to avoid errors in calculations related to prediction of retention times and method transfer. The obtained results may deviate from manufacturer specifications due to a different way of GDV determination. The practical determination of the GDV is described in Section 4.2.9.

What's to know All components of an HPLC instrument upstream from the column inlet contribute to its gradient dwell volume (GDV). The effective value of the total system GDV is crucial for successful method transfer between different instruments and for an objective instrument comparison. It is best practice to individually determine this value with a well-defined and clearly described method.

4.1.2 The Role and Function of the Gradient Mixer

Before further elaboration on gradient mixers, a clear definition of this part is mandatory. The mixer is the part that is required for the homogenization of the blended gradient components and must not be confused with the proportioning valve of a low-pressure mixing gradient pump. It would be much more precise to call the two existing principles low-pressure gradient (LPG) proportioning and high-pressure gradient (HPG) proportioning, but these names have never been successfully established. The mixer is located in both types of gradient generation behind the pumping cylinders in the high-pressure part of the system, but the fluidics in front of the pumping cylinders in the low-pressure mixing gradient pump may also contribute some mixing functionality depending on the design. The technical details and variants of gradient mixers are described in more detail in Section 4.2.12. In order to simplify the naming of the gradient pump type and to avoid confusion due to the word "mixing", this section will from now on use the two well-established names and abbreviations, namely high-pressure gradient (HPG) and low-pressure gradient (LPG). While very common in the literature and communications, these two names have virtually disappeared from HPLC vendors' collateral materials and product names. The reason is that the name low-pressure gradients is perceived to imply that this pump would be less suited to operate at elevated pressures, which is not justified from a technical standpoint. According to the maximum number of contributing eluent components, HPG pumps are normally referred to by the vendors as binary pumps and LPG pumps as quaternary pumps. This naming puts in the spotlight the increased flexibility of the LPG that can combine up to four different solvents in gradient runs. From a technical point of view, these names are sometimes not appropriate to replace LPG or HPG. There are also some binary or ternary LPG systems on the market, and the combination of three or four independently controlled pumps enables ternary or quaternary HPG setups as well, although at a hardly acceptable cost level.

The two different gradient principles are also described in Chapter 2.1 of this book, but are repeatedly illustrated in Figure 4.1 in order to explain the individual specific challenges with respect to mixing. With an HPG or binary pump, the software controls the ratio of the individual flow rates of the two independent pump channels, while an LPG or quaternary pump only features one pumping device and the gradient is generated by a software-controlled proportioning valve in front of this. The technical pumping devices that form the two independent channels in HPG pumps as well as the high-pressure pumping section in LPG pumps will be called pump blocks in the following.

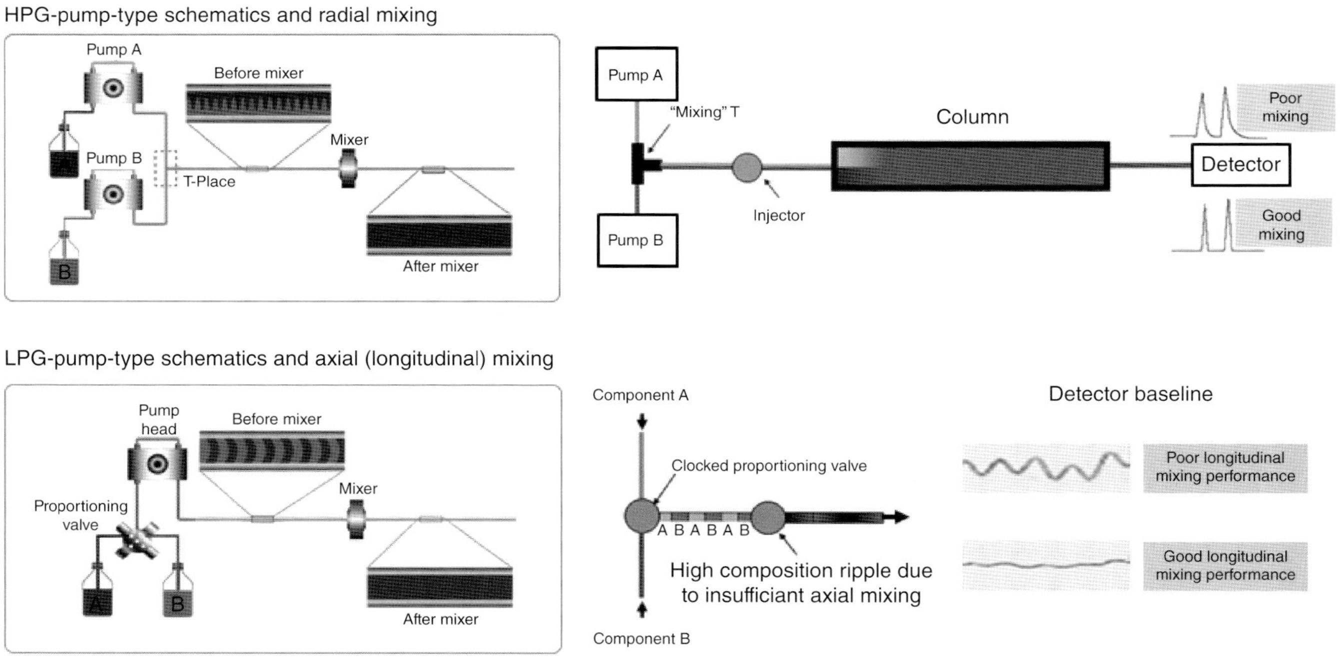

Figure 4.1 Schematic working principle of a high-pressure (often called binary) gradient or HPG pump and a low-pressure (often called quaternary) gradient or LPG pump. The respective function of the mixer and the consequence of incomplete mixing are also shown.

To understand the specific tasks of the mixing device in both gradient pump types, it must be distinguished between radial and axial (also called longitudinal mixing) mixing. The background for both is shown in Figure 4.1. Radial mixing is required when two or more continuous feeds of liquid stream are combined as it is the case with an HPG pump. This radial mixing could also occur quite effectively in the first part of a separation column, but this would negatively impact the chromatographic results. With a radial inhomogeneity at the column inlet, a respective radial inhomogeneity of retention would go along and hence distort the peak shape and negatively affect chromatographic resolution, as can be seen in Figure 4.1. This peak distortion effect would be more pronounced with a short column than with long columns where the eluent mismatch is already leveled off before the substance zone has traveled through a significant portion of the column.

The goal with axial mixing is completely different, as are the root causes that lead to the requirement of axial mixing. Axial mixing means leveling off an inhomogeneity in the direction of the flow rate (or in the time domain), which is the consequence of a discontinuous composition over time due to the technical concept of gradient and flow generation. The clocked consecutive feed of the different liquid segments in a proportioning valve of an LPG is an example where this phenomenon is inherent. Mixing of these liquid segments by diffusion is very limited due to the low diffusivity in the liquid phase at ambient temperature. A certain mixing effect is achieved by changes in the flow direction in the proportioning valve and due to the piston movement in the pump cylinder, but a complete axial mixing of these segments is only achieved in the true mixer located behind the pump cylinder.

The earlier explanations may lead to the conclusion that an HPG pump would only require radial mixing, while an LPG pump only needs axial mixing. For the LPG, this is almost true, although some radial mixing effect will be helpful because the border surface between the segments is neither planar nor orthogonal to the flow direction. This is because of the consecutive feed into the proportioning valve from different directions as well as the formation of a parabolic flow profile in the capillaries. While an ideal pulsation-free HPG pump would really need only radial mixing, the reality is that virtually all pumps exhibit a certain level of pressure (and flow) pulsation. The pulsation in the two individual pump blocks is normally not in sync and will thus translate into a composition ripple, which is represented in Figure 4.1 by the bumpy borderline of the two components in an HPG pump in front of the mixer. While the periods of the liquid segment ripples from an LPG proportioning valve normally fit into a sub-100-μl volume pattern according to the typical LPG proportioning cycles, the ripple in an HPG can have much lower frequencies and volume periods up to many milliliters. This is the case when one of the blocks runs very slowly at the limits of the gradient composition (e.g., <2 or >98%). This implies that effective radial mixing with an HPG that shows a more pronounced pulsation can be quite challenging and may require quite large mixer volumes. A common assumption is that the mixer volume must always be greater than the volume period of the composition ripple. While a large mixer volume is normally helpful, this rule does not strictly apply, and very effective axial mixers can achieve satisfying mixing performance with only a fraction of that volume period. The key to success with smaller mixers is that the amplitude of the ripple generated by the pump block is as low as can be. This is normally achieved with modern high-end UHPLC pumps.

What's to know Gradient mixing addresses the radial homogenization of two constant liquid streams in an HPG pump as well as the axial homogenization of the liquid segments from the proportion valve of an LPG pump and occurs in both cases behind the pumping cylinder. HPG pumps also need some axial mixing due to residual pulsation of the pumps. In practice, the majority of the problems with chromatographic results are due to incomplete axial mixing rather than radial mixing issues.

4.1.3 Deviations from Ideal Behavior of Gradient Generation Resulting from Fundamental Physicochemical Phenomena

In an ideal world, the user could expect from a gradient instrument that it delivers the desired mobile-phase composition at a given time accurately, precisely, pulsation-free, and with the set flow rate to the separation column inlet. In practice, this performance can be affected by several inherent operational flaws, which will be discussed in Section 4.2 in more detail. Besides such technical limitations, there are physicochemical effects that result from excess volumes under solvent mixing as well as solubility limits of eluent additives. Most of the ionic additives (e.g., buffer salts) are far less soluble in the organic eluent component than in water. As these buffer additives are normally only present in the aqueous eluent component, the inherent dilution by mixing with the organic component helps in this respect but may not be sufficient to avoid precipitation. The user can verify off-line in a test glass if the given buffer concentration is still soluble in the organic solvent content used in the gradient. However, neither can this test cover all resulting water–organic ratio situations in a continuous gradient, nor can it absolutely exclude precipitation during pump operation. This is because there are temporary and local compositions at the point of combination of the different eluent components that may differ substantially from the completely mixed eluent regarding both the organic content and the buffer concentration. In such cases, unforeseen precipitation may result, which occurs with LPG pumps in the proportioning valve and with HPG pumps in the mixing-T that combines the two component feeds.

The consequences of precipitation are normally much more dramatic with LPG pumps. Precipitation in LPG instruments can lead to at least partial clogging of the proportioning valve and the whole aspiration tract of the pump, which may result in complete or temporary flow interruptions. Such interruptions, however, can clearly be traced by the system pressure transducer, which makes diagnostics straightforward and even allows for automatic shutdown of the instrument. The much trickier case is that if no pressure breakdown occurs, but the accurate dosage in the proportioning valve is yet compromised. This can strongly affect analyte retention but would not necessarily be detected in the pressure trace as it may be hidden in the normal pulsation behavior. In several methods, there is an influence of the eluent composition on the detector signal, which is not ideal for the trace-level detection performance but is of help for diagnostics. If there are any pulsations, bumps, or steps in the detector baseline that have not been there before and that appear even in a blank gradient without injection, the chromatographer should be alerted of a precipitation problem or similar malfunction of the pump. The first measure would be extensive flushing of all solvent lines with water, then lowering the additive concentration in the mobile and testing the modified method for result compliance and robust operation. If the method then runs flawlessly after this, one can conclude that the original additive concentration caused local precipitation and is not compatible with the given instrument design. This may be a challenge for method transfer between systems in particular in a regulated environment.

HPG pumps are usually less prone to bad chromatographic results as a consequence from precipitation. One reason is that precipitation will hardly affect the composition because the composition is controlled by the flow ratio, which will not react on the precipitation in the mixing-T. Both pump blocks deliver sufficient force to overcome a restriction downstream. Moreover, the force will help to move the precipitate downstream until it is redissolved. The design of most mixers is also sufficiently robust to deal with that situation, and the filtering effect of the mixer will protect the autosampler fluidics. Some modern microfluidic mixers, however, may have serious problems with such precipitation. In any case, the HPG user who applies the methods with elevated additive concentrations should also be alerted by pressure anomalies that may indicate precipitation.

What's to know If the methods use mobile-phase additives and apply solvent combinations beyond the common water/methanol or water/acetonitrile systems, or if the additive concentration exceeds a 20 mmol/l level, the pressure trace and the detector baseline in blind gradients should be monitored carefully over time for any anomalies, as they may indicate additive precipitation. Sometimes, HPG methods cannot be transferred to LPG systems due to precipitation issues. Reduction of additive concentration normally solves this problem but is not always possible in practice.

Another fundamental physicochemical effect that has an impact on gradient elution in HPLC results from the thermodynamic phenomenon of excess volumes in mixed solvents. It is a fact that the total volume of mixtures is not strictly the sum of the partial volumes from the combined solvents of different kinds. When water is mixed with solvents such as methanol or acetonitrile, the resulting volume is lower than the sum of the partial volumes. This volume contraction depends both on the kind of organic solvent and on the mixing ratio, but it can be up to 4% in the case of methanol. The impact of this excess volume on mixing in gradient HPLC differs substantially between HPG and LPG instruments. With LPG pumps, it may also vary with the exact technical implementation of the fluidics and the control of the proportioning valve. With HPG pumps, it is quite straightforward, though of a significant effect, which is often underestimated in practice and not known to all users. In brief, an HPG pump cannot generate a constant flow rate during gradient elution. It is a fact that the flow rate follows the specific excess volume, which is a function of the composition as is depicted in Figure 4.2. What is mostly found in the literature are tables on

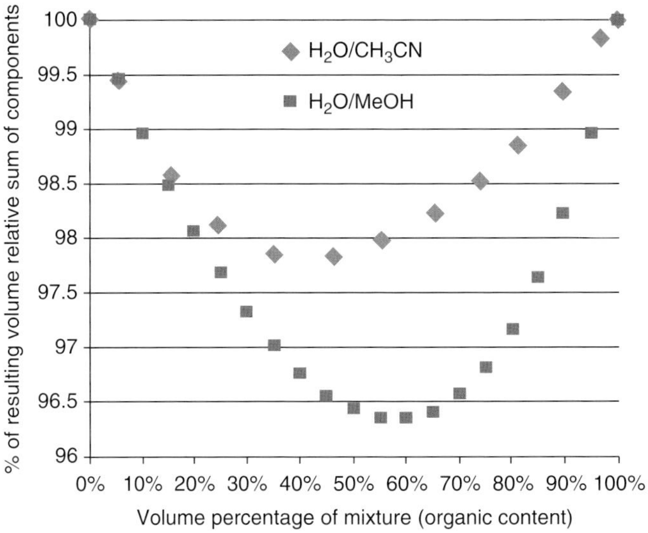

Figure 4.2 Curves of volume contraction due to excess volumes when mixing water with acetonitrile and water with methanol over the entire range. The resulting mixing volume is shown as percentage of the sum of the partial volumes as a function of the volume ratio of the mixture.

the excess volume as a function of the molar solvent ratio [1]. As molar ratios are not very convenient for the HPLC practitioner, the author has calculated the respective volume ratios and plotted them in Figure 4.2. It is very obvious that the volume contraction is much more pronounced in water/methanol than in water/acetonitrile systems. Based on the curve of excess volumes shown in Figure 4.2, it results in, for water/methanol gradients, the flow profile as depicted in Figure 4.3, while the pump flow is constantly set to 1 ml/min. The sum of the partial flow rates of the pump blocks A and B is 1 ml/min at any time, but due to the mixing on the high-pressure side, the resulting volume delivered to the column over time is reduced by excess volume at any specific composition. This leads to the flow rate change curve as shown in Figure 4.3. The clear advantage of an HPG is that it perfectly delivers the programmed composition as long as both blocks work as designed, and this is also illustrated in Figure 4.3. In theory, the resulting excess volumes could be considered in the HPG pump control software, and the flow rate change could be compensated. This would require that the used solvents are exactly known to the pump software (note that they can already be premixed). There are gradient pumps on the market where the user needs to enter manually the composition of the solvents that are filled into the reservoir for the individual gradient components. The control software uses this information for refining the precompression control (see also Section 4.2.3). While this manual approach is less convenient for the user and also more error-prone than a fully automatic compressibility control algorithm, it offers in principle the consideration of the individual solvent characteristics for a flow correction in an HPG. The alternative approaches to automated recognition of the individual solvents

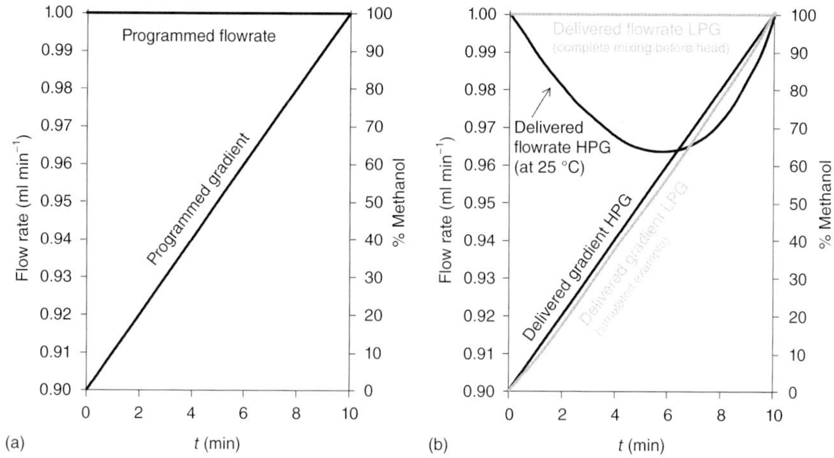

Figure 4.3 Diagram (a) shows the programmed pump parameters of a simple linear gradient running from 100% water to 100% methanol at 1 ml/min flow rate. Diagram (b) shows the real behavior of an HPG relative to an LPG pump with respect to flow rate and eluent composition with this gradient program. Please note that the deviation from composition for the LPG is only schematic and very drastic for better visibility, but its extent does not represent the typical behavior of LPG pumps in the market.

by the pump software or the flow correction by a feedback control are very difficult to implement. To the best knowledge of the author, there was no HPG type at the time of writing that corrects the flow rate for the excess volume effect and provides a constant flow rate in gradients.

The excess-volume-related effects with LPG pumps are substantially different from those with HPG pumps. The fundamental technical difference is that only one pump block controls the flow rate of the already premixed eluent composition. If the aspirated liquid composition is homogeneously mixed when delivered, the flow rate will remain perfectly constant over the entire gradient range. The crucial prerequisite for this constant flow is that the homogeneous mixing of all components must be achieved in the pump cylinder at the latest by the end of the aspiration phase (see also Section 4.2.7). If the mixing is only fully completed in the part behind the pump cylinder during the delivery phase of the piston, there will also be a certain volume contraction, and the result will be a corresponding decrease in flow. The main mixing task in an LPG is the homogenization of the consecutively aspirated segments due to the operational principle of the proportioning valve, and this requires the more challenging axial mixing (see Section 4.1.2). This must be ideally accomplished in front of the pump cylinder and would require an effective axial mixing device at this location, which in return must not affect the correct volume aspiration of the piston in any way. This is not trivial because effective mixing devices normally introduce a relevant flow restriction due to their fluidic design. It is no problem in overcoming this in the high-pressure section, but in front of the pump cylinder, there is only the hydrostatic pressure thanks to the elevated solvent reservoirs available. This is only about 10 mbar (0.14 psi) per 10 cm (4 in.) difference in the solvent level above the pump fluidics and will not be very effective in overcoming the flow restriction. For this reason, all manufacturers have implemented the true mixing device for the best correction of composition ripples in the high-pressure part of their LPG pumps. A dedicated mixing device in front of the pump cylinder is often not present, and the mixing effect from the proportioning valve and in the pump cylinder until the completion of the aspiration phase of the piston is considered sufficient. There are also some LPG pumps on the market that feature an additional mixer in the aspiration part (see Section 4.2.12). It is fair to claim that the premixing effect of all representative commercial LPG pumps is under typical operating conditions sufficient to account for less than 1% flow deviation. Nevertheless, the user should be aware of this effect, which may also distinguish LPG pumps of different vendors with respect to constant flow in gradients and contribute to method transfer challenges, albeit to a minor extent.

The far more relevant effect of LPG pumps is the impact on composition accuracy in gradients, and this results exactly from the fact that a certain extent of liquid segment mixing already occurs in the proportioning valve. As Figure 4.2 shows, on a quantitative basis, there is a volume concentration related to mixing water with methanol or acetonitrile. Figure 4.2 also demonstrates that this effect is not symmetric over the entire range, and in particular, the water/methanol system deviates from the symmetry of the curve. So what is the impact of this on the operation of an LPG pump? Assuming that a given binary gradient starts with pure water delivered from channel A, the proportioning valve will constantly remain in the A position to let the liquid from reservoir A into the proportioning chamber, and the volume of this chamber and all subsequent fluidics will only contain water. As soon as the gradient program starts to increase the content of the organic component delivered by channel B, the proportioning valves will start to switch regularly between the A and the B position for a defined time ratio that corresponds to the desired composition at any point in time. With that, it doses segments of water and segments of organic solvent in a given ratio into the common volume of the fluidics. The prerequisite for accurate dosage of each component is that the incoming liquid volume of each component would exactly correspond to the opening time for each channel. This in return would require that there is strictly no mixing of solvents at the contact surface between the water and the organic phase. The reality is, however, that a certain portion of the component is already mixed during aspiration into the proportioning chamber, and due to the excess volume effect, there is a volume contraction in this mixture. As soon as the valve opens the B position for the first time, a certain portion of the organic solvent disappears virtually in the excess volume, and this leads to an overdose of the organic solvent volume relative to water. Once the valve switches back to the A position and starts to let the water into the chamber, there will also be a certain mixing effect at the next contact surface between the two liquids, and now, some of the aspirated water disappears in the excess volume as well. If this effect was identical (or symmetric) at both contact surfaces, it would be compensated, and no deviation from the programmed composition would result. Due to the substantially different physicochemical properties of water and the organic liquid, the penetration depth will also be different at the two subsequent segment borders. This applies both to mixing effects based on diffusion (anyway, not very pronounced in liquids at ambient temperature) and to mixing by convection (the dominant effect). On top of the different penetration depths adds the difference in the excess volumes between water-rich mixtures as opposed to organic-rich mixtures. Due to the asymmetric curve of water/methanol (see Figure 4.2), the 1.8-fold volume of water will disappear in the dosage of 10% water to 90% methanol as is the case with dosing 10% methanol in 90% water (see in Figure 4.2 that the excess volume is 1.8 higher at the right end for organic-rich mixture). This overemphasizes the water content in the mixed phase and explains why the resulting methanol content is lower than the programmed (see Figure 4.3). This effect adds to the phenomenon that water is more cohesive than methanol, so less methanol penetrates the water segment in contrast to more water penetrating the methanol segment. Moreover, there is a thermal effect as due to the mixing heat of water and methanol, the mixture will expand, and this will result in dosage errors as well. From the sheer number of these contributing effects, it can be deduced that the extent of the composition deviation in an LPG pump cannot be quantitatively predicted. But experience shows that it depends on the technical design of the individual LPG pump; on the solvent system, gradient specifics, and flow rate; on the temperature in the lab; and even on the allocation of the different components at the proportioning valve (it can be different when channels A and B are swapped). One reason for the differences due the proportioning valve channel allocation can be gravity effects on the segment penetration due to differences in solvent density. This will only apply with vertical orientation of the proportioning valve (upward vs. downward orientation of the different inlets), but this is the common orientation to account for the shortest capillary connections from the module front with equal volumes. It can also make a difference whether the two inlets are geometrically opposite or vicinal. All this has to be considered, and the assumption that the choice of the channel would not matter for the resulting chromatogram may sometimes be wrong.

In order to avoid a too big concern from LPG users who read this section, it must be clarified that with a properly functioning instrument, the typical deviations of the composition can be expected to be less than 1%, and these deviations are precisely formed in each repetition of the gradient. Accordingly, deviations between different systems can be expected to be even smaller. So it must be emphasized in Figure 4.3 that the eluent composition curve for the LPG clearly exaggerates the deviation, and it does this only to visibly illustrate the given effect. That being said, it should also be mentioned that composition deviations with the practical application of LPG instruments (mixing water with organic solvents) may be larger than the composition accuracy as specified by the vendor. The reason is that pump operational qualification test is, for good reasons, always carried out by mixing two identical solvents, typically water with water containing a UV marker. Under such conditions, all the excess volume and segment penetration effects described earlier are absent, and solely, the correct mechanical and fluidic operation of the pump is characterized, which is the purpose of the qualification test.

Because LPG pumps are reproducing their deviations from the programmed eluent compositions in a very precise way, all the aforementioned considerations are of no relevance as long as a method is always operated on the same LPG instrument and the allocation of the solvent channels A, B, C, and D is not changed. Things become more challenging, however, if methods are transferred between different types of LPG instruments. This can be either different vendors' instruments or different instrument generations from the same vendor. Some practical guidance and considerations for method transfer are given in the following "What's to know" section.

What's to know The volume contraction by mixing water with acetonitrile or methanol has an impact on gradient generation in HPLC. The consequences are nonconstant flow rates with HPG systems and smaller deviations from programmed composition with LPG systems. These effects result in challenges for method transfer between HPLC systems that exhibit technical differences beyond the contribution of the GDV. For the ideal case of a simple linear solvent gradient and elution of all relevant compounds well within this gradient, the following rules and considerations apply for method transfer:

- For the transfer between different HPG systems, even from different vendors and generations, method transfer should be straightforward if the GDV matches.
- For transfer between LPG systems from different vendors or generations, one can expect changes in retention and selectivity even if the determined GDV matches, but there is a fair chance that methods can be successfully transferred. With nominally equivalent instruments, the allocation of the solvent channels A, B, C, and D should never be changed. With systems that are different in the technical design, the change of the solvent allocation can be an option for method tuning.
- The transfer between HPG and LPG systems or vice versa, even if from the same manufacturer and generation and with perfectly matching GDV, is generally difficult. It always requires method adaptations and may sometimes fail in spite of all endeavors.

All the aforementioned considerations only apply to retention time match and for a simple linear gradient. In more complex multisegment gradients, the elution will also depend on the flush-out characteristics of the individual mixer. The impact of system dispersion and detector flow-cell volumes on peak width and resolution is another important consideration for method transfer. The quality of the detector baselines is not considered either and may be substantially different with methods using trifluoroacetic acid (TFA) as additive (see Section 2.2.14). Finally, the column thermostatting technique may play a significant role, which is not the subject of this section but addressed in Chapter 2.2.

4.1.4 Instrumental Influences on Gradient Elution Outside the Pump

This section describes all instrumental effects outside the pump that may affect retention time, peak shape, or baseline quality but are not attributed to the pump. The respective elements of the instrument are the autosampler fluidics, the connection tubing (mainly those in front of the column) including optional columns prefilters, but also the volume of the detector flow cell and the detector settings to some extent.

It was already mentioned (Section 4.1.1) that the volume of all precolumn system fluidics contributes to the total system GDV. This concerns the following elements outside the pump module:

- Connection capillary from the pump to the injector
- Sample loop of the injector
- Internal volumes of the injection valve
- Needle seat volume and volume of its connection to the injection valve (with flow-through needle autosamplers)
- Connection capillary from the autosampler to the column
- Optional precolumn filters.

Among the listed elements, the sample-loop volume is normally by far the biggest contributor, at least if an integral-loop (flow-through needle) type autosampler is used. It must be considered for the GDV contribution that sample loops are normally labeled according to the maximum injection volume they are compatible with but not according to their true physical volume. Due to the formation of the parabolic flow profile in the loop during the sample aspiration, all loops have a larger volume than the maximum applicable injection volume for partial-loop injection mode (only supported mode for flow-through needle samplers). This additional volume is typically between 30 and 100% of the maximum injection volume and usually at the higher end for smaller sample loops of 10 μl maximum injection volume or less. The factor varies between manufacturers and can hardly be predicted.

A less obvious effect is related to mixing of the sample zone with the surrounding mobile phase prior to its delivery to the column, which may be substantially different between different instruments. Especially with modern UHPLC instruments that are optimized for low extracolumn peak dispersion (see Chapter 3), this mixing cannot be very effective. The possible resulting solvent mismatch on the column can substantially affect the peak shapes and retention times but is even more pronounced in isocratic methods. The reason why it is still mentioned in this section is because many users assume that it could be neglected in gradient elution, but this can be a fatal error. The effect is absent if the solvent of the sample matches the composition at the start of the gradient. If both deviate, it can be that the sample solvent is a weaker eluting composition than the gradient start condition, or it can be a stronger eluting one than the surrounding eluent. If it is weaker, a poor mixing can be of advantage, because the injection zone will even be sharpened at the column inlet (see Figure 4.4a). The user must be aware that this can in cases with high analyte concentration and large injection volumes result in a peak height that overloads the linear detector range. In the opposite case of a stronger elution sample zone than the surrounding starting eluent, the substance can smear with the injection zone through a relevant portion of the column, and the related peak distortion can sometimes not be corrected by the typical peak sharpening effect that is inherent to gradient elution (see Figure 4.4b for a very severe solvent mismatch effect).

It must be emphasized once more that modern optimized UHPLC systems will generally exhibit a very limited mixing capability of the sample zone in front of the column, as mixing always implies the undesired extracolumn dispersion.

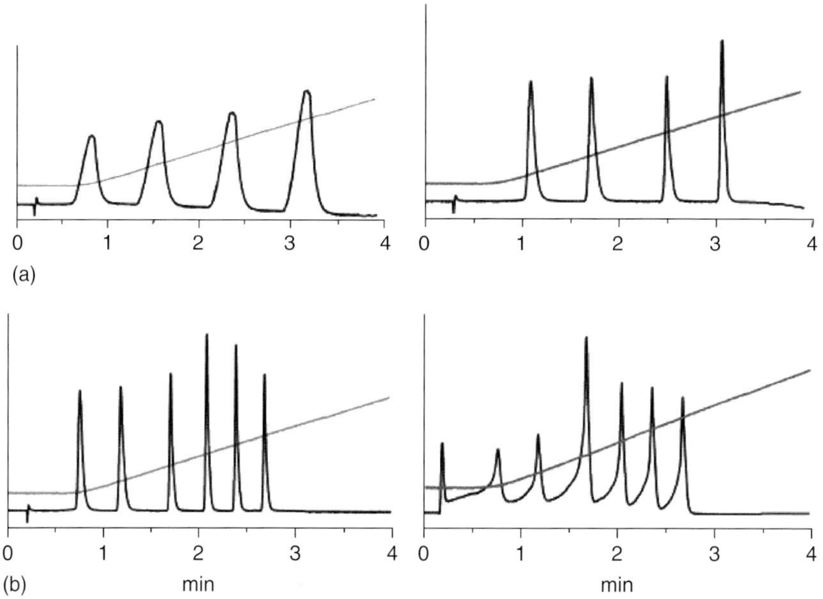

Figure 4.4 Chromatographic effects that occur at the column inlet due to differences in the elution strength of sample solvent and gradient start conditions. (a) These two chromatograms illustrate the focusing effect (right) of a weaker eluting sample solvent to compensate volume overloading (left). (b) This shows the severe peak distortion effect as a consequence of a significantly stronger eluting sample solvent. Both left chromatograms are recorded with the sample dissolved under the start conditions of the gradient.

Another fact is that gradients often start with relatively low elution strengths, and thus, the eluent mismatch due to the higher elution strength of the sample solvent is the more probable case. Moreover, the extent of this effect can change with the changes of the gradient program, in particular changes of the gradient starting composition. To practically test for the absence of such a mismatch effect, one can compare the chromatograms of a standard analyte mix dissolved in the gradient starting conditions versus the same compounds dissolved in the typical sample solvent. If both chromatograms are similar, there is no mismatch effect present. If there is still a not satisfying peak shape, it must have another root cause. If the peak is worse with the stronger sample solvent, the user should account for stronger percolumn mixing. Wider and longer connection capillaries are one option, but a small precolumn mixer is the more effective way to achieve this [2]. Precolumn filters that are optimized for low dispersion are normally not effective in improving precolumn mixing. Last but not least is the reduction of the injection volume, a very effective way to improve the situation, and it can also be combined with instrumental changes for better sample premixing. Very small injection volumes of about 1 μl with modern UHPLC methods are helpful to avoid solvent mismatch effects even with highly optimized precolumn fluidics. If the injection volume of the sample with strong eluting solvent is increased on such an instrument, the adverse effect may come into play.

The last instrumental consideration is related to the detector end. It is a fact that the selected gradient program, in particular the slope and range of the gradient, can have a dramatic influence on the peak volumes eluting from the column. With a steeper gradient, this is normally a positive effect and is referred to as gradient peak compression. The resulting smaller peak volume may, however, require a smaller detector flow cell volume in order to take effect, or the electronic filter (time constant) of the detector may be set inappropriately to detect the sharper peak. If this effect even translates into loss of peak resolution, the user may question the method or the column quality in the first place instead of considering a detector issue as the root cause. To avoid such pitfalls, it is recommended for gradient optimization experiments to work with the smallest detector flow cell available and to set a relatively high data rate and low filter constant (time constant, rise time, response time). Once the separation method has been optimized, the detector settings can be optimized for best signal-to-noise performance in a second step under consideration of the narrowest substance peak that must be quantified.

What's to know With volume optimized UHPLC instruments, peak distortion can occur if the sample solvent elutes stronger than the starting composition of the gradient. This can often be fixed by reduction of the injection volume or dilution of the sample with the weaker eluting component (water, in case of RP). If these measures increase the limits of quantitation too much, there must be a mixing element installed in front the column. If the change of a gradient program increases the peak compression, the detection parameters and the flow-cell volume must be verified to match the reduced peak volume.

4.1.5 The Stress and Wear on Columns in Gradient Methods

The application of a solvent gradient always implies a change in viscosity during the chromatographic run. With the typical flow-constant operation mode (or quasiflow-constant with HPG), this results in pressure changes and thus changes in the forces that act on the column bed. In particular, steep or sometimes called ballistic gradients change these forces very fast. If viscosity changes during the run by more than factor 2 and the maximum pressure is higher than 1000 bar, the resulting stress on the column can be immense. This can both damage the stationary phase particles and affect the structure of the packing. The extent of column wear depends on the individual column type and the method specifics, but several modern UHPLC columns even provide very acceptable endurance under such conditions. An alternative pressure-constant mode where the linear velocity would be constantly adapted to the viscosity change is a possibility, and it is also described in the literature [3, 4]. The key incentive of such a mode is to keep the linear velocity always at the maximum possible, and it will also adapt it to the changing mass transfer that follows the solute diffusivity. This can improve both the speed of analysis and the overall separation efficiency and resolution. At the same time, it eliminates the column stress from the cyclic change of pressure, but it keeps the column permanently under a relatively high pressure. However, there is an important consequence of the constant pressure mode, as it removes the strict relationship between elution volumes and retention times that only exists in the flow-constant mode. This has implications for the chromatographic behavior during column aging. When the column permeability goes down over the lifetime of the column (permeability accounts for the resulting back pressure at constant flow), the retention times will change as a consequence, because the instrument control reduces the flow to keep the pressure constant. The only way to present chromatograms when operating in the pressure-constant mode consistently would be plotting the detector signal over the elution volume rather than over the retention time. This does not easily find acceptance in the user community and is one obvious reason why the pressure-constant mode for gradient elution has not yet gained ground. As a consequence, no vendor has commercialized this mode (by the time of writing), and there is no practical experience with its positive effect on column lifetime. It is also questionable whether the permanent pressure changes are always more adverse than constantly running the column at pressures closer to its stability limit. Moreover, it is known that pressure shocks and sudden pressure dips are very damaging for most columns. These can occur at the point of the injection valve switching, mainly because a non-compressed liquid in the sample loop is switched into the fluidic path. With most instruments, this pressure surge at the injection point affects the column lifetime much more than the smooth cyclic pressure change from gradient elution.

It is fair to claim in general that the wear of a column under routine operation increases with the pressure that the column is exposed to. This fact and other considerations could encourage implementing a pressure-constant mode in a more intelligent way. If this mode would be run at the average pressure of the original flow-constant method rather than at the maximum pressure, the analysis time would be similar, but the column wear should be significantly reduced. Another intelligent way could be to adapt the pressure level to run the system in the van Deemter minimum [5] rather than at a constantly high pressure and further in the C-term. A respective software assistant (beyond the direct gradient slope control with the flow rate change) would be a mandatory prerequisite to bring pressure-constant gradient chromatography in a truly effective and column saving way into routine analysis. It may also enable to solve the problem of changing retention times with changing column permeability.

Finally, it is worthwhile to mention that the maximum pressure in gradient methods is principally higher than the constant pressure in a corresponding isocratic method. It could be concluded that gradient methods always induce more column wear than respective isocratic methods. There is a beneficial effect of gradient for the column lifetime though, namely the periodic cleaning effect of the column surface. This will certainly overcompensate the increased mechanical stress when samples with harsh matrices are analyzed. When isocratic methods are run on gradient instruments, it is therefore common practice to implement a gradient ramp for cleaning after the separation of the analytes and to accept the additional analysis time (mainly due to the required re-equilibration). In the RP mode, such a cleaning gradient will result in a pressure decrease due to the decreasing eluent viscosity with increasing cleaning strength. The related stress on the column packing should therefore be rather negligible, and the cleaning step can be considered good general practice.

What's to know The cyclic pressure change in gradients may negatively affect column lifetime when pressure capabilities of modern UHPLC instruments are exploited. Pressure-constant gradient modes are possible but could not gain ground in practice. The negative effect of pressure surges at the point of sample injection is usually more adverse for column lifetime than the smoother gradient pressure cycling. Isocratic methods always run at constant pressure but do not provide the cleaning effect of a gradient. Therefore, it can increase the column lifetime if a short cleaning gradient is run after the isocratic separation.

4.2 Gradient Elution Technology and How to Systematically Characterize Gradient Instrumentation

This section addresses the detailed technical implementation of gradient elution including the software control, challenges of eluent mixing, and the related physicochemical background. Different technical implementations are explained, and advantages and disadvantages are discussed in an objective way. The main purpose is to enable the user to fully understand the operational principles and resulting phenomena, to objectively and correctly characterize gradient instruments, and to solve occurring problems in gradient applications or transfer of gradient methods between different instruments.

4.2.1 Physicochemical Effects of High Pressure on Liquids

The general functional purpose of a gradient instrument is to precisely and accurately deliver a defined mobile-phase composition at a given time with a precisely controlled flow rate over a wide pressure and flow range, with minimum mixing ripple regardless of the solvents to be mixed and with minimum possible abrasion of wear parts. This sounds already quite complex and is even more challenging when looking into the details in practice. Manufacturers have taken different approaches to meet these requirements, and the respective pros and cons can depend strongly on the specific application requirements. The resulting manufacturing costs and resulting system prices can also deviate significantly. Very important differences in technical requirements have to be considered between LPG and HPG instruments. This will always be acknowledged in this section by discussing the characteristics and effects first for HPG instruments, then for LPG instruments. This can even be in separated subsections if appropriate.

Before the technical implementation can be discussed, some relevant physical and physicochemical basics need brief explanation, in particular for modern UHPLC instruments where even extreme pressures far above 1000 bar can be generated. While liquids change their volume under pressure to a much lower extent than gases, the compressibility of mobile-phase solvents becomes actually relevant under UHPLC conditions. Since the liquid volume changes with pressure, it is inevitable that the resulting flow rate is also a function of the applied pressure. Due to this fact, a very important clarification must be made at this point:

> The flow rate of an HPLC pump, as it is set in the pump control panel, refers to the delivered solvent under atmospheric pressure. This applies to the vast majority of instruments in the market.

The flow rate is controlled by the pump through the aspiration of liquid volume per unit time from the eluent reservoir (at atmospheric pressure), and it can be measured behind the column (or behind the detector) where it is once again under atmospheric pressure. The average flow rate over time is therefore not affected by the mobile-phase solvent compressibility. It is, however, of tremendous relevance for the control of the pump piston movement in order to deliver a pulsation-free flow onto the column. As already discussed in Section 4.1.2, pressure fluctuations of the individual pump blocks of an HPG pump directly translate into mixing ripples and are hence often of high importance for the quality of the resulting chromatography. Figure 4.5 shows the volume changes for the isothermal compression of water, acetonitrile, and methanol at a temperature of 298 K. While water is only compressed by 4% at 1000 bar, this value is twice as high for the two organic solvents, where methanol is even slightly more compressible than acetonitrile. Since the curves are not linear, the compression rate at 2000 bar is not twice that at 1000 bar. Nevertheless, the volume of methanol under 2000 bar is reduced by 14% relative to the atmospheric pressure. This has

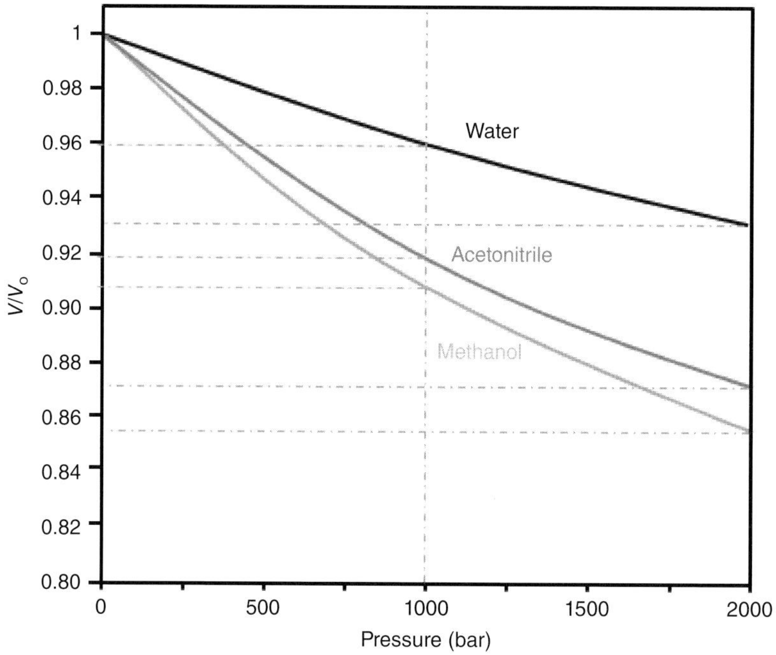

Figure 4.5 Compressibility curves for water, acetonitrile, and methanol at ambient temperature (up to 2000 bar pressure).

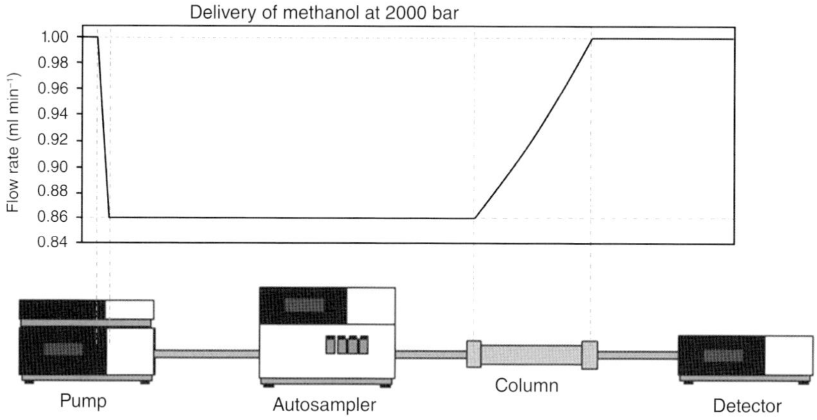

Figure 4.6 Flow rate profile along a UHPLC instrument fluidic path for a method where pure methanol is delivered at 1 ml/min against a column head pressure of 2000 bar (pressure drop in connection tubing is neglected).

significant implications for the effective local flow rate in the UHPLC instrument, as illustrated in Figure 4.6. In this schematic overview, the pressure drop of the liquid along the connection capillaries is neglected for the sake of simplicity. In practical UHPLC, it actually cannot be neglected if a flow of 1 ml/min is pumped through the very narrow tubing of approximately 100 μm internal diameter in the instrument at the given viscosity of methanol at ambient temperature. While the pump aspirates at a flow rate of exactly 1 ml/min, it is very obvious that the flow rate through the autosampler is (due to the methanol compression) only as low as 0.86 ml/min, and it will consequently require 16% more time to flush out the autosampler fluidics than it would take at atmospheric pressure. The most relevant pressure change in the instrument occurs along the column (in this simplified scheme, the only part where the pressure changes). This results in the respective flow acceleration from 0.86 ml/min at the column inlet up to the original flow rate of 1 ml/min at the column outlet. This directly implies a dependency of the column hold-up volume on the applied pressure level. It must clearly be stated that this simple calculation assumed isothermal behavior, which cannot be easily controlled inside the column under such conditions (see also Chapter 2.2). There are many more effects on the column when it comes to extreme pressures. Those will not be further discussed in this chapter, as they are not directly related to the understanding of gradient instrumentation and its operation. For more detailed information on what is going on in the column at extreme pressure, the article by Martin and Guiochon [6] is recommended, from which the data presented in Figures 4.5 and 4.7 are taken. The fundamental physical relationships presented in this section are crucial for a better understanding of the pulsation phenomena and mixing ripples in gradient instruments.

Another physicochemical phenomenon relevant in the presence of extreme pressure is the phase conversion from liquid to solid state. In Figure 4.7, the melting temperature for water and several organic solvents is plotted against the pressure up to a level of 2000 bar (data from [6]). The graph confirms the anomalous behavior of water, which melts under isothermal pressure increase, while

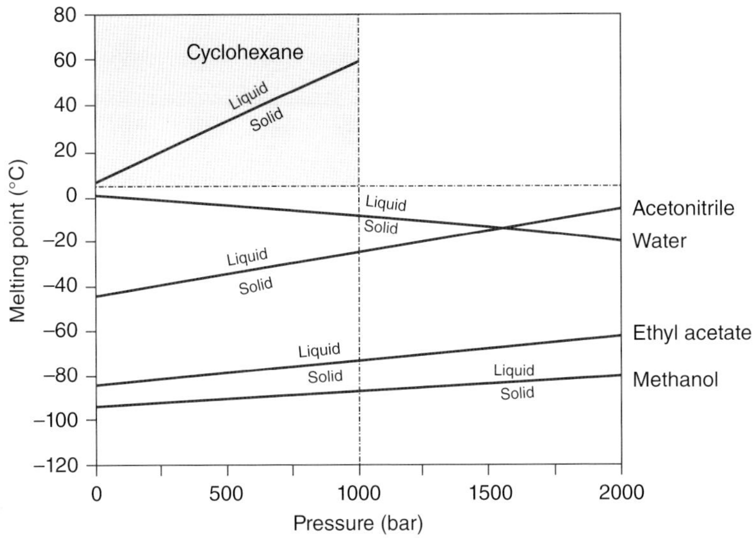

Figure 4.7 Dependence of the melting point of water and several organic solvents as a function of pressure up to a level of 2000 bar.

this curve has an opposite slope for all organic solvents. This behavior of water is the prerequisite for the sliding on ice, which would not be possible on frozen methanol, as this would never melt under the pressure of the sly and not build the respective liquid film for the sliding. Another important information to be deduced from Figure 4.7 is the fact that cyclohexane should never be used in HPLC, because it converts to the solid state at ambient temperature already at a pressure of less than 200 bar, while its melting point under atmospheric pressure is 6.5 °C. The behavior of n-hexane is very different. With its melting point of −95 °C, it is a typical mobile-phase component in normal-phase HPLC. Due to its lower toxicity, n-heptane is normally preferred though. With an extrapolation of the acetonitrile curve to even higher pressures, it can be concluded that this very common solvent would convert to solid state at 3650 bar and ambient temperature. This sets a natural pressure limit to reversed-phase UHPLC, unless the pump fluidics would be heated and the separation also carried out at elevated temperature.

4.2.2 The Need of Solvent Degassing

For all considerations in the previous section, the assumption was to have pure solvents with no oxygen or nitrogen dissolved therein. Nevertheless, the solubility of both nitrogen and oxygen in typical HPLC solvents is in no way negligible [7]. It is fair to say that well-degassed solvents are a must for a flawless operation of HPLC pumps. Even for isocratic pumps where no mixing of solvents has to be accomplished, the dissolved gas can disturb the delivered flow as a consequence of outgassing. The extent of this effect is a function of the amount of dissolved gas, the pump inlet fluidic geometry, and the applied flow rate. The higher the flow rate and the narrower the fluidics, the more likely the occurrence of outgassing effects. Outgassing leads to formation of a gas bubble in the working cylinder of the pump during the drawing phase, and this will still be present when the delivery phase starts, and it will displace a certain volume of solvent in the cylinder. Under the steep pressure increase, the bubble will often quickly redissolve, but the extremely high compressibility of the temporary two-phase system in the cylinder will result in pulsation and flow deviation. The latter cannot mostly be compensated even in pumps with automatic compressibility adaption. With an HPG, it is not likely that this effect occurs to the same extent in both pump blocks. As a consequence, it will not only lead to flow pulsation and deviation but also affect the composition accuracy and continuity. In the initial phase of the flow ramp-up of a pump, it can also occur that a gas bubble sticks in the check valve, prevents this from proper operation, and prevents the pump from starting flow delivery. But even if the pump control manages to dissolve small gas bubbles properly and without relevant flow and pressure pulsation at the outlet site, the outgassing can occur after the liquid expansion behind the column, which will lead to heavy disturbances in the detector baseline. It is recommended to watch in a transparent outlet capillary behind the detector cell that gas bubbles are the cause of the baseline noise. Often, however, these bubbles may be just too small to be visually observed. The resulting detector baseline problem can be solved with a restriction capillary mounted behind the detector flow cell to simply suppress the outgassing effect, but it is mandatory to consider the maximum back pressure that the flow cell tolerates when taking this measure. The user should control the pressure at the pump without this post flow-cell restrictor first. After mounting the restrictor, it is recommended to ramp up the flow rate only slowly while making sure that the incremental pressure is clearly below the maximum flow-cell pressure specification. Such countermeasures may be very helpful in basic isocratic instruments in the absence of devices for eluent degassing.

With gradient instruments, the proper degassing of the mobile-phase components is indispensable for a perfect flow and composition control. This is even more critical with LPG-type pumps. If a gas bubble is formed in one of the solvent channels in front of the proportioning valve, this will directly lead to a significant proportioning error, which can in no way be corrected downstream. If gas bubbles of smaller size can be redissolved and the pulsation can be kept low, thanks to automated compressibility adaption, the user will not spot this at the pressure sensor, but still the eluent composition and thus the gradient profile will be changed and affect the retention times. This will never enable to meet the retention time precision expectation to a modern high-end gradient pump and absolutely must be avoided through proper solvent degassing.

It is also worthwhile to note that the amount of dissolved gas even in the absence of outgassing has an impact on the compressibility of a solvent. This is also relevant for the solvent precompression control, which will be discussed in more detail in Section 4.2.7.

What's to know Thorough on-line eluent degassing is mandatory for LPG systems to account for robust precise and accurate gradient formation. While this is somewhat less critical for HPG systems, it is fair to suggest that all gradient instruments should feature an effective on-line degasser.

4.2.3 The Different Types of Pump Technology (Serial, Parallel, Cam Drive, Linear Drive) and Their Specific Properties and Requirements

HPLC pumps work according to a discontinuous principle in each cylinder, where a loading and a delivery phase occur alternately with a given frequency. Single-piston (or single-cylinder) pumps interrupt the flow for a short period of time to quickly refill the cylinder. They are not common anymore as their flow delivery performance and pulsation specification do not match those of modern HPLC requirements, even when combining them with very effective pulse dampers. All modern HPLC and UHPLC pumps follow either the concept of a serial 1½-cylinder pump or the parallel 2-cylinder pump. Their different functional principles are explained in this section.

The concept of a serial 1½-cylinder pump is explained at first and can be seen in Figure 4.8 in the variant with one motor and transmission via cams coupled with a cam shaft. This pump principle is very cost-effective to build and also requires a relatively low number of critical wear parts. The motor power transmission to the cam shaft is usually accomplished with a belt. This belt sets the cam shaft into rotation, and this rotation is transmitted into translation movement of the pistons by the cams. Due to the coupling of both cams to one shaft and one motor, the piston movement in the working cylinder and in the equilibration cylinder is strictly coupled, and hence, no individual control is possible. A closer look at the position of the inlet and outlet check valves of the working cylinder shows the first-in-first-out principle, where the eluent enters at the opposite end of the cylinder than at the outlet. Depending on the proportioning cycle (see Section 4.2.8), this can be very advantageous to minimize composition fluctuation in the gradient. It can also be seen that the cam of the working cylinder is significantly larger than the one for the equilibration cylinder, because it has to generate approximately twice the flow rate (at the same frequency). This is due to the functional principle of serial pumps where during its delivery phase (piston moves to the left in Figure 4.8), the working cylinder has to use a portion of its generated flow delivery to refill the equilibration cylinder, while the other portion equals the flow rate delivered to the column. When the working cylinder is in the loading phase (piston moves to the right in Figure 4.8), the equilibration cylinder takes over the flow delivery to the column. As can be concluded from Figure 4.8, only the working cylinder requires check valves with this pump type. During the loading phase, its inlet check valve must open and the outlet check valve must close so that the equilibration cylinder can build up the required pressure to deliver mobile phase to the column. As soon as the mobile phase in the working cylinder is compressed to the existing column head pressure, the outlet check valve opens, and its delivery phase begins simultaneously with the loading phase for the

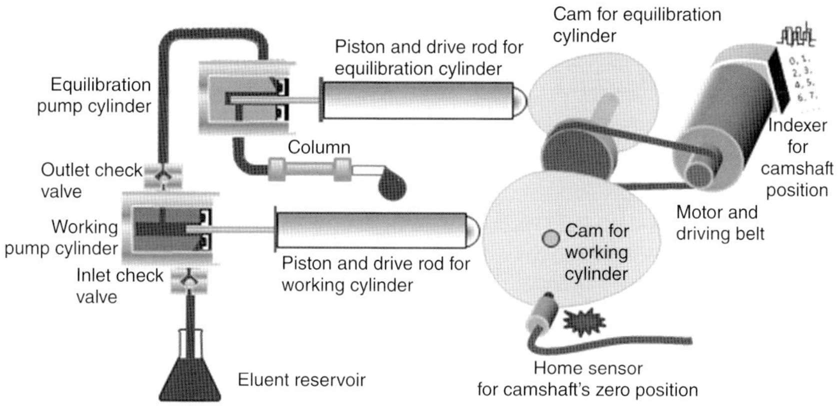

Figure 4.8 Technical scheme of a cam-driven serial 1 ½-cylinder pump. This pump technology can be applied for both HPG pumps (by combining with a second pump and aligned flow control of both) as well as LPG pumps (by integrating a time-controlled proportioning valve in front of the pump).

equilibration cylinder. Consequently, the equilibration cylinder requires no check valves. The outlet valve of working cylinder that closes during the delivery phase of the equilibration cylinder acts simultaneously as the inlet valve of the second cylinder. No outlet valve is required for the equilibration cylinder because mobile phase is delivered continuously through its outlet toward the column. Both pistons alternately deliver mobile phase. For smooth and pulsation-free delivery, it is mandatory that the handover between both cylinders occurs with virtually no flow interruption. Due to the compressibility of liquids as discussed in Section 4.2.1, the transition from the loading (or aspiration) phase to the delivery phase requires a so-called precompression phase. During this phase, the pistons move into the cylinder to compress the solvent until the required pressure to open the outlet check valve is reached, but no flow is generated during this phase. The challenge with the pump control in this phase is that the required stroke length for the precompression depends on the solvent compressibility and the system pressure level at which the solvents need to be compressed. This in turn is a function of the solvent type and, to a certain extent, of the solvent temperature and the amount of dissolved gases. An intelligent control of the piston movement can certainly consider all of this, but the pump control algorithm will require certain information on the solvent properties. If the user needs to know these and has to enter them into the control software for a specific method, this is a possible source of error. Certain pumps on the market measure the compressibility directly from the pressure rise during the compression phase, calculate the true compressibility on the fly, and consider this for the proper piston movement. This does not only avoid possible user errors, but is also a way to automatically adopt the precompression control to changing solvent compressibility. This adoption is crucial for LPG pumps where the composition and thus compressibility in the pump heads continuously change when the gradient evolves.

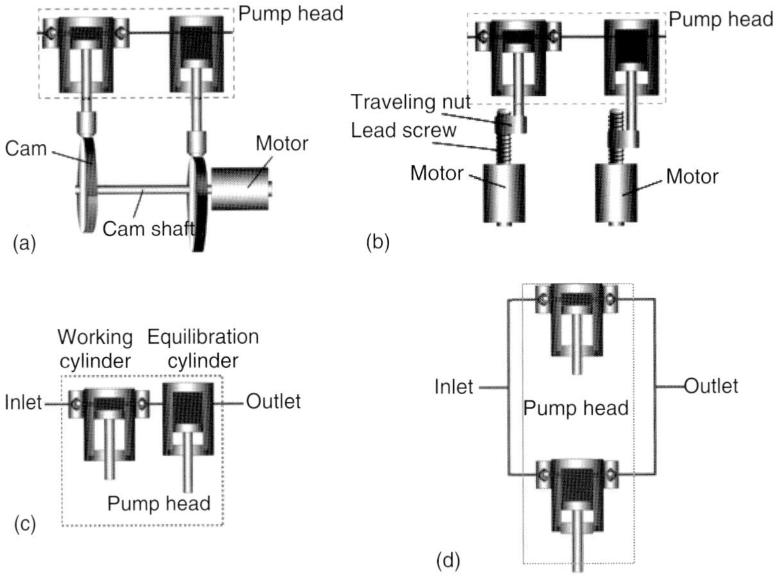

Figure 4.9 Schematics of the two common HPLC pump drive principles, cam (a) and lead screw (b), as well as the two common solvent delivery concepts, serial (c) and parallel (d). It is not mandatory that cam-driven pumps have only one motor resulting in coupled movement, but most cam pumps on the market follow this cost-effective principle.

Figure 4.9 shows in direct comparison the main variants for technical solutions of the pump drive, namely (a) cam or (b) linear drive with a lead screw. Next to this, it shows the two typical solvent delivery variants that are (c) serial and (d) parallel. The advantage of the serial principle is the reduced number of check valves, while with a parallel pump, both cylinders with absolutely equivalent functionality require inlet and outlet check valves. Advantage of a parallel pump in addition to several specifics that address a pulsation-free flow delivery is a significantly reduced wear at the piston seals. In fact, a parallel pump exactly distributes the mechanical workload between the two alternating cylinders, while in a serial pump, the working cylinder needs to refill the equilibration cylinder. With that, its frequency at given flow rate and stroke volume must be twice as high, resulting in twice the workload and stress on the wear parts. This also applies to the check valves. While their number is reduced, they need to accomplish twice the operation cycles, so no savings result in cost of operation from the smaller number of valves.

The advantage of the coupled cam drive type relative to the more advanced alternative concept is mainly the economical factor. The fine mechanical demands for the related devices are less delicate, but the main difference is the need for one motor only. This relatively basic pump principle in comparison to the concept of completely independent linear drives implies a reduction factor of 2–3 in both the manufacturing costs and the maintenance costs. The clear disadvantage is the strict coupling of the piston movement in both cylinders. While an intelligent movement control in the compressibility phase of the working cylinder is possible, it is obvious that any change in the speed of the piston movement will also apply to the second piston in the equilibration cylinder that is still in charge of the flow delivery during the working cylinder compression phase. The shape of the cams and the advanced control algorithms account for this to the best possible extent and can make even a pump of this simple principle competitive for today's requirements on minimized pulsation. Nevertheless, here, it must clearly be stated that such a pump can never achieve the smoothness of flow delivery similarly to a pump with independent motors and linear drives, which is in turn much more expensive. Considering the complexity for both principles with respect to the mechanical details and the control algorithms, it can be concluded that even with one type of pump, there can be pronounced differences in the quality of flow delivery between different manufacturers and even with different tiers of instruments from one manufacturer.

Another limitation of the common implementation of cam-driven pumps (that do not change the direction of motor movement) is the fix stroke volume of the pump cylinders. This implies that the stroke volume cannot be adapted to a specific flow rate, which can be of advantage if a wide range must be covered. Exactly, this is the case when HPG pumps run in gradient mode. If the stroke volume of a pump cylinder equals, for example, 50 μl, the pump must execute 20 cycles/min to generate a flow rate of 1 ml/min. This relates to serial pump where the mentioned stroke volume applies for the working cylinder while a parallel pump with that stroke volume would only execute 10 cycles/min on each pump cylinder. For a reduced stress on the wear parts, on the one hand, and for reduced challenge on the mixing cycle requirements (see Section 4.2.12 and following), on the other hand, it is of great advantage to adapt the stroke volume and cycle to the required flow rate, so that the pump never runs at extremely high or low frequencies. This can best be accomplished with a linear drive pump. A variable stroke volume can also be advantageous for LPG pumps, as the stroke volume also contributes to the GDV with this type of gradient instruments. In this respect, it is worthwhile to mention that the flush-out volume of a pump cylinder is significantly larger than its stroke volume (often by more than factor 2), so the relative effect on the GDV reduction will be less pronounced than one might actually assume.

What's to know The rule "you get what you pay for" also mostly applies to the functional principles of HPLC pumps. An advanced parallel pump with independent motors and intelligently controlled linear drives enabling to specifically adapt the stroke volume will always be far superior in performance relative to a serial pump with coupled cams driven by one motor (just to distinguish between the two extreme cases). The suitability of a pump will always depend on the specific application demands. This can be reflected in the detector baseline quality (especially for TFA methods) and retention precision (especially for shallow gradients starting at 0%).

4.2.4 The Specific Gradient Pump Type and Its Implications for Practical Operation

The discontinuous operation of HPLC pumps as explained in the foregoing section may have a significant influence on the detector baseline in certain applications. Baseline effects are, however, not the only effect on chromatography. It is a logical consequence from the discontinuous flow rate or even discontinuous eluent composition over time that the movement of analytes through the column does not occur at a constant pace. It may even result in a "stop and go" motion leading to chopped peaks in protein chromatography. This very special effect is described in more detail in Section 4.2.15. While such effects are rather rare, a discontinuous movement of analytes almost always has adverse effects on retention time precision. The baseline problem being the more visible consequence is mostly challenging with applications where TFA is used as an eluent additive. Hence, a separate section is devoted to this problem (Section 4.2.14). It is a prerequisite for a good understanding of the related challenges to know the specifics and the role of gradient mixers in full detail, and this will be covered in Section 4.2.12.

We start this section with the commonly less known consequences of the technical flow and gradient generation on the resulting retention time precision. Unlike the behavior of proteins (see Section 4.2.15), the extent of discontinuous movement as a result of eluent composition ripples does not translate into clearly visible peak shape changes with small molecule analytes and the vast majority of applications. The retention time is basically controlled by the average velocity of the substances inside the column during the separation process. Even if there is discontinuous motion, its effect on the average velocity is often negligible. This assumption can be made as long as the number of these "speed up and speed down" cycles is sufficiently high, so that the exact phase of this pattern relative to the injection and elution points is of no significant effect for the average speed. This essentially depends on the ratio of the pump stroke volume relative to the applied flow rate. Hence, it depends on the number of strokes until the elution of a given analyte. For example, a conventional HPLC column of internal diameter of 4 mm or wider, which is typically run at flow rates of 1 ml/min or higher, is used. Assuming a 4.6 mm column run at 1 ml/min flow rate and a typical pump stroke volume of 50 µl, one can calculate approximately 50 stroke cycles until the apex of a peak with 2.5 min retention time is eluted, because the pump has to perform 20 cycles/min at this flow rate. Transferring this method to a 2.1 mm i.d. column at the same linear velocity, the flow rate must be reduced by a factor of 4.8 to a level of 0.21 ml/min. Consequently, the number of strokes until the elution at 2.5 min is reduced by the same factor to a level of approximately 10. It is a valid point that modern 2.1 mm UHPLC columns are mostly run at two- to fourfold higher flow rates. At the same time, the retention times are significantly reduced due to the combination of high linear velocities applied on short columns. This is the essential idea behind method speed-up in UHPLC. If the corresponding peak in such a UHPLC method now elutes at a time of 0.3 min, even at a flow rate of 0.8 ml/min, the pump with 50 µl stroke volume only performs four to five cycles in this time period, which is a very low number indeed. With a single-digit number of strokes until the elution point, any discontinuity in the delivery absolutely matters for the retention time precision, unless it is intelligently compensated. In fact, even severe discontinuity is of no consequence on the retention time as long as the fluctuation pattern is exactly copied from run to run, and thus, the average speed of the analyte zones is exactly identical. This can only be achieved through an exact match of the sample injection point to the working cycle of the pump. The related challenge is, however, substantially different between HPG- and LPG-type pumps and will be discussed separately for these two types in the following sections.

4.2.5 HPG Pumps and How Discontinuous Pump Cycles Resulting from Pressure Pulsation Impact Retention Time Precision in Practice

HPG pumps are generally perceived to be superior in performance relative to LPG pumps. This is the reason why users are willing to pay a higher price for the related technology that requires a larger number of expensive parts to build them. Regarding the accuracy of composition control and the ability to accurately generate very fast composition changes, this is a valid perception, and it is rounded up with the favorable smaller GDV relative to an LPG. At the same time, the exact match of the pump cycle phase from run to run is far more challenging with HPG pumps than with their LPG equivalent. This is particularly critical at high percentages of one component combined with low flow rates, as explained at the end of the foregoing section. In reaction to this, different vendors have developed technologies to deal with it, and many of these are also patent-protected. The vendor-specific situation and the related intellectual properties will not be the subject of this section. The goal is in fact to educate the user on a fundamental understanding of the problematic and to indicate ways how the resulting effects on retention time precision can be minimized in practical applications.

It is important to first understand how the stroke cycles of an HPG pump occur while a gradient is formed. We assume a fixed stroke volume during the gradient formation regardless of the individual flow rates. This is applies to the majority of pumps on the market. As both components of the binary gradient from an HPG are generated by a flow ratio change from two independent pump blocks, there is no fix ratio between the flow rate and the frequency. In fact, the frequency of the piston cycle in both pump blocks changes throughout the gradient. If an HPG pump delivers 100% of one component, only one of the two blocks actually works, and it performs 20 cycles/min if the stroke volume is 50 μl and the flow rate is 1 ml/min. For a mixture of 50/50, both blocks need to deliver 0.5 ml/min. So both will perform 10 cycles/min, but they will do so with a random phase offset, because there is no easy way to bring the two blocks in sync. If we consider what happens when the pump has to mix only 1% of component B at a total flow rate of 1 ml/min, the B-block will then run at 10 μl/min, which implies a cycle time of 5 min or 0.2 cycles/min. If at this mixing ratio, the flow rate is additionally reduced to 0.2 ml/min, the cycle time of the B-block is increased to 25 min and the frequency reduced to 0.04 cycles/min. These extreme ratios that can occur in HPG pumps together with the fact that the two pump blocks work independently can make it very difficult to start the two pump blocks from a defined position at the point of sample injection when the chromatographic run begins. It is obvious from the aforementioned consideration that it can take 20 min or more to bring the piston of the slowly pumping block in a defined position. At the same time, the piston of the A-block moves on independently with a relatively fast speed and will never be in the same position as at the start of the foregoing run at the time when the B-block is in the same position. If a method starts with 100% component A, there is no way to bring the piston of block B in a defined position once the starting conditions of the gradient are reached, just because the piston in block B stands still. The only way out would be an intelligent control of the pump during the backward gradient at the end of the run and during the re-equilibration phase. This approach could only work accurately if the piston control algorithm also considers the duration of the sample injection cycle. The sample draw phase in a gradient program is usually not controlled according to fix time stamps in the program, but as a series of commands where, for instance, the syringe draw time is a result of the programmed draw speed and injection volume. It is not impossible to consider all these factors, but it is definitely very elaborate and will also result in the consequence that the duration of the equilibration phase varies from run to run. On top of all these challenges comes the respective patent situation that does not give all vendors freedom of operation in this respect. As a consequence, the injection synchronization with the piston cycles of an HPG that runs at constant stroke volumes is quite uncommon in practice. The situation would be different with an HPG with variable stroke volume where the pistons of the two blocks can be run in synchronized phase but with changing stroke volume during the gradient. If such a pump needs to form a 99% component A mixture and has a maximum stroke of 100 μl, block A would apply a stroke volume of 99 μl while block B moves the piston only to the extent corresponding to 1 μl. Such small stroke volumes together with all the precompression requirements are difficult to control accurately. This challenge is the reason why this concept is also very rarely applied in practice, and the patent situation also contributes in this respect. Moreover, it must be taken into account that with an HPG pump of synchronized piston cycles, both blocks are running all the time at the maximum frequency to generate a certain flow rate. This implies an increased stress to all wear parts relative to a pump that slows down the frequency of the block that needs to deliver only a minor percentage of its component and thus a low flow rate.

The most common solution to achieve excellent retention time precision in gradient runs with HPG instruments even without synchronization of the injection to the pump cycle is to reduce the discontinuance of the composition over time (also known as mixing ripple) to a negligible level. The most effective way to achieve this is a very advanced linear piston drive with independent motors for the two pistons of each pump block. This will inevitably result in significantly higher costs relative to those for a pump using cams connected by a cam shaft and simply driven by the same motor for both pistons. If the technical level of the pump drives cannot account for pulsation-free piston control, it is possible to level off the resulting composition discontinuity by very effective longitudinal mixers that commonly exhibit volumes of several hundred microliters. If an HPG pump is equipped with a large volume mixer, it actually abandons one of its big advantages with respect to the lower GDV relative to an LPG.

A common assumption is that a pulse damper would be a very effective tool to suppress mixing ripples of an HPG pump. While such a device can reduce the total pressure pulsation that occurs at the column inlet (that can also be transmitted to the detector flow cell), a central pulse damper behind the mixing point of the two components can never selectively suppress the individual pulsation of the respective pump blocks that actually generate the undesired mixing ripple. Two independent pulse dampers for each block that are separated by a restriction capillary from the following mixing T to minimize cross talk could better address this challenge but are not very common in practice due to resulting challenges with the pump control and the general effort involved. Moreover, there are two fundamental difficulties with pulse dampers in gradient instruments. The first is due to the fact that the pulse damper volume contribution to the pump GDV changes with the pressure level, as its flush-out volume changes accordingly. Hence, the GDV of the instrument will also change when a column is aging and increasing its back pressure under operation by partial clogging. The other issue is even more severe and applies to pumps with an automated control of the solvent precompression to react to the different compressibilities of solvents (Section 4.2.1, Figure 4.5). The control algorithm of such pumps absolutely depends on the prerequisite that it is not disturbed by any other elastic element other than the solvent. If a pulse damper is mounted behind such a pump, the major elasticity actually results from this element and comes into effect at the point when the outlet check valve opens. This can even result in resonances, and the precompression control may fail completely. Therefore, a pulse damper must NEVER be mounted to a pump with automatic precompression control and is anyway of limited use to reduce mixing ripples of gradient pumps.

There are two other HPG-specific effects that are hardly known to users, though they can have a tremendous influence on retention time precision. As they have no impact on the detector baseline, at least in the part of the chromatogram where analyte peaks have to be integrated, they are often overlooked. The first effect only occurs when a gradient is started from 0% of one of the components. In such methods, the respective pump block is standing still in the preinjection phase and only starts its motion at the time of sample injection with a very low albeit increasing velocity. With many HPG pumps, however, this piston movement does not translate into solvent delivery until the solvent in the respective cylinder is compressed to the column head pressure level and the outlet check valve opens. If a stroke volume of 50 μl is assumed and by chance the piston would be at the return point after the aspiration phase, the implication of starting a delivery of methanol against a column head pressure of 1000 bar needs attention. For a 10-min-long gradient running from 0 to 100% of component B at a flow rate 500 μl/min, one can calculate an acceleration rate of 50 μl/min^2 for this pump block. As discussed in Section 4.1.3, methanol is so compressible that its volume shrinks by 4% under a pressure of 1000 bar. Hence, the piston must displace 2 μl in a 50 μl cylinder to compress the solvent to a 1000 bar level. This is even a simplified assumption, because there is additional dead volume in cylinder that also needs to be compressed. To stick to the assumption of 2 μl compression volume and considering the given acceleration rate of 50 μl/min^2, the piston displaces 25 μl in the first minute. To displace the required 2 μl, it needs 2/25 min, which is almost as long as 5 s. If by chance the other extreme happens and the piston is very close to the return point at the end of the delivery phase, the compression volume might be as low as several nanoliters, and thus, the compression to 1000 bar occurs within several milliseconds.

As a consequence, the timepoint when the methanol delivery actually starts may vary within a range close to 5 s. With that, the time when the gradient arrives at the column also varies by 5 s, which translates into retention time change in the order of 5 s. A retention time change of 5 s means a relative retention time change of 8% for a peak that elutes at 1 min, which is not an uncommon elution time in UHPLC methods.

A possible solution to correct for this first effect is to start the gradient at the lowest percentage of component B that the gradient control allows for, usually 0.1%. The resulting retention time changes between the gradient start at 0% and 0.1% will be negligible for most methods. Indeed, this trick is at the same time a diagnostic tool to identify if the precompression challenge of the B-block is the root cause for poor retention time precision as well as a possible way to cure it. The precompression challenge may also be a trap where a gradient method user with the best intention may fall into. It is another good practice to premix mobile phases for gradient elution rather than letting the pump mix pure solvents or, even worse, pure organic solvent with highly concentrated aqueous buffers. This may result in baseline ripples up to pump malfunction due to buffer precipitation. If a gradient starts with 10% organic component, the user may be inclined to premix the 10% organic level on channel A but start the program with 0% B. While the mixing challenges may have been cured now, this practice when applied to an HPG may promptly lead to the discussed retention time precision issue because block B starts from zero movement and needs to precompress prior to delivery. So it would have been better to premix only 5% organic component into channel A but start the gradient from 5% B (more exactly 5.3%) instead of 10%. There are HPG pumps on the market that can address the challenge of starting the gradient precisely from 0% (e.g., dynamically adapted compression volume with linear drive, keeping the cylinder precompressed all time, special precompression algorithm for start from 0%, etc.). Nevertheless, the user cannot rely on such features and should carefully consider the aforementioned recommendations.

The second effect to be mentioned here can occur with any gradient method, and its impact increases with increasing pressure level during the sample injection or, more exactly, at the switching time of the injection valve. This effect can even occur in isocratic methods when both pump blocks are active to automatically generate the desired mobile-phase composition (dial a mix). The effect is due to the randomly different volumes of the compressed solvents in the different pump blocks A and B. They will expand these compressed volumes into the sample loop that hosts liquid (sample solution and mobile phase) at atmospheric pressure at the point in time when the autosampler valve is switched to the inject position. Actually, the compressed liquid from both ends (column and pump) streams into the sample loop in order to balance the pressure levels. Due to the relatively strong flow restriction in the column packing, the majority of compressed liquid comes from the pump side. With HPG-type pumps, the volume ratio of compressed organic to compressed aqueous solvent depends on the relative position of the pistons in the active cylinder. With a serial pump, the compressed volume also depends on whether the working cylinder or the much smaller equilibration cylinder is active when the injection valve is switched.

Depending on this arbitrary volume ratio, the composition of the mobile will randomly change between the mixing point and a possibly substantial part of the GDV. This random composition at the start of the gradient together with the flow interruption due to the sudden pressure dip can seriously affect the retention time precision. The extent of retention time variation increases with the column head pressure at gradient start, but also with increasing stroke volumes of the pump, increasing compressibility of the used solvents, and increasing size of the sample loop, including metering device (if present in the autosampler). Once the user observes a systematic drop of retention time precision with increasing pressure, which can be checked by varying the flow rate in a given method, it is very likely that this effect is the main root cause. The user has limited options to counteract this, but two hints can be given. If the mounted sample loop is much bigger than is needed for the applied methods, a reduction of the loop size can reduce the effect. If the gradient window is significantly smaller than 100%, an appropriate premixing can correspondingly reduce the difference in solvent composition in the two pump blocks and with that the resulting extent of composition variation. As mentioned earlier, this premix approach should not encourage to start the B-block from 0%, unless the user can be sure that an intelligent pump control takes care of the resulting random compression delay.

The instrument manufacturer can also implement features that reduce or eliminate the effect of random solvent compositions expanding into the sample loop. One solution would be to bring the active pump pistons in a defined position at the time of sample injection. The difficulty with this was discussed earlier in this section. Another way to solve this issue would be a precompression of the sample loop prior to switching the autosampler valve. Such a loop precompression will not only remove the described issue of random solvent composition after injection but also remove the flow interruption and the pressure shock that is quite adverse for the moving parts in the instrument and will also reduce the column lifetime. The technical implementation of the loop precompression is outside the scope of this section, but the effect on the recorded pressure trace and the resulting retention time precision are demonstrated in Figure 4.10. The fact that no visible pressure dip occurs with the activated precompression translates into a marked improvement of the retention time precision in a representative gradient UHPLC separation of alkyl phenones even at a relatively moderate maximum column pressure of 300 bar. The improvement on retention precision for seven consecutive injections from 0.045% RSD to 0.007% RSD for the benzophenone peak is arguably significant and shows the importance of this precompression possibility to eliminate the effects of flow interruption at the injection point and the random solvent composition by the expanding volumes from the two pump blocks.

What's to know If the retention time precision in the gradient method run with an HPG pump is not satisfying, it may have several root causes. Discontinuous eluent composition due to mixing ripples can be one reason, but this normally translates more into detector baseline problems than retention time effects. A synchronization of the pump cycles with the injection solves the problem but is difficult to implement

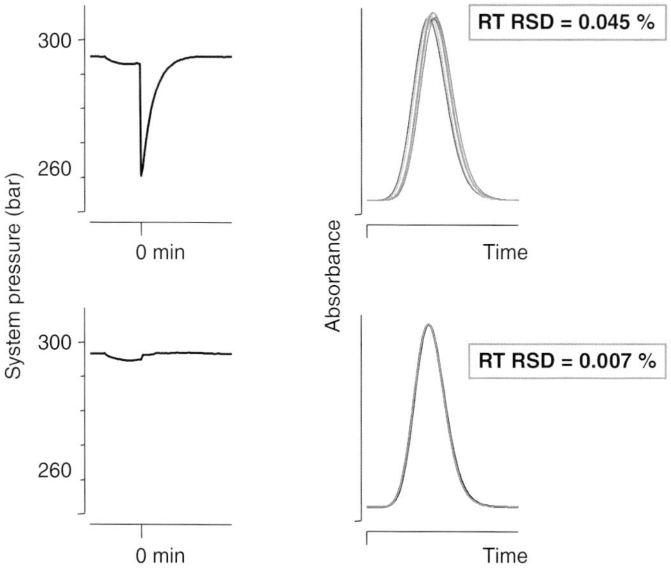

Figure 4.10 Influence of the sample-loop precompression on the recorded pressure trace during the run and more important on the retention time precision on a selected peak (benzophenone). Method: phenone separation in a gradient in 3.6 min from 40 to 100% acetonitrile and a Thermo Scientific Hypersil GOLD 1.9 μm 2.1 × 30 mm column. Flow rate: 0.26 ml/min, temperature: 25 °C, injection volume: 1 μl, detection: UV at 254 nm.

with an HPG. The user can increase the volume of the mixer, if the corresponding GDV increase is acceptable. If a gradient starts at 0% B, the retention time precision may suffer from the random position of the B-piston that needs to precompress the solvent prior to delivery. The user can modify the method to start the gradient at 0.1% B instead. Also, the random expansion of the solvents in the two pump blocks into the noncompressed sample loop can also be a source of poor retention time stability. The reduction of the sample-loop volume and a premixing of solvents (provided the gradient does not run from 0 to 100%) can significantly improve in this case. Care should be taken that the premixing does not result in a gradient start from 0% B and hence introducing the related problem of random pump precompression time in block B. All described effects usually become more severe with increasing operating pressure but play no noticeable role in conventional HPLC at pressures below 300 bar. The extent of all described effects significantly depends on the technical possibilities of pump and autosampler control, and their absence reflects the quality of a modern UHPLC system.

4.2.6 LPG Pumps and How Their Immanent Discontinuous Generation of Gradient Composition May Impact Retention Time Precision in Practice

While an ideal and thus pulsation-free HPG pump does not generate any longitudinal discontinuance in the mobile composition, such discontinuance is system-immanent with LPG pump due to the timed solvent segment aspiration in the low-pressure proportioning valve. This valve generates sequences of different solvent zones, which must be transformed into a more continuous composition by an effective longitudinal mixing device. At the same time, the maximum size of these segments is clearly defined by the operation cycles of the proportioning valve. Proportioning valve cycles are usually matched with the aspiration and delivery cycles of the pump piston that simply follows the proportioning unit. The exact implementation of the proportioning cycles and the respective volumes is vendor-specific and may change markedly between different LGP pumps. A quite common mode of operation is to match one proportioning sequence exactly to one stroke volume of the pump piston and to match the phase accordingly under consideration of the volume delay between the proportioning valve and the pump cylinder. To give an example, we consider once again a pump with 50 μl stroke volume. This pump is meant to generate a flow rate of 500 μl/min and to form a composition of 50% A, 25% B, 25% C, and 0% D. One pump stroke will last 6 s at this flow rate, and this time period must fit the entire proportioning cycle. To do this, the valve may dispense component A in for 3 s, followed by component B for 1.5 s, and lastly, component C also for 1.5 s. Another pump may for specific reasons (e.g., to deal with the mixing challenge) work as follows: 1.5 s A, 1.5 s B, 1.5 s C, and again 1.5 s A. Whatever the exact sequence will be with a pump that matches the proportioning cycle to the stroke cycle, the volume phase of the repeating segment blocks will always be 50 μl with this specific stroke volume, regardless of the respective mixing composition of the two to four components. This implies that the subsequent mixer always has to mix volume packages of a defined size of these 50 μl, no matter what the required composition will be. This is completely different to an HPG with residual pump pulsation that translates into corresponding mixing ripples.

Alternative to the matching of one proportioning cycle to one stroke cycle, one proportioning cycle can also be synchronized with multiple stroke cycles. This is a common practice with pumps that feature small stroke volumes, and this mode may only be activated once a certain flow rate is exceeded. The motivation for this concept is to avoid too short proportioning intervals in the range of 10 ms or even lower. This also relates to the general challenge for an LPG to proportion a small percentage of one solvent component accurately. As soon as the proportioning cycle is extended, the mixer volume must be sufficiently large to still match this (see Section 4.2.13). This describes the key challenge for an LPG to work over a wide flow rate range with acceptable GDV.

It should be mentioned that there are also a few LPG pumps that work according to a different proportioning concept, namely without the strict match of proportioning cycle with the stroke cycle. In these pumps, the proportioning cycle will be matched to the respective flow rate in a way that the timing of the switching cycles for a given composition will not change significantly when the flow rate changes. This is another way to get rid of extremely short proportioning cycles at elevated flow rates, while it automatically adapts the volume packages to a smaller size at lower flow rates. This has the massive advantage that a smaller mixer can be used for smaller flow rates to achieve a matching GDV. Practical experience shows, however, that an LPG pump without the synchronization between the proportioning valve and the pump block does not achieve an equivalent proportioning accuracy and precision. Advanced pumps with a variable stroke volume can match everything and adapt it to the flow rate and mixing volume accordingly. The overall control requirements for such a pump are very demanding, and its technical implementation and manufacturing are cost-intensive.

In spite of all the attempts to match proportioning with pumping and mixing, it is inevitable that LPG-type pumps exhibit a residual composition discontinuance cycle, albeit the amplitudes of this will be very small with high-end LPG pumps. In order to achieve the best possible retention time precision, it is mandatory that these cycles are timed with perfect repeatability from run to run. This can be achieved with synchronization of the proportioning cycle with the injection time point. Normally, the injection occurs at the start of a cycle. This can be done relatively easily with an LPG where the cycle is only controlled by one pump block, and the cycle time is relatively short. Given the example of the pump that delivers 500 μl/min flow with a stroke volume of 50 μl in a one-to-one volume match of proportioning and delivery, the proportioning cycle time results in 6 s. This equals the maximum delay time until the next injection can occur, which can be easily tolerated with analysis cycle time of several minutes or more. For extremely fast analysis cycles of less than 1 min, a wait time of 6 s is already significantly impacting the analysis throughput. For this and for other reasons, such fast analyses should always be accomplished with HPG-type pumps. As mentioned earlier, different vendors handle the match between proportioning and pumping differently, and the same applies for the way the injection synchronization of the pump is carried out. Nevertheless, this synchronization is essential to achieve the best possible retention time precision with LPGs. Some vendors offer the possibility to switch this synchronization on or off, because some users expect more disadvantages than benefits with the sync on. The decision will always depend on the exact application, and also other specifics may play a role here, such as a delayed injection relative to the gradient start. Every user who can influence the sync may be encouraged to test both variants for each method or to consult experts from its vendor on this. Nevertheless, the described relationships should be well reflected, and in doubt, the sync should rather be set on than off.

What's to know LPG pumps exhibit concept-immanent solvent discontinuance due to the proportioning cycles. To achieve the best possible retention time precision in spite of this, the sample injection should be synchronized with these pump cycles. When all the cycles are well matched, an appropriate mixer is installed, and when the injection sync is active, modern LPG systems can deliver similar precision similarly to HPGs, at least for the methods with simple linear gradients at moderate speed (but yet gradient cycles <5 min).

4.2.7 Thermal Effects in Gradient Pumps and How Intelligent Instrument Control Can Minimize the Consequences on Chromatography

The fact that liquids are compressible was already discussed in detail in Section 4.2.1. In addition to the volume change with pressure increase come the thermal effects. In fact, the compression of the liquid in the pump cylinders occurs much closer to an adiabatic than an isothermal mode, because the precompression phase is urged to be relatively short and will not leave sufficient time for thermal equilibration. The extent of the resulting compression heat depends in a relatively complicated way on the molar volume (or density), the thermal expansion coefficient, and the heat capacity of the solvent, while all these parameters depend themselves on the actual pressure [6, 8]. It would be outside the scope of this section to explain this in all physicochemical details. What is important though is that the user should have an idea of the extent of heat-up effects and how they differ with the solvent, which makes it much more challenging to deal with them from the instrument control perspective. Water only heats up by 2–3 °C under a compression from ambient pressure to 1000 bar, while the temperature of methanol increases by 14 °C under the same conditions [8]. Therefore, the effect cannot easily be generally compensated based on the pressure level alone but requires more intelligence in the control algorithm.

Why does the heat-up play such an important role for the pump control in order to generate a smooth and pulsation-free flow? The temperature increase is immediately effective once the precompression has been accomplished. After that, the temperature will not remain constant, as it is now higher than the ambient temperature. So if no compensation algorithm is implemented, the pressure continuously decreases in the respective piston chamber during the inevitable cooldown, hence causing a significant reduction below the original pressure that was generated by the precompression. As soon as the flow is meant to be delivered by the respective piston (handover point, where the delivery of the other piston has stopped), the system pressure will drop until the continuous piston movement has brought the respective system volume back to the pressure at the end of the delivery phase of the other piston. Only after that has been completed, the flow rate comes back to normal. The user will observe a pressure dip as a consequence of this phenomenon, and over time, this creates a pulsation with the frequency of the pump cycle. This pulsation would be very serious at elevated operating pressures of 1000 bar and higher.

Thermal effect compensation is an intelligent technical implementation including the respective pump control algorithm to minimize pulsation due to compression heat effects. It is a crucial prerequisite for smooth detector baselines to account for low detection limits, as well as precise retention times and peak areas resulting from pulsation-free flow delivery and eluent composition. As an example for an intelligent compensation algorithm (that is well known to the author), the adaptive thermal effect compensation (ATEC) of the Thermo Scientific Vanquish H pump will be discussed in more detail with the help of the curves in Figure 4.11.

To facilitate the understanding of the technical background of ATEC, the piston movement is subdivided into five different phases:

1) *Aspiration phase*: Liquid is drawn in from the reservoir at constant atmospheric pressure and constant ambient temperature.
2) *Precompression phase*: Liquid is compressed up to a certain value (x) below the system pressure (comparison of pressure sensor in the pump head with system

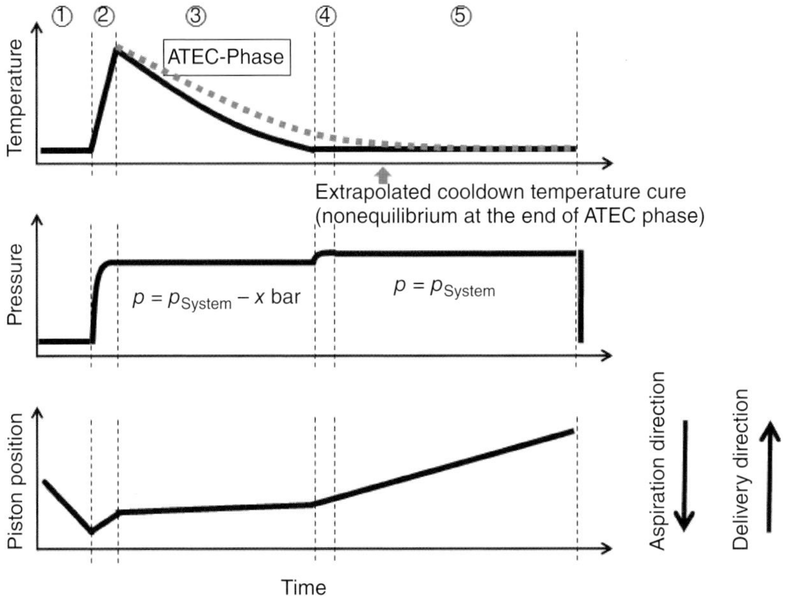

Figure 4.11 The technical and physical steps of solvent aspiration, solvent precompression, and solvent delivery in an HPLC pump operating at elevated pressure level. The scheme illustrates that related thermal effects in the graph on the top. The gray dotted line shows the scenario where the thermal equilibrium is not achieved at the end of the ATEC phase. The ATEC algorithm will then extrapolate the curve and consider the effect for the piston movement in the subsequent delivery phase.

pressure sensor), and the liquid heats up as a consequence of this adiabatic process (see temperature profile in the graph on top)

3) *Eluent heat dissipation and initial ATEC phase*: While the heat in the pump head starts to dissipate, the solvent temperature decreases inside, and pressure would decrease accordingly (without adequate pressure compensation). During this phase, the pressure decrease is monitored and maintained at the precompression level by ultraprecise movement of the piston. The cooldown may be incomplete at the end of this phase, but this will be considered by the pump control software, which is able to extrapolate the cooldown curve. With this information, even the required compensation in the delivery phase can be accomplished (see step 5).
4) *Compress to actual system pressure phase*: The next step is the compression to the full actual system pressure level. Reaching the system pressure level is a prerequisite for the outlet check valve to open.
5) *Solvent delivery phase*: At the end of the fourth phase, the check valve opens and the delivery of eluent to the column starts. It is part of the ATEC algorithm that the piston movement in this phase is not necessarily at constant speed but may still compensate for incomplete heat dissipation in the third phase. This will avoid any pulsation such as pressure drop on the column by this thermal effect and account for the smoothest flow possible.

A **parallel** dual-piston design is an important prerequisite for the very effective ATEC control in this special pump. The complete decoupling of the two alternately working pump heads together with the relatively large stroke volume of 120 μl enables an extended heat dissipation phase (~3 s at 1 ml/min), which is mandatory for the best possible ATEC control. Moreover, the delivery phase is only influenced by the movement of **one** piston. Figure 4.8 explains the operational principle of a serial pump with two coupled camshafts driven by one motor. This design does not decouple the piston movement for eluent delivery between the two contributing pump heads, because the piston of the working cylinder needs to fill the equilibration cylinder at the same time while it delivers the eluent to the column. In this phase, the movement of both pistons has an impact on the pulsation behavior of the pump, which makes it much more difficult to control than with a parallel design pump.

ATEC is an essential contributor to the excellent retention time precision that can be achieved with such a pump. It is also a key prerequisite for obtaining the low baseline ripple in applications using TFA as eluent additive (see Section 2.2.14). The primary effect of the ATEC algorithm is presented in Figure 4.12. A zoom into the pressure trace shows the significant reduction in pump pulsation. With a quite typical but highly compressible eluent such as H_2O/MeCN 5/95 v/v, ATEC is able to reduce the pressure ripple by factor 4 under the given flow rate and pressure conditions and achieve a pulsation level of only 0.25% of the total pressure.

What's to know The heat-up of solvents under a compression to 1000 bar and the consecutive volume contraction when this heat is dissipated may lead to substantial pressure pulsation if it is not adequately compensated in the pump control. An effective compensation of thermal effects is a major quality criterion for high-end UHPLC pumps. It is very relevant in HPG pumps for minimization of detector baselines ripples resulting from UV-active mobile-phase additives (in particular TFA).

4.2.8 Ultrafast Methods with Very Steep or Ballistic Gradients

Extremely fast analysis cycles down to significantly below 1 min are one of the use cases of modern UHPLC instruments. If such analyses are performed with gradient elution, there is a high demand on the pump technique that must be able to change the eluent composition accurately and precisely in a very short time. Such gradients are often referred to as ballistic gradients. The demand is that the pump potentially needs to change the composition over a wide range in only a few seconds. Gradient times of 10–30 s are not uncommon for ballistic gradients. An HPG pump provides clear advantages for ballistic gradients and is actually the only option to perform them adequately. The only limitation with an HPG pump is the maximum positive and negative acceleration rate of the two pump blocks, because the gradient is formed only by the individual flow rate changes of the two pumps. The key requirement is a sufficient time resolution for the pump control, which can be limited by the driver, firmware, and

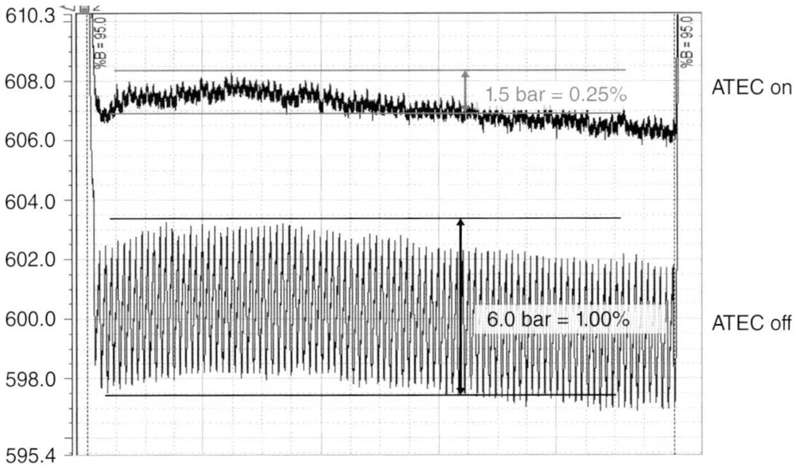

Figure 4.12 Damping of the residual pulsation resulting from compression heat effects to demonstrate the benefit of the automated thermal effect control (ATEC). Pump: Vanquish H Pump (Thermo Scientific), mobile phase: MeCN/H_2O 95/5 v/v, and flow rate: 2.0 ml/min delivered against a restriction capillary at 600 bar.

internal electronics. The mechanical inertia can also be a limiting factor, in particular when the flow has to be quickly ramped up against very high backpressures. Modern UHPLCs are typically powerful enough, and the maximum steepness is only limited by the control. To generate a smooth and ripple-free flow ramp, the software should allow control at every percent step (better every 0.1% step) accurately at the respective timepoint in the gradient program. Modern UHPLC-HPG pumps normally update the flow rate at a frequency of 100 Hz or higher. At a control rate of 100 Hz, even a gradient from 0 to 100% in only 10 s would be resolved into 1000 separate steps, each step corresponding to 0.1% composition change. This fully meets the requirements defined earlier and makes such a pump ideally suitable for ballistic gradients.

Due to the completely different operational principle of an LPG pump, the considerations and challenges to generate ultrasteep gradients are entirely different. In an LPG pump that matches the proportioning cycle to the stroke cycle, the resolution of setting the individual composition changes over time can never be higher than the given stroke frequency. Every new stroke is thus a new mobile-phase composition, and a linear gradient is in fact only a sequence of step gradients with a certain resolution. Within one stroke, no controlled composition change can be accomplished. As the steps are defined by the stroke volume and correspond to volume packages, the obvious consequence is that the flow rate also matters for the maximum steepness of LPG gradients, which is not the case with an HPG. The number of steps or packages and the % change in composition can be calculated with the following equations:

$$n_{Steps} = \frac{t_G \cdot F}{V_{Stroke}}$$

$$\Delta\%_{Step} = \frac{\Delta\%_G \cdot V_{Stroke}}{t_G \cdot F}$$

The second equation allows calculation after inserting typical values such as a gradient range $\Delta\%_G$ of 50%, the often referenced stroke volume V_{Stroke} of 50 µl, a flow rate F of 600 µl/min (common for fast runs on 2.1 mm columns), and a gradient time t_G of 1 min that the %−age change per step $\Delta\%_{Step}$ equals 4.2%. If a full-range gradient ($\Delta\%_G = 100\%$) is demanded, each step would even be $\Delta\%_{Step} = 8.4\%$. With this, it becomes very obvious that the demanded maximum of 1% change per step for a smooth and well-controlled gradient can never be achieved for such steep gradients with an LPG pump of this design. Resolving the second equation for the gradient time t_G and inserting the earlier used values of $F = 600$ µl/min, $V_{Stroke} = 50$ µl, $\Delta\%_G = 50\%$, and the demanded $\Delta\%_{Step} = 1\%$ yields a minimum gradient time of 4.2 min to fulfill the 1%-step requirement. The duration of 4.2 min cannot really be considered a fast gradient, and if the gradient range needs to be increased to more than 50%, it becomes even worse.

To elucidate the possibilities and limitations of an LPG with 50 µl stroke volume, the schematic representation in Figure 4.13 should be useful. For this graph, the representative values of $F = 600$ µl/min, $V_{Stroke} = 50$ µl, and Δ%G = 50% are once again considered. The graph illustrates the shape of the gradient for the four different scenarios of $n_{Steps} = 64$, 32, 16, and 8 and lists the resulting step heights $\Delta_{\%Step}$ and gradient times t_G. With the given size and resolution of the graph,

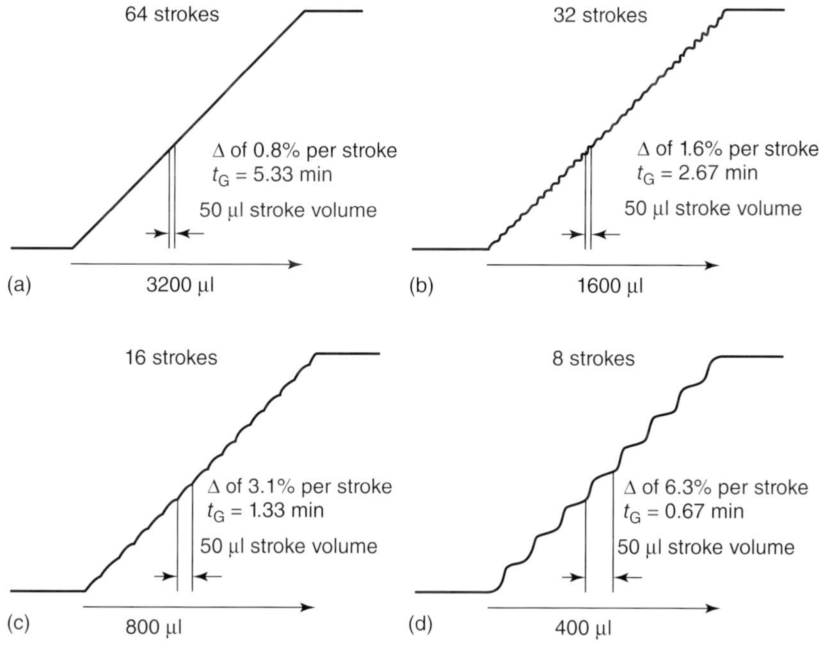

Figure 4.13 Relationship between the resulting waviness of a linear gradient formed on an LPG and the resulting gradient volume at a given flow rate as a function of the gradient slope. Conditions: F = 600 µl/min, proportioning cycle synchronized to 50 µl stroke, V_G/t_G of (a) 3200 µl/5.3 min, (b) 1600 µl/2.67 min, (c) 800 µl/1.33 min, and (d) 400 µl/0.67 min.

the gradient with 64 steps already appears perfectly linear, while all other profiles seem to be more or less waving. The waved gradients may well be perfectly repeatable from run to run and not impact the retention time precision. Nevertheless, the separation will deviate from that generated with an HPG that does not generate these waves. It will strongly depend on the individual application and how much these gradient waves matter for the peak resolution. Experience tells that deviations in the chromatogram become obvious in most applications once the step height exceeds 2% composition change. From this rule, it can be concluded that for typical UHPLC flow rates of 0.5–0.8 ml/min, stroke volumes in the order of 50 μl, and a gradient range of 50% or even more, the shortest possible gradient time is between 2 and 3 min. As the minimum reasonable gradient time correlates with the stroke volume, a reduced stroke volume would be a possible way to make LPG pumps compatible with steeper gradients. Cam-driven pumps do not have this degree of freedom, and they must be constructed and built for smaller stroke volumes from the scratch, or a drive control that allows a direction change must be implemented. The latter is not available in any of the HPLC pumps on the market at this point in time. A pump with linear drives provides this freedom, but with the reduction of the stroke volume, the next challenge arises, namely the precise control of the proportioning valve working at a higher frequency. Given the example assuming 600 μl/min flow and 50 μl stroke volume, one stroke lasts 5 s. So for the proportioning of 1% of a certain component, the valve must open the respective channel for a time of 50 ms. This is already almost mechanically impossible, but the reduction of the stroke volume with the related increase of the pump frequency would even reduce this dispense time further. It could be concluded that the challenge of LPGs for steep gradients could easily be tackled at lower flow rates, for example, 100 μl/min for a fast run on 1 mm columns. It must be considered though that the best GDV-optimized LPG systems still exhibit dwell volumes of 250 μl and more. At a flow rate of 100 μl/min, it would already take more than 2 min until the gradient arrives at the column. So it must be concluded that whatever the method conditions and optimization attempts would be, LPG systems are not capable of running gradients shorter than 2–3 min at flow rates significantly below 1 ml/min. If such fast method should be run on LPGs with acceptable gradient delay times, the user must change to column internal diameters, of minimum 4 mm and accept the high solvent consumption.

The challenge of the system GDV in the context of ultrafast gradient methods has been mentioned now. Before the determination of the GDV will be discussed in detail in Sections 4.2.9–4.2.11, the implications of a too large GDV are briefly described here and solutions are discussed. If in the course of a method transfer, the ratio of V_G to the instrument GDV changes substantially, it can result in changes of relative retention up to inversion of the elution order. This can only be generally avoided if the GDV is kept in strict proportion to the gradient volume, or it must be more than factor 10 smaller than V_G. This is, however, hardly possible given the small gradient volumes applied in ballistic gradients. In several cases, it can even happen that some substances already elute before the gradient arrives at the column. This is often the case when the retention capacity of column is relatively low, so that already the elution strength at the gradient start can elute the weaker retained substances. An extreme example of this is shown

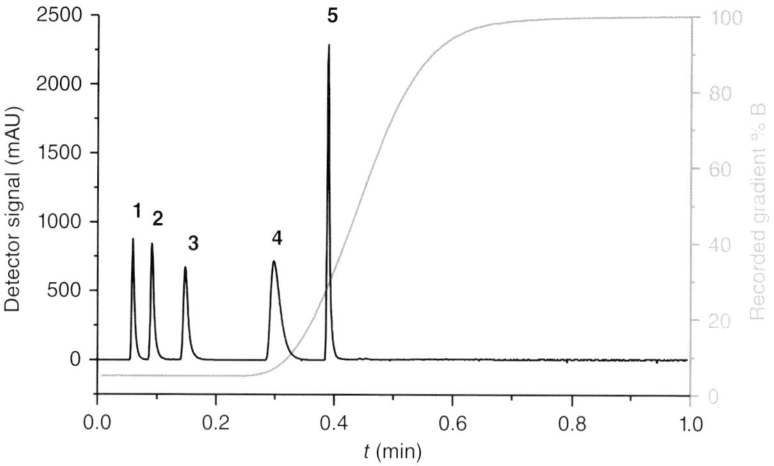

Figure 4.14 Example of a chromatogram with elution of the majority of analytes prior to arrival of the gradient at the column, due to inappropriate GDV. Column: MICRA NPS 1.5 µm ODS 1; 33 × 4.6 mm, flow rate: 4.0 ml/min, gradient: H2O/MeCN 5–100% linear in 0.3 min, GDV: 1200 µl, and peaks: thiourea (1), methylparaben (2), ethylparaben (3), propylparaben (4), and butylparaben (5).

in Figure 4.14. The retention capacity of the column for small molecules such as parabens is relatively moderate because it is a nonporous material with very low specific surface area. The gradient arrives only at 0.3 min at the column, which is so late that solely butyl paraben is eluted within the gradient while the other three parabens elute isocratically prior to gradient arrival. This is also reflected in the peak shape. The peak width markedly increases for the first three compounds, and the substance eluted in the gradient shows a sharp peak due to the gradient peak compression effect.

If the GDV of an instrument is too large for a certain application with steep gradient or if even the effect shown in Figure 4.14 appears, the user has several options to counteract this:

- Reduction of the physical GDV by swapping the mixer or sample loop
- Reduction of the effective GDV by switching the injector back to load after the sample is flushed out of the loop (loop bypass)
- Delay of the sample injection relative to the gradient start.

A size reduction or even removal of the gradient mixer to achieve a smaller GDV definitely requires caution. It was intensively discussed in the foregoing sections how important thorough longitudinal mixing may be for the continuity of mobile-phase composition. A complete removal of the longitudinal mixer is only recommended with HPG instruments that have an advanced control algorithm for low pulsation. Details on the technical implementation of mixers will be discussed in Section 4.2.12. In several cases, depending on the pump design and individual application demands, it may well be possible to exchange the default mixer against another one with smaller volume. After the change, it is highly recommended to requalify the detector baseline as well as the peak resolution and retention time precision, as all these criteria may depend on the mixing volume. If they still meet the requirements, the smaller mixer can be used for the given application.

The early switch of the injection valve to the load position (sample-loop bypass) may help to significantly reduce the GDV of an HPLC system. In systems with highly optimized HPG pumps but flow-through needle samplers with extended volume range (e.g., up to 100 µl or more), the relative GDV reduction will be substantial. If the sample dosage is based on a metering device, meaning that the high-pressure compatible syringe is also part of the sample loop, the GDV contribution of the injector may well be three to four times higher than that of the pump. While there may be a very beneficial effect of sample-loop bypass, this measure also requires caution. It is a prerequisite that the sample is sufficiently flushed out of the loop, at the same time, the gradient start must not yet have arrived at the sample loop. If the switch to LOAD occurs too early before the sample is completely flushed out, the peak area precision may be negatively affected, and severe sample carryover to the next analysis may be observed. If the switch occurs too late when the gradient start already has entered the loop, the first part of the gradient will be cut off, and the elution strength increases with a step at the start of the separation. This may strongly affect the early part of the chromatogram. In addition, a portion of mobile phase being stronger than the composition at gradient start will be stored in the sample loop and may be switched onto the column with the next injection, if the loop is not flushed before a small volume of the next sample is drawn. This will further change the eluting conditions at the start of the gradient. As a rule of thumb, it is strongly recommended to flush out at least the threefold injection volume before the switch to the LOAD position. This leads to the prerequisite that the maximum injection volume equals 1/3 of the GDV contribution of the pump alone. Modern HPG pumps can have a GDV as low as 30 µl, which results in a maximum injection volume of 10 µl. For modern UHPLC methods, this limitation is usually acceptable, but it must always be considered. If the sample volume is anyway low, the user may also check if the sample loop can be changed to a smaller volume and reduce the GDV contribution of the autosampler by this, instead of the early switch back to LOAD. The loop bypass also requires that the control software allows the user to switch the injection valve at any time in the program. Ideally, the software fully automates this by taking into consideration the injection volume and presample-loop GDV. However, this volume would be changed by the user once a mixer is swapped. So, automation routines for the switch back also require some caution and should be well documented in the software, in particular for regulatory compliance. In any case, the user should always thoroughly check the detector baseline, retention time precision, and sample carryover before such a sample-loop bypass method is implemented for routine use.

The delayed injection to wait for the arrival of the gradient is the third option, which must be assessed only by an apparent GDV reduction rather than a physical one. For this procedure, the time program to launch the gradient generation in the pump must start before the point in time of injection valve switch. The gradient program will be timed as usual, but the injection time will not be at zero but shifted by a defined time. This time shift will be determined based on the requirement of how much the gradient delay has to be shortened. The procedure requires that the sample is already drawn into the loop when the injection command occurs and that this entire process is supported by the control software. It should also be avoided that the sample stays already a longer time (several minutes) in the loop, because it will spread by diffusion, which may negatively affect separation efficiency. This negative influence can be very relevant in highly optimized UHPLC methods with small injection volumes ($\leq$1 µl), while conventional methods on a larger volume scale are much less prone to it. A full software automation of the injection delay with appropriate timing of all required steps in the autosampler is the ideal scenario. The timing for the sample draw will significantly depend on the sample volume, draw speed, equilibration times, and the applied needle wash procedure. Eventually, the resulting total time demand must be matched with the gradient start, so there is a lot to calculate and consider for the control software. The sample draw and needle wash cycle may also well be longer than the time of injection delay, so the gradient start should occur during the sample draw (and needle wash) cycle. If all this is well implemented in the software, the delayed injection may be a very effective method to avoid a late gradient arrival after elution of the first analytes. It may also facilitate transfer of a method to an instrument with a larger GDV than the original one. Unlike the two procedures described earlier (physical GDV reduction by mixer or sample-loop change and sample-loop bypass), the delayed injection will not or hardly reduce the analysis cycle time. The chromatogram will well be shorter, which may lead to a misconception, but an equivalently longer time elapses before the analysis actually starts, namely the injection delay time. At least for instruments and methods where the autosampler cycle time is shorter than the injection delay, there will be absolutely no gain in the analysis throughput.

What's to know To run fast analyses using steep gradients and relying on a small GDV, an HPG-based instrument is generally recommended. Modern highly optimized LPG instruments may also accomplish gradient times down to 2–3 min without significant composition waving from the proportioning cycle but cannot go below that without any impact on the chromatogram. A negative influence of a too large GDV can be addressed by reduction of mixer volume, sample-loop volume, or sample-loop bypass by switch back to LOAD position. For all these changes of the instrumental setup or method, the user must verify that there is no negative impact on the chromatographic result (check detector baseline, retention time and peak area precision, sample carryover, peak shape, and resolution). A delayed injection can also mimic a reduced gradient delay volume but will not help to increase throughput, as the analysis cycle time will virtually stay the same.

4.2.9 Fundamental Considerations on the Determination of a Gradient Delay Volume

Before going into detail on the practical determination of GDV, there are several general explanations required for a fundamental understanding of the related challenges. This will help to avoid misconceptions and resulting errors but also nurture the understanding of deviations in the results due to different ways of GDV determination. It must be taken into account that most methods depend on the addition of UV-active compounds that are dissolved in one of the mobile-phase components to form the gradient. The chromatographic and spectral properties of these compounds in the respective solutions can have a marked influence on the experimental results. To begin with, there is a set of fundamental rules to follow:

- GDV determination using UV-active marker substances must always be accomplished without a column, as the inevitable retention of the marker compound on the column would influence the result.
- So the column must be replaced by a restriction capillary that must generate the specified minimum operating pressure of the pump under the selected flow rate and mobile-phase viscosity.
- Unlike the normal operation mode, all pump channels must contain the same solvent (mostly water) at least for the standard Dolan tests, because spectral shifts of the marker compound under a changing solvent will affect the result by absorbance change that is nonproportional to the gradient composition change.
- All fluidic elements behind the column connector are not part of the GDV, and their contribution to the test result must be subtracted. It is often overlooked that detector flow cells may exhibit substantial heat exchanger volumes in front of the irradiated detection channel that also contribute to the delay.
- Evaporation of the marker compound or the solvent in the vacuum degasser in front of the pump may alter the effective concentration of the marker solution and with that the experimental result (at least for Dolan tests). This effect is more pronounced at low flow rates and after the solution has been standing in the degasser cartridge for a longer time before the flow on this channel is initiated. In practice, this effect will lead to a dependence of the result on the applied flow rate.
- Due to the compressibility of solvents, the observed time to overcome a given physical GDV is generally a function of pressure and the properties of the solvent. The delay time close to atmospheric pressure at a given flow rate is the smallest possible. It increases with increasing pressure, and this effect is stronger with highly compressible solvents. As described in Section 4.2.1, the observed gradient delay with methanol as mobile phase will be 10% longer at 1000 bar than it is at atmospheric pressure. This effect will not impact the proper GDV determination at pressures below 100 bar and with water as solvent. It explains, however, why the gradient may arrive with more delay under other conditions. This rule applies to the vast majority of pumps, where the set flowrate refers to the aspirated volume at atmospheric pressure.
- If the gradient pump is equipped with a compressible pulse damper, then even the physical GDV of the pump is a function of the applied pressure beyond the compressibility of the solvent. This phenomenon has similar consequences than the solvent compressibility alone, but it is usually of a much bigger extent and is equally strong with water as mobile phase. An approximate value of 1 μl GDV change per bar can be found in the literature for a commercial

instrument [9]. In pumps with a pulse damper, it is always recommended to perform the GDV determination with different restrictors to cover the entire pressure range at which the instrument is operated. The differences in the result will give an impression on the amount of the GDV deviation.

All these points must be taken into consideration by the user for planning of the GDV determination experiments and for the interpretation of the results. Some of the effects will be discussed in more detail in the following sections. Before explaining two methods based on a UV marker and without the use of a column, it should be mentioned that there are also ways to calculate the GDV from chromatographic results using a column. This is commonly based on the elution behavior of homologous series and requires advanced calculations but is reported to result in a superior accuracy across different instruments and conditions. Magee *et al.* described and evaluated such an approach, and the article also provides access to software for performing the required calculations [10].

What's to know There is hardly any method for GDV determination that can reliably deliver absolutely accurate and precise results. Users have to consider all experimental details thoroughly and should understand the possible systematic error resulting from deviations. It is therefore mandatory to describe the experimental details exactly and comprehensively and always call the method that was used for GDV determination together with the specified value.

4.2.10 The Marker Pulse Method as a Quick Way for GDV Determination

With this method, an instrumental setup with replacing the column by an appropriate restrictor capillary is applied. Normally, both channels are filled with water, but a UV-marker compound (e.g., acetone, caffeine, benzylic alcohol, etc.) is added to the water in the second channel. Alternatively, a typical HPLC solvent with some UV absorption (e.g., methanol at wavelengths <215 nm) can be used instead of a solution with a marker compound. A method is programmed that uses a relatively low flow rate (100–250 μl/min) depending on the pump type and expected GDV. The method comprises a short pulse of the marker-containing component shortly after the start of the data acquisition. To generate that pulse, a gradient step with a very large change in composition is programmed, and then the high content of the component with marker is kept for several seconds and is followed by a step back to the initial gradient composition. This initially rectangular concentration profile of the marker will now travel through the gradient mixer and autosampler and undergo the normal dispersion phenomena that turn it into a more Gaussian-like concentration profile. The profile can be recorded by the UV detector and integrated in the same way as a chromatographic peak in order to automatically determine its elution time. This elution time must now only be multiplied by the flow rate, and the resulting volume is equal to the GDV. To correct for the delay and width of the marker pulse in the gradient program, the time that elapses from data acquisition start to the center of the rectangular pulse should be subtracted

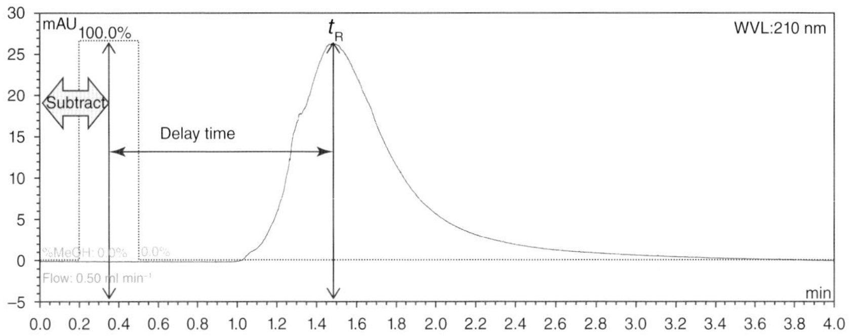

Figure 4.15 Application of the marker pulse method for GDV determination by direct detection of the pure component B (methanol). Instrument: UltiMate 3000 Quaternary SD System (Thermo Scientific), pulse between 0 and 100% from 0.2 and 0.5 min (water to methanol), flow rate: 0.5 ml/min, and detection: UV at 210 nm.

(see "subtract" in Figure 4.15). The result should also be corrected for the contributing postcolumn volume (that can only be accurately determined with a second detector in line). The relatively low flow rate is selected to minimize the influence of the marker pulse timing precision. In spite of the application of low flow rates, this method is much faster at executing the well-established Dolan tests (see Section 4.2.11). The biggest advantage over the Dolan test is the fully automated processing of the results by standard peak integration. Short run times and automated data processing easily allow for 6–10 repetitions to assess the precision of the method. There are several possible sources of systematic errors to be considered with the marker pulse method:

- It must be excluded that the UV marker is retained at the wetted inner surfaces of the instrument that can be composed of materials with substantially different properties (see also Chapter 7). While this cannot easily be proven, it already helps if different marker compounds (e.g., acetone, caffeine, benzylic alcohol) result in the same elution time. This allows the conclusion that hardly any retention can be assumed, but the test requires additional effort. If the results are different, the user should be warned and also change the carrier solvent for additional experiments. The smallest elution time observed should be considered for the calculation. The risk of marker retention can be considered minimal if pure methanol is used to generate the pulse (see Figure 4.15).
- The pulse time must be short enough that the zone arrives at the detector with a sharp maximum, but it must also be long enough to allow the pump for an appropriate control of the timing, and it can never be shorter than the smallest time increment that the gradient editor enables. A duration of the marker pulse of 0.1 or 0.2 min proved to be effective In practice. If the resulting peak shows no sharp maximum, the time of the pulse should be shortened. If the height of the peak is too low to distinguish from noise, or the precision of the elution time is not appropriate (RSD > 3%), the duration should be extended. The marker concentration and the detection wavelength are other effective ways to influence the resulting peak height.

- With an HPG-type pump, and in particular when running this at low flow rate, the baseline for the experiment should not be at 0% B. Starting the B-block from zero introduces uncertainty due to the solvent compression lag (see Section 4.2.5). It is recommended to switch the HPG between 20 and 80% for the marker pulse. With that, both blocks continuously run at reasonable flow, but still the pulse is significant.
- As in many other respects, the situation is completely different with an LPG-type pump. Here it is highly recommended that the marker pulse is set between 0 and 100% so that the proportioning always opens only one channel and does not need to accomplish more complicated mixing pattern. It is also relevant in this context that the switching sequence of the proportioning valve is synchronized with the piston stroke of the LPG. Therefore, the flow rate should not be selected too small relative to the stroke volume, ideally not less than one stroke during the pulse duration, which is hardly possible at flow rates significantly below 1 ml/min. In any case, the synchronization of the stroke cycle with the injection must be activated. To profit from this in order to attain the proportioning cycles and thus the timing of the marker pulse in sync from run to run, there must be an injection applied with this test. The smallest possible volume (<1 μl) of the pure solvent (normally water) from channel A should be injected. The possible baseline disturbance by this injection will arrive so much earlier at the detector that it should not affect the integration of the marker pulse peak.
- The technical design of the mixer (see Section 4.2.12) can also have an influence on the shape of the marker compound peak. Small-volume longitudinal mixers that operate based on discrete channels of different lengths (e.g., Agilent Jet Weaver [11] or Knauer SmartMix [12]) may show even split peaks under certain conditions. If this is the case, the pulse duration should be increased until unimodal peak shape, even if with not perfectly sharp summit, is obtained.

If all listed points are taken into consideration, the marker pulse method normally provides very reasonable GDV results and is relatively easy to accomplish. The disadvantage is that it is hardly described in the literature if at all and is therefore not commonly accepted as a reliable and well-transferable procedure for GDV determination. A big advantage of the marker pulse method with the switch from water to another solvent (such as methanol) is that the pulse can also be detected, by most nebulizer-based detectors (evaporative light scattering detector, charged aerosol detector, mass selective detector, etc.) and is thus not limited to the presence of a UV detector.

What's to know Marker pulse methods especially by simply applying a methanol pulse with the gradient pump and detecting its arrival at the detector by absorption at low wavelength are a very fast and fully automated method for GDV determination, although not standardized and commonly accepted. The pulse should be as short as the gradient control allows for but not shorter than 1 s. It should be applied with HPGs between 20 and 80%. With LPGs, it should be performed between 0 and 100% with injection synchronization activated and with applying a blind injection (small volume of solvent from channel A). Low flow rates are recommended, but pump specifications must be considered.

4.2.11 The Dolan Test as the Classical Method for GDV Optimization

The Dolan method [13] essentially applies the same experimental setup as the marker pulse method with a restrictor instead of column and a detectable solution in channel B. The principle is well described in the literature and can be found in many textbooks on HPLC [14] and is also often cited in lab journals [15]. It is briefly described in Chapter 9 of this book but is addressed in more detail in this section. To repeat a very brief description, a linear gradient is normally programmed from 0 to 100% with a slope of 10–20 Δ%/min at typical flow rates of 0.5–1.0 ml/min. This gradient is typically run from pure water to water that is spiked with less than 1% acetone, and the gradient ramp is detected with UV mostly at 254 nm. The determination of the offset relative to the programmed gradient is usually done manually on the printout according to Figure 4.16. The time offset is also multiplied with the flow rate to calculate the GDV, while a correction for the postcolumn volume should also be applied to be accurate.

Next to its very broad acceptance for GDV determination, it allows to completely trace the characteristics of a linear gradient profile including the re-equilibration phase. This provides a good estimation on the gradient formation quality and the flush-out behavior of the mixer. The single biggest disadvantage of this test is the required experimental time, especially if a statistically relevant number of repetitions are performed. The time can be shortened by running a higher gradient slope (30–50 Δ%/min). While this is already a deviation from the typical protocol, it is also important that the slope does not become steeper than that the pump can form at the given flow rate with appropriate linearity. The practical quality indications are the width and smoothness of the resulting linear ranges and the precision of the calculated GDV values. The extreme variant of a step gradient from 0 to 100% in zero time is also often discussed for GDV determination but has nothing in common with what is referenced as Dolan test. In most instruments, the step gradient

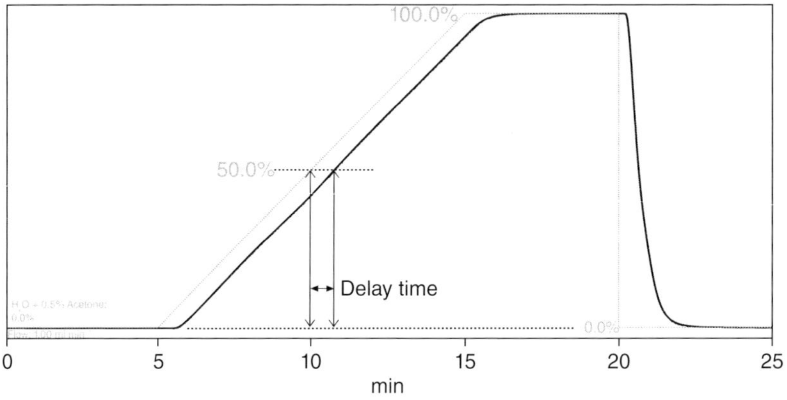

Figure 4.16 Example for a typical Dolan test result. The round edges in the part of the re-equilibration back to initial conditions are also indicative of the gradient mixer characteristics. Deviations from perfect linearity can also be detected but may sometimes be a bias from the determination method (degasser effects).

does not yield any linear part where the distance to the programmed step can be reliably determined. Step gradients can be very effective to characterize a system for its mixing behavior but are not at all recommended for GDV determination.

Even when closely following the standard protocol of a Dolan test, there may be substantial deviations from the true gradient formation characteristics of the instrument resulting from evaporation effects in the degasser. With the typically applied and quite volatile marker acetone and depending on the degasser vacuum performance, the acetone concentration may drop to virtually zero in the complete volume of the degasser cartridge, which can vary from 0.5 ml to several ml. With that, the recorded gradient is delayed by the degasser volume after the acetone solution stands for a certain time period in the degasser. It will actually normally not result in a straight delay, but in a waving profile, because the volume in the solvent line behind the degasser is not affected and arrives first at the detector. When a relatively large degasser volume, a relatively low flow rate, and a relatively steep gradient come together, a worst-case scenario occurs where the majority of the gradient ramp may deviate from linearity. It is needless to mention that this will affect the accuracy of the GDV determination. If severe deviations from linearity are observed, degasser effects should always be considered. There are three possible ways to modify the Dolan test in order to correct for degasser effects:

1) Use a less volatile marker instead of acetone (e.g., caffeine, benzyl alcohol). It may be necessary to adapt the concentration and/or detection wavelength in order to keep the absorbance at 100% B-level safely in the linear range of the detector. With compounds that are less volatile than water, the opposite effect may occur in the degasser, and the marker will be enriched due to water evaporation. This effect will also result in deviations from a proper tracing of the gradient ramp, but due to the much lower volatility of water relative to acetone, it will be much less pronounced.
2) The gradient is not started from 0%, but from a level of 5 or 10%, in order to avoid that the solution is standing in the degasser channel prior to the start of measurement.
3) The gradient is run into the opposite direction from 100% to 0%, which implies that the channel with marker is running at full flow prior to the start of the experiment. If the marker is added to channel A instead of channel B, this principle can even be applied to characterize the typical direction of the gradient.

Each of these workarounds to address the degasser effects implies a deviation from the standard Dolan process, which will certainly affect its common acceptance. The second option does in addition not characterize the complete gradient range. It is also a reality that none of these measures can completely eliminate all degasser effects, in particular when the degasser is very effective and the flow rate is low. A combination of the first action with either the second or the third should normally suppress the degasser effects to negligible minimum.

What's to know The Dolan test is the best defined and most accepted method to determine system GDV and is also considered valuable for a comprehensive characterization of the gradient linearity and mixer flush-out behavior. Both the results and the recorded gradient shape with the test can be biased by evaporation effects in the solvent degassers. Degasser effects can be minimized by increasing the flow rate and decreasing the volatility of the marker compound (e.g., caffeine instead of acetone).

4.2.12 Designs of Mixers and Their Effectiveness Relative to Their GDV Contribution

The basic requirements to a gradient mixer and the distinction between radial and longitudinal mixing have already been elucidated in Section 4.1.2. This section gives a deeper insight into the possible designs and classes of gradient mixers and their individual characteristics.

The most known classification is between the two fundamentally different types of static and dynamic mixers. Dynamic mixers work on the basis of a moving element inside the mixing chamber. This is typically a stirrer that can be accomplished even at a moderate scale by a small magnetic stir bar. Dynamic mixers can be very effective for both radial and axial mixing, while the properties for axial mixing depend markedly on their size, similar to the behavior of most static mixers. While the mixing is generally very effective with respect to homogenization, dynamic mixers are very difficult to miniaturize below a volume range of 200–300 μl. Once the magnetic stir bar falls below a certain size, the magnetic forces become too weak to account for a solid coupling especially at high frequencies and elevated eluent viscosities. A small size inevitably leads to high linear velocities at elevated flow rates, and this may also disturb the rotation of the stir bar. The stir frequency is essentially an adjustable parameter to optimize the effectiveness of a dynamic mixer for the given requirements that may even change with the mobile-phase composition during a gradient. As the best mixing effect cannot easily be experimentally determined under the variation of the stir frequency, it is not trivial for the user to optimize this parameter. This is the main reason why the stir frequency cannot be changed with many commercial dynamic mixers but is optimized by the manufacturer for a specified flow rate range. In addition to the relatively large volume, dynamic mixers exhibit other disadvantages. The mechanical friction of the stir bar may lead to abrasion, and small particles may be released to frits and sieves in the instrument of column or, in the worst case, to the autosampler fluidics where they can result in clogging problems or contamination. There is also a residual risk that the magnetic bar decouples from the rotating stirrer, and the device will at least temporarily drop its performance completely. For all these reasons, dynamic mixers are not very widespread in modern analytical HPLC, and they are noncompatible with UHPLC where the risk of clogging from the stir bar abrasion is very high due to miniaturized fluidics and sieves and frits with small mesh sizes.

Static mixers are much more common, they are the only mixers used in UHPLC instruments and can exhibit very different technical and geometrical designs.

The classical static mixer is the so-called mixing column, normally an HPLC column packed with nonporous beads with inert surface, commonly glass beads with a size of 100–200 μm [16]. A key requirement to the packing material of these columns is that the particles have no capacity to selectively retain a solvent component, which may disturb the eluting gradient profile. The particle size must be small enough to form a stable packing that generates effective radial mixing convection but large enough to avoid a significant back-pressure generation. Large particles generate, according to the van Deemter theory, a pronounced eddy dispersion (A-term), and this equals a longitudinal mixing effect. Although 100–200 μm range proves to be relatively effective in this respect, it is a reality that mixing columns are not very powerful longitudinal mixers relative to their volume. Diameters below 2 mm can hardly be packed with such large particles, and a regular packing is, by definition, impossible once the particle diameter exceeds 1/10 of the column diameter. Reduction of the mixing column length reduces the effectiveness markedly so that lengths below 10 cm are relatively useless but would even at a small diameter of 2 mm create a volume of 120–180 μl depending on the packing density. Packed columns are never used as gradient mixers in modern UHPLC instruments.

A relatively common design that is also partially used in modern UHPLC equipment is the metal-frit-based mixers of cylindrical geometry. The frit can be made of stainless steel or other iron-free alloys in the so-called biocompatible pumps (see also Chapter 7). They exist in sizes from 10 μl up to several ml, but the smallest versions act mainly as filter frits and hardly perform as mixing devices. If the morphology of the frit material is appropriate and the geometry for inlet and outlet is also well designed, they can act as both radial and longitudinal mixers, but the longitudinal properties are normally superior. A relevant longitudinal mixing effect that meets the needs in HPLC pumps is, however, hardly achievable below a frit volume of 100 μl, regardless of the design. The truly effective frit mixers have volumes in the range of 200–800 μl.

For the radial mixing effect in an HPG pump, even the simple T-piece is already quite effective, at least as long as both partial flow rates are in the same order of magnitude so that the mutual penetration of the solvents based on inertia is well balanced. This is obviously not fulfilled when only 1 or 2% of one component needs to be effectively mixed with 99 or 98% of the other. At the same, it is a known phenomenon that at equal flow rates (between 48 and 52% mixing ratio), there often occur resonance effects that rather generate oscillation than homogeneous mixing. A classical, but still relatively common, radial mixing device is based on an array of intelligently designed fine structures inside the central volume of the mixing-T, the so-called spin chambers. These structures practically resemble a labyrinth on the disk with channels that lead into cyclone chambers to induce the spin. A stack of small disks with these cyclone chambers is accomplished that generates effective local spins that spread in radial direction from layer to layer to distribute over the entire width of the chamber. Radial mixers based on that principle are commercially available down to the mixing volumes of 10 μl but also exist in the dimension of 200–300 μl. Their very effective radial mixing capabilities also qualify them as mixing chambers in reaction detectors since the early days of HPLC [17]. A commercially available T-piece-like

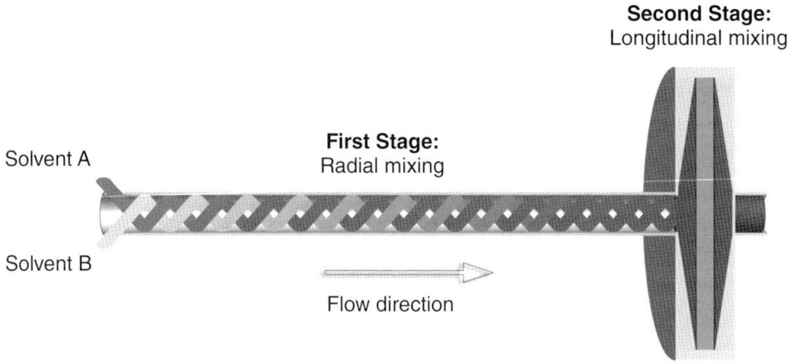

Figure 4.17 Schematic representation of the technical concept of the SpinFlow™ mixer (Thermo Scientific) [19]. It is a two-stage mixer that combines a radial mixing capillary with a helix structure inside (first stage) with cylindrical or disk-like frit mixer for subsequent longitudinal mixing (second stage).

mixer with such internal spin chambers is the so-called Visco Jet™ micromixer (Lee Scientific) [18].

Excellent radial mixing performance at relatively low mixing volume can also be achieved in mixing capillaries that contain an internal helix structure to set the liquid streams into a rotation around a longitudinal axis. For an HPG pump with pulsation-free flow generation, such a mixing capillary mounted behind a simple T-piece is the ideal mixing device, as no additional longitudinal mixing is required. If additional longitudinal mixing in an HPG is desired, the radial mixing capillary can be connected to a subsequent frit mixer. This concept of a two-stage mixer is depicted in Figure 4.17. In the first stage, the capillary with the internal helix accounts for the radial mixing of the two components, while a subsequent cylindrical frit mixer with optimized aspect ratio accounts for longitudinal mixing.

A critical requirement for an effective longitudinal mixing is to provide the liquid a multitude of paths with different lengths that are taken with an equivalent probability. This generates different run times to overcome a certain length of the mixer bed and thus longitudinal mass exchange once the different streams are subsequently combined to one common volume. The smaller the diameter of the frit, the closer will be the different path lengths and the less effective over a given length is the longitudinal mixing effect. If the diameter becomes too wide, it is not equally probable that the liquid streams through the outer channels. When increasing the diameter, the longitudinal mixing performance does not evolve proportionally to the volume and the flush-out characteristics become more and more asymmetrical. All frit-based mixers provide a multitude of random paths of different lengths similar to a packed column but no discrete channels.

The approach to split the liquid stream into different partial streams, guide them along defined paths of different lengths, and subsequently combine them has led to the microfluidic mixers. A relatively early version of a microfluidic mixer is the so-called SmartMix™ device (Knauer) that can be used for HPG- and LPG-type pumps depending on the operation mode and plumbing [12].

A further development of microfluidic mixers, the so-called Jet Weaver (Agilent Technologies) [11], was already described in Section 2.1.2. Well-designed microfluidic devices exhibit excellent mixing properties relative to their volume and can generate effective longitudinal mixing already at mixing volumes below 100 μl. When these devices are used, it is absolutely critical that no particles can enter the channels as this would lead to partial clogging and strongly impact on the mixing effect. Therefore, an appropriate in-line filter frit must always be mounted in front of a microfluidic mixer. Moreover, microfluidic mixers are relatively complicated and expensive to manufacture, and it is not straightforward to scale them to different volumes with respective scaling mixing capabilities. This scalability is much less difficult with frit mixers, and this is the reason why vendors that rely on a frit mixer design can much easily offer a wide range of different mixer volumes. This facilitates to physically adapt the mixing volume and with that the GDV to individual demands.

In the context of mixer changes for such adaptations, it is also important how the mixer is mounted in a pump, how it is accessible, and to what extent an exchange is supported. It is also crucial that the individual mixing demand in connection with the position of mixer in the pump fluidics is well understood by the user. This is another matter where HPG and LPG pumps can be substantially different. It was discussed in Section 4.1.3 how important the thorough mixing of the segments from the proportioning valve of an LPG is and that it must occur at the latest in the pump cylinder. If the mixing is still incomplete by the end of the aspiration phase, a volume contraction will occur in or after the delivery phase and will result in a nonconstant flow rate. It depends on the exact fluidic design of the proportioning valve and the pump cylinder as well as the fluidic connection between both and how well this mixing is effectuated. The mixing quality will additionally depend on the solvent properties, the flow rate, and even the lab temperature. To account for a complete mixing by the end of the aspiration phase, the majority of the mixing effect must be achieved prior to the pump cylinder. This is best possible if an effective mixing device is mounted between the proportioning valve and the pump cylinder. An example for such an LPG-specific premixer is the so-called inlet-weaver (Agilent Technologies). Such a design is ideal to account for perfect flow-constant delivery of an LPG pump. The use of premixer in an LPG provokes the question of why such LPGs usually have a second mixer behind the pump cylinder. To understand this, it must be distinguished between the mixing quality that is sufficient to keep the volume of the mixing at a given pressure constant and the mixing quality that is required for perfect longitudinal continuity of the composition. To meet the constant-flow requirement, the demands for mixing homogeneity are much lower than for a perfect continuous mixing as is required for a smooth baseline in a method with TFA addition (Section 4.2.14). It is still a fair assumption that the second mixer behind the flow delivery device could be spared out as long as the mixing in front of it is already sufficient. Therefore, it might appear more sensible to just design the premixer right and get rid of the postmixer completely. There is, however, a very important limitation for the premixer to be taken into account. The available pressure to drive the viscous liquid through the premixer is at a maximum of 1.1 bar. This is generated at 1000 bar atmospheric pressure and with the solvent bottle level 1 m above the mixer inlet and ignoring all fluidic friction in front of this mixer. In practical operation, the pump piston must not evacuate the liquid in the pumper cylinder below a value of several 100 mbar, because otherwise serious problems with cavities due to organic solvent evaporation and gas bubble formation would occur. On the other hand, it is not technically feasible to design a very effective mixer with relatively low GDV contribution without implementing relatively narrow fluidic channels. This leads inevitably to a mixer design with limited permeability that cannot be operated in the aspiration part of the pump. Behind the flow delivery cylinder, however, it does not matter if the mixing devices create a back pressure of several bars, and in modern UHPLC instruments, even 10–20 bar would not matter, as the pressure drop in narrow capillary connections is even higher at the given flow rates. This is the technical and physical rationale behind the design of combining a premixer and a postmixer in an advanced LPG pump. The second (postmixing) device can be reduced in size or even removed if the mixing properties prove to be sufficient for the given application. The premixer must never be removed. If it is accessible for the user, it is only to replace it after clogging. A thorough filtering of solvents for eluent preparation and the mandatory use of inlet frits at the solvent lines should help to avoid clogging of expensive microfluidic-based premixers.

With HPG pumps, only mixing behind the cylinders of the two pump blocks is required, as only one gradient component streams into the pump cylinder. The mixer permeability problem therefore does not exist with an HPG. Where the mixer in an HPG is located and how it is mounted will also depend strongly on whether it is an integrated module or a computer-controlled combination of two isocratic pumps. In the latter case, both pumps have their individual purge valves and the mixer is an independent part that has two inlets and one outlet. It is thus usually a combination of a T-piece and the actual main radial mixing device. If the mixing volume should be increased or in particular the longitudinal mixing performance improved, it is common practice to combine a second longitudinal mixer serially behind this device. With both pump blocks combined into one module, the mixing device usually is an integrated part of this. Such pumps usually combine the purge valve into the T-piece that connects the two pump blocks and sometimes integrates even the actual mixer as well. In the majority of HPG pumps on the market, the mixer is a separate part, regardless of its technical and functional design. It can thus be exchanged with one with a different volume or in some cases even be removed. This requires easy access and fluidic connections that the user can deplumb and replumb with standard devices, ideally even without tools. Before the decision to completely remove such a mixer can be made, several requirements must prove to be fulfilled. The chromatographic integrity (reflected in retention time precision, peak resolution, detector baseline quality) is only one criterion. It is important to know that frit mixers also act as crucial filtering devices behind the mechanically stressed parts of a pump. If they are removed, there is a high risk that abrasion particles enter the autosampler fluidics and may clog or even damage them. This cannot occur if the purge block already contains an outlet frit, but this is mostly not the case. If there is no outlet frit in the purge block and the user wants to run a test with a completely removed mixer, it is mandatory that at least a narrow mesh (≤2 μm) filter frit cartridge (10 μl volume) is mounted instead of the frit mixer.

What's to know Modern UHPLC instruments only use static mixers, because dynamic mixers introduce a too big GDV contribution and create problems with abrasion and particle release. Static mixers are very often frit mixers or the more expensive modern variants of microfluidic mixers. With an LPG pump, a sufficient mixing in the aspiration part is crucial for constant-flow delivery. Therefore, some LPG pumps feature a dedicated premixer. It is always desirable to easily exchange the actual mixer in the high-pressure part of a gradient pump (LPG or HPG) in order to adapt to the individual mixing performance and GDV contribution demands. Before a frit mixer can be completely removed, it must be checked if there is still an effective filtering device present in front of the autosampler. If this is not the case, the mixer must be not just removed but replaced by a small-volume filter device.

4.2.13 Systematic Characterization of the Mixing Efficiency and Gradient Formation of a Pump

Even with relatively ineffective mixer technology, the mixing result can normally be improved by increasing the mixer volume or by connecting several mixers in line. This measure will not only increase the system GDV but also commonly change the flush-out characteristic. It results in more rounded edges in the recorded profile when the gradient slope is changed at a certain point, and these curved deviations from the programmed shape will be increasingly pronounced, the stronger the change in slope is. Even a simple linear gradient with short hold step at the highest elution strength will look different at the transition to and from the linear part (start and end). This is what has already been discussed in Section 4.2.11 with the Dolan test. It can be observed in Figure 4.16 and is most obvious at the point of return to the initial conditions for re-equilibration. The rounding can be considered a composition averaging over a certain time interval. As long as there is a steady change of composition over this time interval as is the case in the linear part with constant slope, there is no effect of the mixer flush-out on the gradient shape. Once the slope changes, the averaging generates curvature instead of a sharp edge, which becomes more pronounced, the more significant the slope change is and the larger the averaged or longitudinally mixed time window is (the amount of longitudinally mixed volume and specific flush-out characteristic). This discussion already shows that longitudinal mixing and smearing edges at gradient slope changes are related to each other, but there are still differences in the behavior of different mixer types. A good longitudinal mixer is characterized by very effective mixing of relatively small volume increments rather than mixing over a large volume increment with mediocre performance. In other words, a good mixer turns a bumpy baseline into a smooth one without needing a big mixing volume to achieve this. At the same time, such good and effective mixers will create less rounding in the recorded gradient profile at slope changes, for example, in a Dolan test.

These rounded edges may well be relevant for the chromatographic result. They cause the gradient to start earlier after the initial isocratic phase, but with a constantly increasing slope until the actual linear range is eventually reached instead of the programmed constant slope. Obviously, this deviation can change the chromatographic results markedly on the peaks that elute in this range of the gradient. This can be a concern for a method transfer from an instrument with equivalent GDV but much sharper flush-out characteristics of the mixer or the autosampler fluidics. If methods are to be transferred from an instrument that generates rounded gradient edges to one with a sharp characteristic at gradient slope changes, the rounded gradient may be even mimicked by programming it on the instrument that can sharply control the profile. This principle is applied in software for automated method transfer but always requires a pump that can shape a gradient without significant flush-out effects. It should be mentioned in this context that it is not enough to program a longer isocratic hold if a modern pump with low GDV is meant to mimic a seasoned instrument with a large GDV. The larger GDV will always go along with rounded gradient start, and this must be mimicked as well. To do this either by a programmed curved gradient segment or, in the more conventional way, by increasing the mixer volume, it should always be assisted by characterization of the gradient characteristics that will help to interpret the related chromatographic results.

It has been mentioned several times that the experimental setup for the Dolan test (Section 4.2.11) allows this. The characterization of rounding at slope change can be combined with the determination of the gradient composition accuracy and precision, as it is done in the typical gradient pump qualification procedures. For this purpose, a multistep gradient is programmed with sufficiently long hold steps at different levels to determine the relative composition from the measured absorbance even under severe rounding effects. Figure 4.18 shows an example of such a test gradient that even characterizes small composition changes of 1% at different levels of the total gradient range. The experiment is conducted with an advanced UHPLC-HPG that exhibits very sharp edges as a result of relatively small GDV of 35 µl and a mixer that is optimized for radial mixing. The applied UV marker is acetone in this test, but the gradient steps are run from the highest percentage of the acetone-containing channel downward to eliminate the influence of degasser effects. The relatively elevated flow rate of 2 ml/min also helps in this respect. Figure 4.18 shows the gradient shaping of this pump at its maximum flow rate and also relatively nearby its maximum pressure. A complete characterization of the pump performance requires recording such a profile at several combinations of flow rate and pressure across its entire operational range.

If this test is used to quantitatively determine the composition accuracy, it is mandatory that both channels contain the same solvent. Ideally, water with a non-volatile UV marker is run against pure water to avoid concentration changes in the degasser that may affect the result. If the pump would be running a practically relevant gradient as from water with UV marker to pure acetone, the influence of the solvent change on the spectral properties of the UV marker would make a direct correlation of the composition with the absorbance impossible. This is clearly a disadvantage of the test because it does not characterize the pump under the real live conditions where changes in viscosity and compressibility together

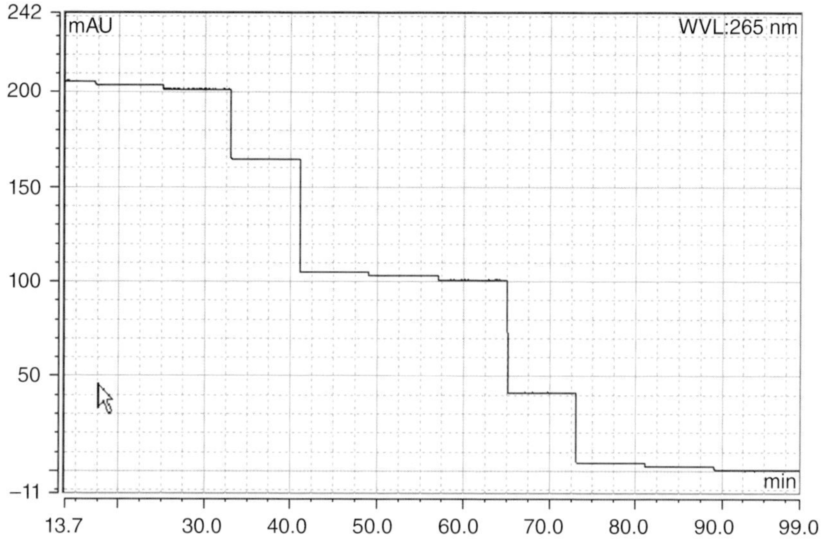

Figure 4.18 Characterization of the mixing accuracy and the fluidic behavior (sharpness of edges) of a gradient pump. Instrument: Vanquish H pump (Thermo Scientific), solvents: water and water spiked with acetone, flow rate: 2 ml/min, and pressure (generated against restrictor): 1200 bar.

with the excess volumes from mixing put additional challenges in particular on LPG pumps. The test is therefore not suited to characterize the advantage of an HPG over an LPG with respect to composition accuracy (see Section 4.1.3, Figure 4.3). Nevertheless, this test is the commonly agreed way to measure and specify composition accuracy and precision, albeit under conditions of very limited practical relevance. There are marker substances that show at selected wavelengths relatively small variation of the spectral absorbance, with changes in the water-to-acetonitrile or water-to-methanol ratio. These procedures will not be discussed in further detail here, as they are not recommended for quantitative gradient accuracy determination in mixtures of water and organic solvents. The only way to objectively characterize the accuracy under such conditions is to calibrate the detector independently for all the composition levels in between 0 and 100% that are mixed in the step gradient, for example, nine different levels (99, 98, 80, 51, 50, 49, 20, 2, and 1%) as shown in Figure 4.18. To do this, all nine water/acetonitrile (with marker) compositions must be manually prepared with the best possible accuracy and subsequently delivered to the detector flow cell. The only way to do this in an automated mode is the use of a 10-position solvent selector valve in front of one pump channel. By calling each position and delivering the respective composition to 100% by one channel to the detector, the absorbance for the different levels can be calibrated. Once the expected absorbance values are determined, the respective step gradient is run, and the accuracy at these nine steps is calculated by comparison to the calibration points. The reader may be encouraged to take this effort if there is a sincere interest in the determination of an LPG composition accuracy.

As discussed earlier, every real mixer represents a certain compromise between its longitudinal mixing efficiency and its gradient edge sharpness. The quality of longitudinal mixing can much easily be characterized than the radial mixing quality as it is expressed in composition changes over time that the detector can record. Radial mixing can only be technically characterized with dedicated fluidic setups that will not be discussed here. The often-mentioned setup for the Dolan test is, however, also suited to characterize the performance of longitudinal mixing device under defined conditions. These conditions can be established with a well-controlled standard HPG pump. A well-defined condition to characterize a longitudinal mixer is the generation of a sinusoidal composition change with a given amplitude and period (or frequency). A well-controlled HPG can generate such sinusoidal pattern for periods down to 1 s or less. It was discussed in Section 4.2.8 that modern HPGs can be controlled with a resolution of 100 Hz or more, which allows to shape these fast periods even down to the sub-1-s level. Figure 4.19 shows such a sinusoidal pattern with a time period of 1.2 s translating into a volume period of 20 μl at a flow rate of 1 ml/min. The gray line shows by Dolan-test-like tracking in the UV detector what pattern the pump releases without any mixer but just from the T-piece. The black line shows the resulting pattern to which a 100 μl (nominal value) SpinFlow™ mixer (see Figure 4.17) damps the original pattern from the pump. The change in amplitude between both lines provides a quantitative estimation for the damping effect of a given mixer at a given volume period of a composition fluctuation. Such a setup allows comparison of different mixers under strictly defined conditions.

It can be deduced from the result in Figure 4.19 that the mixer of a nominal volume of 100 μl damps the fluctuation with the 20 μl period very effectively.

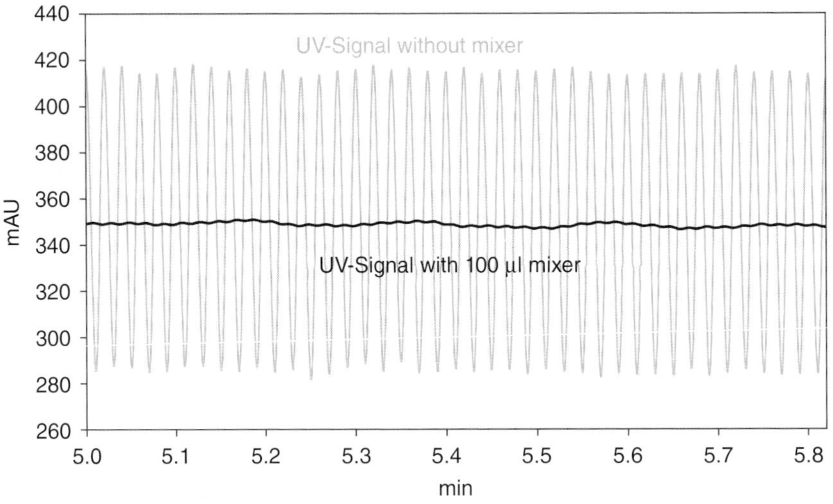

Figure 4.19 Programmed sinusoidal composition pattern generated with an HPG at time period of 1.2 s over the full composition range from 0 to 100% (corresponds to a volume period of 20 μl at the given flow rate of 1 ml/min). The sinusoidal composition change from water to water with marker is traced with a UV detector. The relative change in the amplitude provides a quantitative measure for longitudinal mixing efficiency at the given volume period.

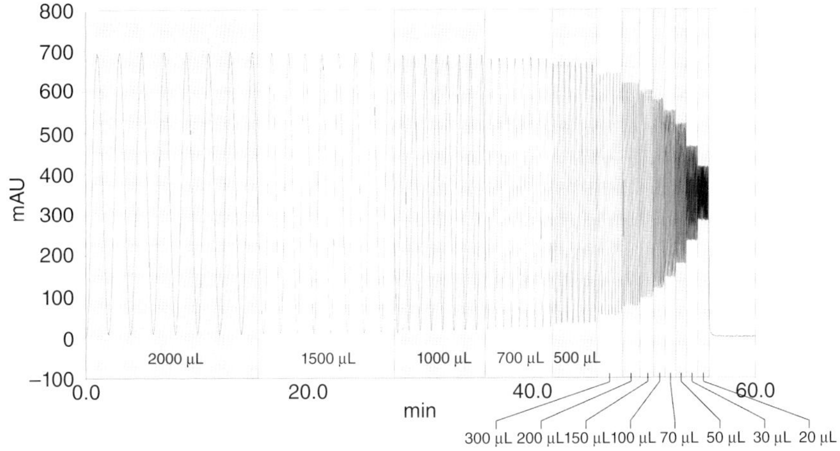

Figure 4.20 Programmed sinusoidal composition pattern generated with an HPG pump with 13 different volume periods between 2000 and 20 μl. The patterns are detected without mixer with the same setup as shown by the gray line in Figure 4.19.

A closer look reveals that the original pattern from the pump showed a certain fluctuation of its amplitude but with a much higher volume period. The black line where the 20 μl fluctuation amplitude is down to 2% of the original value also shows a longer period wave with an even higher amplitude and prone to stem from a pattern with similar period in the original data. This leads to the obvious conclusion that the damping effect strongly depends on the volume period of the composition fluctuation that is to be mixed.

The employed process allows the variation of the volume period of the pattern in a defined way to characterize different mixers. Figure 4.20 shows a sinusoidal pattern over a wide range of volume periods from 2000 μl down to 20 μl (as shown in Figure 4.19) but without a mixer behind the T-piece in the purge block of the HPG. Even the largest volume periods are of relevance when the residual pulsation of an HPG block at low gradient percentage comes into effect. It can also clearly be deduced that the inertia in the fluidics of the pump already damps the amplitude even without a mixer. This effect becomes significant at volume periods below 500 μl. The phenomenon will be of no influence for the mixer characterization, since at each volume period, the damping effect is calculated relative to the amplitude without a mixer but at the same period.

The damping effectiveness of different mixers is demonstrated in Figure 4.21 for two different volume periods of 20 and 200 μl. The graph shows the damping effect as relative residual pulsation amplitude in percentage plotted against the experimentally determined dwell volume of different mixers with determined volumes between 40 and 1500 μl. The Thermo Scientific SpinFlow mixers span this wide volume range, but they all follow the same technical concept as depicted in Figure 4.17. From the results with the 200 μl period, it can be deduced that the residual pulsation decreases exponentially with increasing dwell volume of the mixer. This decay of the amplitude is much steeper for the small period of 20 μl,

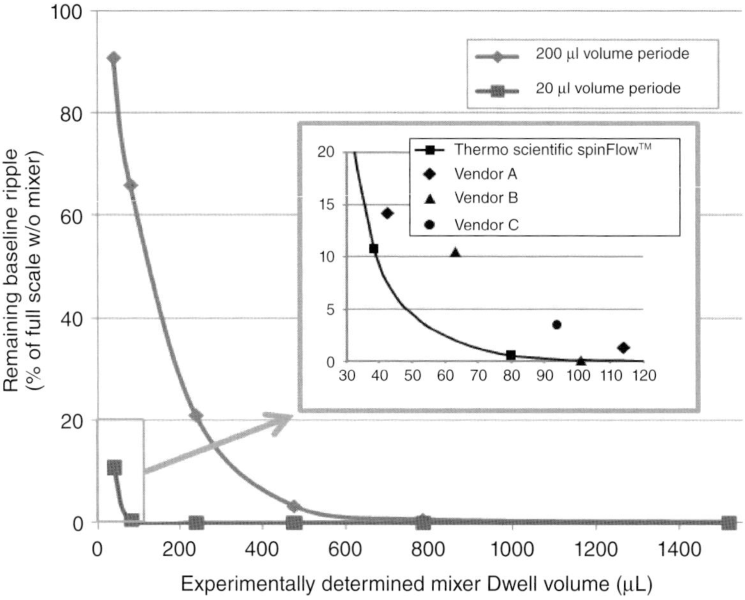

Figure 4.21 Residual pulsation amplitude as a function of mixer dwell volume measured at two different volume periods of pulsation (20 and 200 μl). Zoomed range compares mixers of different vendors at 20 μl period.

as can be deduced from the other curve. One can also state that a smaller mixing volume is required to mix smaller volume periods effectively, a conclusion that is absolutely intuitive. The SpinFlow mixer with a dwell volume of 80 μl could almost fully damp the 20 μl period, while the respective 40 μl mixer removed 90% of the original pulsation. This can be best seen in the view on the right of Figure 4.21 that is zoomed to the range of mixers with volumes up to 120 μl. In this view, mixers of other vendors, following other technical concepts also are shown, some of them based on microfluidic technology. One can see that the damping effect relative to the dwell volume varies over a wide range and even mixers from the same vendor and using similar technology behave differently at different dwell volumes. The mixer from vendor B with a dwell volume of 65 μl showed four to five times the residual pulsation relative to the extrapolated behavior of a 65 μl SpinFlow mixer, while the 100 μl mixer from vendor B could completely damp the 20 μl period pulsation similarly to the SpinFlow mixer at this volume level.

The picture of relatively inconsistent behavior across vendors and techniques was backed up by tests at other volume periods (not shown). Interesting to discuss is the curve measured for the different SpinFlow mixers at the 200 μl volume period. A dwell volume of about 500 μl is required to damp that pulsation period below a 5% level, but a mixer with 80 μl could already remove 35% of the original pulsation, and even 40 μl showed a measurable effect of almost 10% reduction.

Based on the characteristics, the wide range of different dwell volumes, but similar mixer design, the following conclusions can be drawn from the experiments:

- The amplitude is damped to roughly 50% if the dwell volume is in the same range as the volume period, but even mixers with dwell volumes smaller than the volume period could markedly damp the pulsation.
- With a dwell volume of twice the volume period, roughly 90% of the original pulsation could be removed.
- To almost completely remove the pulsation, a dwell volume of factor 3–4 larger than the volume period is required.

These conclusions together with the discussions in the foregoing sections help to understand the longitudinal mixing challenge that occurs with HPG pumps. If we consider the often-cited stroke volume of 50 μl and a pump that generates 80% water to 20% organic solvent mixture, the slower piston that delivers the organic solvent might show a ripple with four times the stroke volume (50 μl) as it only moves once in four cycles of the piston that delivers water. A pulsation period of 200 μl would be expected from this, and this would already need a 400 μl dwell volume SpinFlow mixer to reduce this ripple to 10%. At 5% or less organic content, the resulting period would need more than 1 ml dwell volume in the mixer for the same effect. This shows that it is very beneficial to design HPG pumps for minimized residual pulsation if the user wants to apply them in applications that are sensitive to mixing ripples. The findings from Figure 4.21 clearly indicate that mixing volume increase will always help to cure ripples but will, at the same time, remove one of the key strengths of HPG pumps, namely the capability to accomplish a small GDV.

What's to know The instrumental setup of the Dolan test allows for characterization of many more parameters than solely the GDV determination. If appropriate gradient patterns are programmed, the test helps to characterize composition accuracy as well as flush-out behavior and effectiveness of the gradient mixer. The flush-out effect is determined by the rounding of edges in step gradients and the composition accuracy from the absorption level at different steps. Systematic characterization of mixers is possible by programming continuous composition waving at defined volume periods. All these characterizations are very helpful in method transfer between different instruments, as the goal is to obtain the gradient pattern between two systems as similar as possible by programming the gradient differently or changing the physical mixer. Effective longitudinal mixing to obtain smooth detector baselines in challenging applications requires mixer volumes that are at least factor 2 larger than the volume period of the original pulsation. If a detector baseline shows a periodical ripple, its volume period can be calculated by multiplying its time period in the chromatogram with the applied flow rate. A ripple with a time period of 0.2 min observed at a flow rate of 500 μl/min indicates a volume period of 100 μl. To remove 90% of this ripple, even a very effective longitudinal mixer needs a dwell volume of 200 μl or more.

4.2.14 Optimizing the Mixing Volume in Dependence of Pump Type and Flow Rate for Demanding Applications Such as TFA Gradients

It was already mentioned in the foregoing sections that the impact of the longitudinal continuity of mobile-phase composition heavily depends on the application and method specifics. The term longitudinal continuity is intentionally used here, although it may look somewhat cumbersome. A more common, but actually misleading expression is mixing effectiveness or mixing quality. It is demonstrated in this section that the mixing effectiveness is only one of the contributors to the longitudinal continuity in composition. It is in fact strongly dependent on the delivery characteristics of the pumping and/or proportioning devices and what mixing capabilities are actually required to achieve a certain or ideally a perfect continuity without observable ripples. To what extent a given composition ripple is observed will very strongly depend on the application and, to some extent, on the susceptibility of the detector to dynamic refractive index changes. In this section, the extraordinary strong impact of composition ripples on the UV-detector baseline in applications with TFA as mobile-phase additive is discussed in detail.

Solutions of TFA as prepared by a mixture of water and acetonitrile have characteristic UV spectra that show a strong dependence on the water/acetonitrile ratio [20]. This translates into a strong dependence on the extent of baseline drift in gradients that will vary with the detection wavelength. The TFA spectra in varying solutions show an isosbestic point at 215 nm where the absorbance at distinct molar concentrations of TFA is independent of the respective solvent ratio (water to acetonitrile). This implies that the baseline drift in TFA gradients can be minimized by detection at 215 nm. The observed residual drift will, however, strongly depend on the spectral bandwidth of the detector, and also the mixing volume effects will lead to a nonconstant TFA concentration over the gradient. So there will hardly ever be complete absence of drift, but when working at 215 nm detection wavelength with spectral bandwidth not larger than 8 nm, and at a concentration of 0.1 vol-% TFA addition in both the aqueous and the acetonitrile phase, the drift can be kept at an acceptable level. Since the absorbance of TFA increases with increasing acetonitrile content at wavelengths above 215 nm, it can be a smart solution to reduce the TFA content in the organic phase to 0.09 or 0.08% (relative to 0.1% in water), in order to keep the baseline drift as low as possible for detection wavelengths between 220 and 250 nm. To clear out a common misconception in the context of TFA addition, it must be mentioned that this spectral shift as a consequence of composition discontinuity originating from the gradient pump is only of very minor contribution to the detector baseline ripples in TFA methods.

To grasp the true root cause, a good understanding why TFA is used at all as eluent additive will help. TFA is a strong acid that fully dissociates in aqueous solution. At the same time, it is a very hydrophobic compound due the CF_3 group that even in the dissociated state interacts very strongly to the surface of a reversed phase. This property of TFA helps to achieve decent retention for very hydrophilic basic or amphiphilic compounds, for example, hydrophilic amines or peptides. The retention can be either understood as dynamic cation exchange with TFA acting as the agent to form a cation exchanger by modifying the RP surface or ion pair retention with TFA as ion pairing reagent for protonated basic compounds. Regardless of the retention mechanism remains the fact of a strong retention of TFA at the reversed phase, which is in turn strongly modulated by the acetonitrile concentration as it is the case for every compound in reversed-phase chromatography. If the composition of water/acetonitrile is not constant over time (and over space in the column), a corresponding discontinuity of the TFA concentration over time and space will promptly result. This will happen with the pulsation period of the pump. So the TFA is enriched on and desorbed from the stationary phase in respective cycles. Lower acetonitrile implies enrichment, and high acetonitrile results in mobilization of the formerly enriched TFA. Once an enriched zone is eluted from the column end, it travels as a TFA concentration peak to the UV detector and creates the respective ripple. This is then normally referred to as mixing ripple, but its origin is even more complicated when TFA is involved. Obviously, this effect cannot be cured by detecting at 215 nm, as it is not the spectral shift causing the ripple but the tremendous change in TFA concentration as a result of even only minor fluctuations in the eluent composition.

The relatively chaotic pattern that results from simply mixing 5% acetonitrile with 95% water and 0.1% TFA in both channels can be seen in Figure 4.22. For this experiment, a coupled cam-driven serial HPG pump with relatively small mixer was used, which produces a relatively pronounced composition discontinuity just by design. The graph shows the resulting mixing ripple when pumping this along a standard C18 column, in direct comparison to the baseline of the same eluent delivery but after replacing the column by a restriction capillary to create equivalent back pressure. The result shows in all clarity that the disturbing TFA baselines only originate from the TFA interaction with a column. For this reason, it is completely pointless to characterize an HPLC system for its

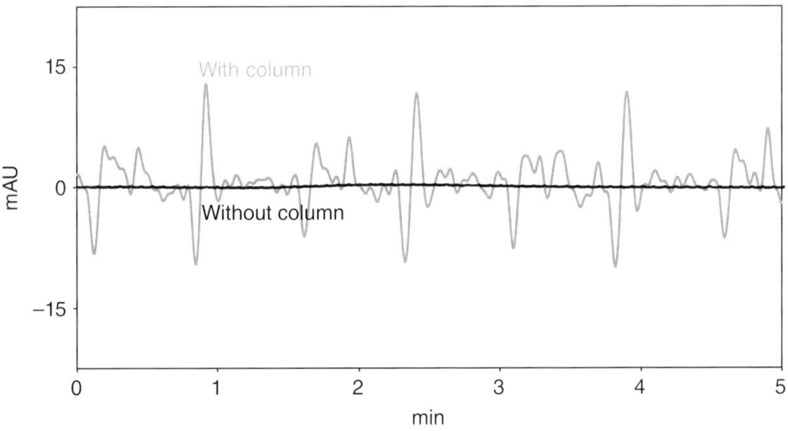

Figure 4.22 Amplitude comparison of baseline ripples with TFA to demonstrate the retention effect of the column, here under isocratic conditions of 5% eluent B, but mixed with an HPG pump at 35 µl mixer. Conditions: water/acetonitrile 99:1 v/v with 0.1% TFA (eluent A), acetonitrile with 0.1% TFA (eluent B), flow rate: 1 ml/min, temperature: 35 °C, and column: Acclaim C18, 3 µm, 250 × 3.0 mm.

TFA-friendliness without employing a column in the experimental setup. This may sometimes lead to very bad surprises in real applications later on. For the same reason, the properties of the HPLC column are crucial for the generated baseline ripples. As a rule of thumb, it can be assumed that the ripple increases with the stationary-phase hydrophobicity and column loading capacity. To compare the TFA behavior between two different HPLC instruments, it is very important that all the parameters including the column are kept constant. Ideally, both instruments are fed from the same eluent bottles, and the two columns employed should be cross-changed during the experiment to exclude minor differences between two nominally identical columns. Even the smallest difference in TFA concentration, reagent purity, and column characteristics can matter and lead the experimenter to wrong conclusions. There is hardly any other field in HPLC where experimental mistakes and misconceptions are so common as with TFA applications.

To practically characterize an instrument for the suitability to run a certain TFA methods, the best way is to exactly run this very method but with a blind gradient without injection of a sample. The results of such experiments for such TFA gradient runs with a given instrument, but with two different mixers, are shown in Figure 4.23. The method applies relatively typical but demanding conditions for the pumping system. The gradient runs from 0 to 30% acetonitrile with 0.1% TFA in both solutions, and detection is at 220 nm. The employed pump is again a cam-driven serial HPG with two mixers of different sizes (400 and 1550 µl) but based on the same technical concept. With the help of the almost fourfold larger mixer, it was possible to reduce the amplitude of the baseline ripple by a factor of close to 9. At the same time, the visible period of the waving was markedly enlarged, which is very favorable for the integration of low peak heights to quantify small concentrations.

With this respective setup and the TFA gradient, all available mixer volumes that had been already employed for the experiments shown in Figure 4.21 were tested . In Figure 4.24, the results of the six gradient experiments are plotted in the same way, and the exponential drop of the residual mixing ripple with increasing

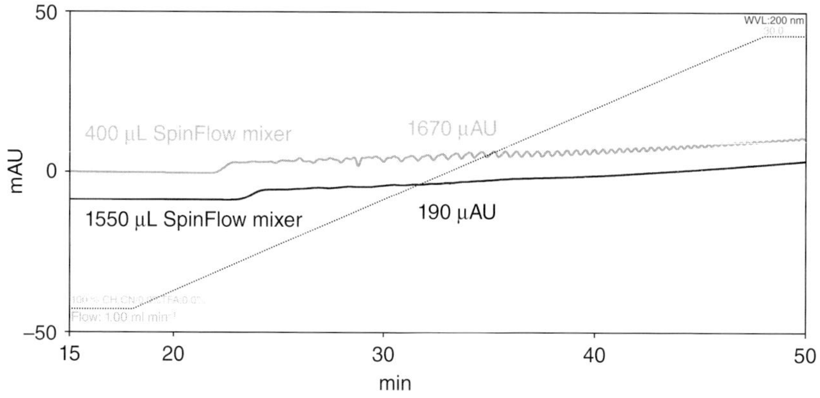

Figure 4.23 Comparison of detector baseline with two different mixer volumes (same mixer type) in a typical TFA gradient.

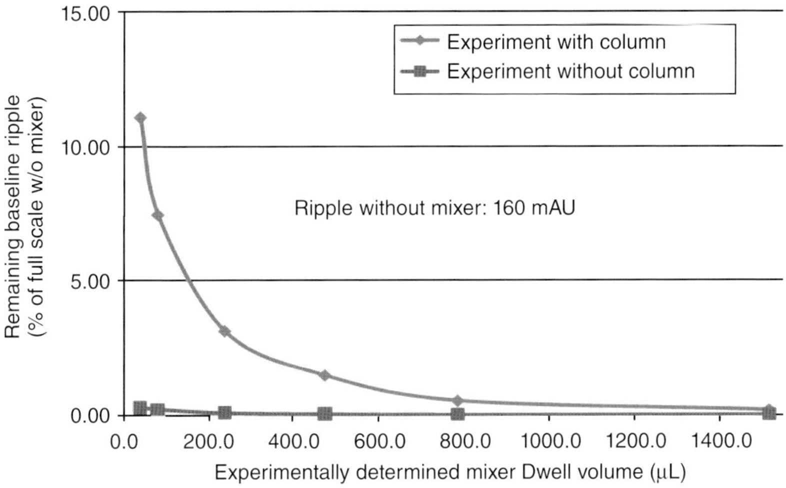

Figure 4.24 Residual baseline ripple amplitude in a TFS gradient with and without column in dependence on the mixer dwell volume.

mixing volume could be reproduced. This proves that the systematic relationship between mixing volume and baseline waving as determined with the experiment from Figure 4.21 also applies to demanding TFA gradients with even varying pump ripple period. Figure 4.24 additionally shows the baseline ripple amplitude with restriction capillary instead of column. Once the column disappears, the ripple in the full gradient disappears as well.

All experiments shown in Figures 4.22–4.24 were performed with an HPG pump that generates a certain residual pulsation in the two pump blocks due to its coupled cam-driven and more economical design. It was demonstrated that the baseline can be significantly improved by employing a large mixing volume. As discussed in Sections 4.2.3 and 4.2.7, an advanced pump technology can effectively minimize residual pulsation, which can be very detrimental with HPG pumps. To build an HPG pump for truly smooth TFA-gradient delivery, there is considerable effort required in the mechanical piston drive technology and the intelligence of the control algorithm. This will inevitably also be reflected in the investment cost for a respective pump. An appropriate consideration for comparison is an HPG pump that employs parallel piston design with four independent motors, linear high-precision drives, and advanced piston control, including ATEC (see Section 4.2.7). The beauty of the performed comparison experiment is that the mixers behind this advanced pump were exactly the same as those used behind the more economical HPG pump with only two motors and cam drives. Under identical experimental settings, the two pumps were compared with either a 35 or a 200 μl, while the 35 μl variant exhibited mainly radial mixing capabilities that can hardly damp baseline ripple root causes. The experimental conditions were to dial a mix of 15% acetonitrile isocratically with 0.1% TFA in both the aqueous and organic components and detection at 200 nm. Figure 4.25 depicts the observed baseline of the less advanced cam-driven serial pump (gray line) in direct comparison with that of the highly advanced parallel pump with linear

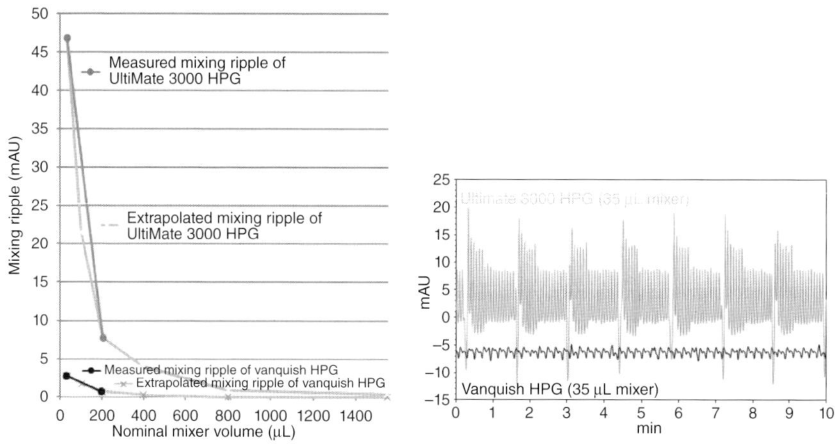

Figure 4.25 Comparison of the absolute baseline ripple from isocratic mixing of 15% acetonitrile with 0.1% TFA to aqueous 0.1% TFA solution for two different HPG pumps, but equipped with identical mixers. The ripple amplitudes with nominal 35 and 200 μl mixers have been experimentally determined, the amplitudes at the other nominal mixer volumes are extrapolated according to the curves determined in Figure 4.24.

drives (black line), both using the identical mixer with the small volume of 35 μl. From the advanced flow delivery technology of the high-end pump alone, the observed mixing ripple with TFA could be ninefold reduced although the mixing devices were identical.

Figure 4.25 also shows the plot of the ripple amplitude of both pump types as a function of the mixing volume for the two tested mixers (35 and 200 μl), as well as the extrapolation to the other available mixing volumes using the characteristic function determined from the experiment shown in Figure 4.24. Comparison of the two curves shows that the advanced pump generates with the 35 μl mixer less than half the ripple that the economical pump still generates with the 200 μl mixer, which increases its GDV contribution almost sixfold. This clearly demonstrates that the crucial contributor to TFA mixing ripples in an HPG is the quality of the flow delivery technology while the downstream mixing device can only have a corrective function. The better the continuity of composition behind the mixing-T of an HPG, the less longitudinal mixing efficiency is needed to suit the advanced demands for TFA applications. Increased mixing volume can compensate for less smooth flow delivery in the individual pump blocks but comes at the expense of increased GDV. If TFA compatibility is demanded at a pump GDV of 100–200 μl, the user just has to accept the cost level of an advanced HPG instrument.

Another practical consideration is the already mentioned premixing of water and acetonitrile (Section 4.2.5), which is only possible with gradients that do not start from 0%. It has been discussed in Section 4.2.5 that this may lead to another challenge for an HPG, namely to start the run with a noncompressed liquid in the B-block. This can be circumvented by complicated calculations for the premixing to account for a start of the gradient at 1% or even below. This effort is hardly accepted in laboratory practice, and therefore, advanced HPG pumps with the lowest possible residual pulsation are very favorable for TFA applications when a low GDV is required.

The TFA mixing behavior with LPG pumps is again substantially different. Since every LPG pump employs a respective mixing volume to longitudinally mix the inherent composition discontinuity originating from the proportioning valve. This longitudinal mixing performance automatically accounts for a decent TFA suitability in the default configuration of the pump. This reflects once again the advantage of LPG pumps that produce composition mixing with very strong discontinuity, but this discontinuity has relatively short volume periods and can be corrected with a relatively small mixer volume. A less advanced flow pulsation control has no impact on the TFA performance with an LPG. This is another advantage of the LPG concept besides the numerous disadvantages when it comes to fast gradients. If we consider an HPG and an LPG pump based on equivalent stroke volumes and an economical serially coupled cam-drive design, it can be expected that for the LPG, a mixer with smaller volume will suffice to achieve a similar TFA-baseline quality. This does, however, not imply that the LPG profits from this advantage with respect to GDV, because a significant GDV contribution of an LPG is generated prior to the mixer in the pump cylinders and proportioning device. As a rule of thumb, it can be assumed that a less advanced HPG as was used to generate the gray baseline in Figure 4.24 will require the additional size of the mixer that is equivalent to the GDV contribution of the LPG outside the (smaller) mixer. The practical consequence is that such a lower cost HPG will generate a similar GDV as the respective LPG, once the appropriate mixer for TFA applications is installed. With that, the economical HPG completely loses its GDV advantage over a respective LPG when it comes to demanding TFA applications.

What's to know The use of TFA as eluent additive to better retain hydrophilic basic compounds in reversed-phase chromatography puts extraordinary demands on the gradient regarding the impact on the detector baseline. The following key rules should be considered for best practice.

1) The quality of UV baselines in TFA gradients is a consequence of discontinuity of eluent composition originating from the pump and translating into discontinuity of TFA concentration in the detector that is only generated by the interaction of TFA with the stationary phase.
2) The suitability of an instrument for TFA applications can only be characterized with the respective column, as the ripple will completely disappear once the column is left out. The retention behavior of the column for TFA can vary. More hydrophobic columns produce worse baselines under otherwise identical conditions.
3) The baseline ripple from HPG pumps is not only a consequence of poorer efficiency of the mixer but mainly a consequence of pronounced residual pulsation of the individual pump blocks. A costly high-end HPG with very low pressure pulsation is able to achieve the same baseline performance with substantially smaller mixer volumes.
4) If an HPG exhibits pronounced baseline ripples resulting from elevated pressure pulsation, the increase of the mixing volume can significantly damp that but at the

expense of the GDV increase. HPGs from mid-to-low price segments will usually need mixer volumes between 500 and 2000 μl for acceptable TFA baselines.

5) LPG pumps from mid-to-low price segments can generate similar baseline quality as respective HPG pumps, but with significantly smaller mixers (200–500 μl depending on stroke volumes), thanks to the much shorter periods of the mixing ripples. This will not generally give them a GDV advantage over the HPG, as an LPG generates substantial GDV contribution outside the mixer. Still it can provide similar TFA performance at similar GDV but with lower investment for the pump.
6) As practical hints for the method optimization, it will help to keep the TFA concentration as low as possible (0.1% is normally sufficient), account for the best possible TFA purity, and select detection wavelengths close to 215 nm if low wavelengths are required. In order to avoid pronounced baseline drift at detection wavelengths between 220 and 250 nm, it is recommended to reduce the TFA concentration in the acetonitrile reservoir to 80% of that applied in water reservoir. If the analyte shows sufficient absorbance at elevated wavelengths, it may pay off with instruments that have problems with TFA baselines to detect above 260 nm where the TFA absorption has almost vanished.

4.2.15 Exceptional Elution Behavior of Proteins with Mobile-Phase Mixing Ripples

In this last section, a relatively rare effect resulting from mixing ripples is briefly addressed. In HPLC with small-molecule analytes, even a severe composition fluctuation will never result in other effects than detector baseline ripples or reduced retention time and peak area precision but will never affect the integrity of peak shapes. The elution mechanism of a protein in reversed-phase HPLC is different, and this difference is even more pronounced with a large protein such as lysozyme. Proteins elute from a given RP column in a very sharp window of the eluent composition, which is reflected in an almost infinite slope of the plot of ln k versus % organic component in the mobile phase. Below that specific eluent composition, the protein stays immobilized at the stationary phase, and once the composition is achieved, it moves without retention through the column. The pronounced gradient peak compression effect resulting from this is the reason why longer columns hardly improve resolution in protein separation but only increase the sample loading capacity.

If insufficient longitudinal mixing creates discontinuous eluent composition over time, it may occur that after the elution composition for the protein was already achieved, the composition may again drop below that in the next pulsation cycle and its elution pauses. If the next upward ripple arrives, the protein will be mobilized again, and this process can repeat several times at a relatively high ripple frequency (result of combination of stroke volume and flow rate). The result will be a chopped peak as can be seen in Figure 4.26 when using the small 35 μl mixer. A look into the baseline before and after the peak clearly shows that the TFA effect is not the main reason for this chopped peak with the small mixer, as the baseline is much smoother than the peak profile. The second

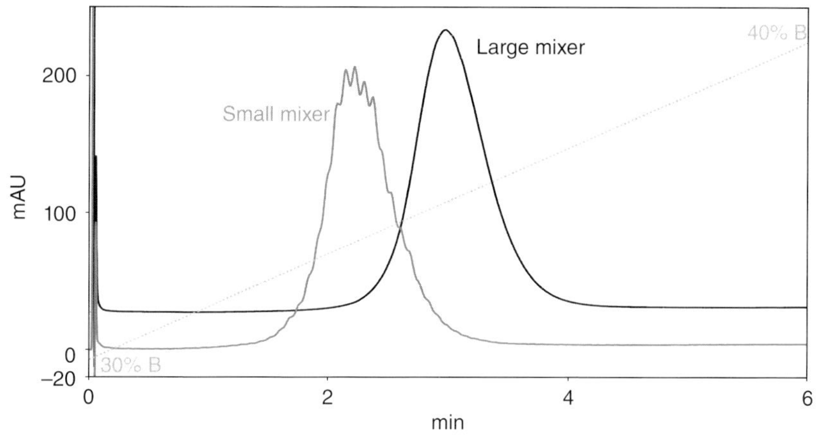

Figure 4.26 Chopped protein peak as a consequence of longitudinal discontinuity of mobile-phase composition and demonstration of influence of the gradient mixer volume on this effect. Sample: lysozyme, instrument: Thermo Scientific UltiMate 3000 binary RS system, small mixer: 35 μl SpinFlow, large mixer: 400 μl SpinFlow, flow rate: 1 ml/min, temperature: 30 °C, gradient water with 0.12% TFA to acetonitrile with 0.12% TFA, and "column": Phenomenex SecurityGuard C18, 4.0 × 3.0 mm.

chromatogram proves that a larger mixer to remove the composition fluctuation results in a clean and unimodal peak.

What's to know Due to a special elution mechanism of proteins, eluent mixing ripples can result in chopped peaks, in particular with large proteins. It is very important to exclude such a peak chopping effect resulting from mobile-phase composition ripples as they may lead to the misconception that the chopped pattern is a partial separation of protein isoforms. If such a pattern occurs with proteins, one should always check the volume of the gradient mixer and possibly repeat the experiment with a larger mixer to surely exclude the peak chopping effect and account for correct interpretation of the chromatographic result.

References

1 Van Meurs, N. and Somsen, G. (1993) *J. Solution Chem.*, **22** (**5**), 427–436.
2 Thermo Scientific (2012) Thermo Scientific Application Note 1023: Determination of Sudan Dyes I–IV in Curry Paste, Sunnyvale, CA, http://www.dionex.com/en-us/webdocs/113854-AN1023-LC-2D-Sudan-Dyes-Curry-Paste-AN70212_E.pdf (accessed 23 June 2016).
3 Broeckhoven, K., Verstraeten, M., Choiket, K., Dittmann, M., Witt, K., and Desmet, G. (2011) *J. Chromatogr. A*, **1218**, 1153–1169.
4 Gritti, F., Stankovich, J.J., and Guiochon, G. (2012) *J. Chromatogr. A*, **1263**, 51–60.

5 Petersson, P., Frank, A., Heaton, J., and Euerby, M.R. (2008) *J. Sep. Sci.*, **31** (**13**), 2346–2357.
6 Martin, M. and Guiochon, G. (2005) *J. Chromatogr. A*, **1090**, 16–38.
7 Battino, R., Rettich, T.R., and Tominaga, T. (1984) *J. Phys. Chem. Ref. Data*, **13**, 563–600.
8 Gilpin, R.K. and Zhou, W. (2008) *J. Chromatogr. Sci.*, **46**, 248–253.
9 Rogatsky, E., Zheng, Z., and Stein, D. (2010) *J. Sep. Sci.*, **33** (**10**), 1513–1517.
10 Magee, M.H., Manulik, J.C., Barnes, B.B., Abate-Pella, D., Hewitt, J.T., and Boswella, P.G. (2014) *J. Chromatogr. A*, **136**, 73–82.
11 Agilent Technologie Technical Overview, Performance Characteristics of the Agilent 1290 Infinity Binary Pump, http://www.agilent.com/cs/library/technicaloverviews/Public/5990-4536EN.pdf (accessed 23 June 2016).
12 Schmidt, C., Knauer, H., and Földi, P. (2009) Mischkammereinheit zur Verwendung in einer HPLC Anlage. German Patent DE 102006058026 A1, Dec. 2009.
13 Dolan, J.W. and Snyder, L.R. (1998) *J. Chromatogr. A*, **799**, 21–24.
14 Guillarme, D. and Veuthey, J.L. (2012) *UHPLC in Life Sciences*, RCS Chromatography Monographs, RCS Publishing, pp. 43–49.
15 Dolan, J.W. (2015) http://www.sepscience.com/Techniques/LC/Articles/219-/HPLC-Solutions-86-Measuring-Dwell-Volume (accessed 23 June 2016).
16 Schwartz, H.E. and Brownlee, R.G. (1985) *J. Chromatogr. Sci.*, **23** (**9**), 402–406.
17 Kobayashi, S. and Imai, K. (1980) *Anal. Chem.*, **52** (**9**), 1548–1549.
18 Med Device Visco-Jet Micro-Mixer, The Lee Company, http://www.meddeviceonline.com/doc/visco-jet-micro-mixer-0001 (accessed 23 June 2016).
19 Wiechers, J. (2015) Mixing device for liquid chromatography. US Patent 8,979,358 B2, Mar. 17, 2015.
20 Choikhet, K., Glatz, B., and Rozing, G. (2003) LCGC Europe, pp. 2–9 http://images.alfresco.advanstar.com/alfresco_images/pharma/2014/08/22/ec00494d-f2e4-4243-92b1-4450653ac623/article-45019.pdf (accessed 23 June 2016).

5

Requirements of LC-Hardware for the Coupling of Different Mass Spectrometers

Terence Hetzel, Thorsten Teutenberg, Christoph Portner, and Jochen Tuerk

5.1 Introduction

Mass spectrometry (MS) is of increasing importance in the field of routine analysis. In particular, the advantage of mass-related detection in combination with highly efficient (U)HPLC systems seem to promote this detection technique. Today, even more comprehensive analyses of complex samples in the shortest possible time are required. Especially for the analysis of complex samples, the hyphenation of liquid chromatography (LC) with mass spectrometry is advantageous compared to spectroscopic methods. The general benefits of mass spectrometry are not further discussed in this chapter. The reader is strongly encouraged to consult part one of this series, where an overview is given about the general hyphenation of liquid chromatography with mass spectrometry [1]. The focus of this chapter is therefore mainly laid on explaining the problem to obtain a sufficient number of data points across a chromatographic peak. It is often assumed that the latest versions of mass spectrometers are fast scanning and hence ideally suited for hyphenation with UHPLC systems. We will give some practical hints for a proper data acquisition on the basis of general applications in the field of small-molecule analysis.

At the end of this chapter, an outlook with respect to the opportunities of miniaturized separation techniques in combination with mass spectrometry is given in order to demonstrate the potential of this hyphenation.

5.2 From Target Analysis to Screening Approaches

5.2.1 Target Analysis

In general, there is a differentiation between target and screening approaches. For target analysis, usually substances that have to be quantified in a sample are defined. At the beginning of method development, reference standards are used

The HPLC Expert II: Find and Optimize the Benefits of your HPLC/UHPLC, First Edition.
Edited by Stavros Kromidas.

in order to achieve the most sensitive method with low limits of detection (LOD). At first, syringe pump experiments are done to determine compound-specific MS parameters followed by flow injection experiments to adjust temperature and gas streams for the ion source. The careful optimization of these source parameters exerts a strong influence on the ionization of the analytes, and these are important factors for the overall signal intensity. For a further understanding of this optimization procedure, the reader is referred to the corresponding literature [1–4].

5.2.2 Suspected-Target Screening

In suspected-target screening, chromatograms and the mass spectrometric information from unit or high-resolution mass spectrometers (HRMS) were analyzed for suspected compounds of interest. In contrast to target analysis, in suspected-target screening, the comparison with a reference standard is only done to create a list of identification criteria such as retention time, precursor ion, sum formula, and/or MS/MS spectra. This information can be stored in a self-created as well as open-source or commercially available database. Afterward, the samples are measured and compared against a reference database. The higher the number of identification criteria, the higher the probability that the identity of the detected compound can be confirmed [5, 6]. A comparison to a reference standard can be done for an unequivocal identification of the suspected compound.

5.2.3 Nontarget Screening

In true non-target screening, where no initial information on the analytes is available, automated peak detection and spectra deconvolution algorithms are applied, which typically reveal several thousands of peaks in a complex sample. Using statistical evaluation procedures such as multivariate data analysis (Principal Component Analysis, PCA) enables to identify differences between samples. The first criterion for an unequivocal identification of a substance is its accurate mass. In addition, using the isotopic pattern allows for a determination of the elemental composition or sum formula. The next step, the transformation of a sum formula into a structural formula, is a challenging task. Therefore, in non-target screening analysis, it is absolutely necessary to measure fragmentation spectra of the detected compounds. For this reason, several HRMS, such as quadrupole time-of-flight mass spectrometer (Q-TOF) or Orbitrap, have been established, because these systems offer the possibility to combine different acquisition modes such as high-resolution MS full scan and MS/MS fragmentation. With such an instrumental setup, a specific precursor ion can be selected and afterward fragmented by Collision-Induced Dissociation (CID) or Higher-Energy Collisional Dissociation (HCD) to detect the formed product ions. These MS/MS spectra allow to draw conclusions about the structure of the substances. The integration of meta-information such as exact mass, sum formula, solubility, REACH-Data, and literature data support the process to unambiguously align a structural formula to the accurate mass and/or sum

Table 5.1 Overview of the different LC-MS strategies and workflows.

Target analysis	Suspected-target screening	Non-target screening
• Known substances • Direct comparison with reference substance • Identification using retention time, at least two specific mass transitions, precursor ion, and/or fragmentation pattern • Quantification.	• Suspected substances • Known sum formula and/or fragmentation pattern • Verification by comparison of the precursor ion, sum formula, and/or MS/MS spectra with self-created, commercially, or freely available databases such as CheLIST, ChemSpider, DAIOS, DrugBank, HMDB, mzCloud, Norman MassBank, MetFusion, METLIN, PPDB, Stoff-Ident, TOXNET [8–19].	• Nonexpected or unknown substances • No *a priori* information about the elemental composition • Determination of the sum formula • Containment of possible chemical structures by the evaluation of the fragmentation pattern • Additional verification using *in silico* fragmentation[a)] and/or transformation/metabolism by special software [20] • Statistical evaluation by multivariate data analysis (PCA).

a) The experimentally determined sum formula and product ion spectra are compared to simulated fragmentation pattern for further characterization of unknown substances.

formula. At this point, it should be mentioned that non-target screening is extremely complex and error-prone. The interested reader is referred to the respective literature to learn more about general workflows in non-target screening [5–7].

The most important and increasingly discussed acquisition modes are summarized in Table 5.1. The databases listed in the second column of Table 5.1 not only contain MS/MS spectra. Using ChemSpider facilitates the pooling of information, which is important for a characterization of an unknown compound. Furthermore, the listed databases are not known to many users. That is why the reader is invited to visit the internet links to get a general idea of the included information [8–14].

5.3 What Should Be Considered for UHPLC/MS Hyphenation?

5.3.1 The Interface and the Optimum Flow Rate

Although there are several different ionization techniques commercially available, usually electrospray ionization (ESI) is used. A much smaller part of the applications is based on chemical ionization at atmospheric pressure (APCI).

Even though the MS manufacturers specify flow rates up to 3 ml min^{-1} in their product catalogs, the flow rate should be set below 1 ml min^{-1}. This is especially true for ESI, because the ionization efficiency is usually higher when the flow rate is adjusted between 300 and 500 µl min^{-1}. In contrast, the optimal flow rate range for chemical ionization at atmospheric pressure is between 500 and 1000 µl min^{-1}. Very high or low flow rates, which are around the limit of specification, are usually unfavorable.

5.3.2 Optimization of MS Parameters

The first step in LC-MS method development is optimization of the MS parameters. When analyzing large molecules (e.g., in bioanalysis), often multiply charged precursor ions ($[M+nH]^{n+}$) are formed. In contrast, for the analysis of small molecules (<1000 g mol^{-1}), singly charged, protonated, or deprotonated precursor ions ($[M+H]^+$ or $[M-H]^-$) are generated. Furthermore, adduct ions (e.g., $[M+Na]^+$, $[M+K]^+$, $[M+NH_4]^+$, $[M+HCOO]^-$, $[M+CH_3COO]^-$) might be formed. Especially, the formation of adduct ions depends on the purity and quality of solvents, buffer concentration, or ionization additives, the matrix as well as on the ion source itself. The optimal settings are evaluated by syringe pump experiments followed by flow injection analysis (FIA). Within a short period of time, the influence of ionization additives such as formic acid or ammonium formate on the signal intensity or adduct formation can be investigated. A discussion of these optimization procedures is omitted here for reasons of clarity. The interested reader should consult the available literature [1–4, 21].

5.3.3 Optimization of the Chromatographic Parameters

Since a mass spectrometer is able to separate coeluting compounds on the basis of their specific mass-to-charge ratio (m/z), it should be critically asked whether a baseline separation of all target analytes is really necessary. Analyzing 20 or more substances, which need to be separated within a single run, can already be very challenging. If the method includes 100 and more components, a baseline separation is usually impossible. Even using two-dimensional separation methods, there are still areas in the chromatogram where a complete separation cannot be obtained. In general, a chromatographic separation is only necessary if isomeric or isobaric substances with similar fragmentation pattern are included in the sample. Complex gradients with several steps for the optimal separation are usually not required if a mass spectrometric separation can be achieved. Another important chromatographic parameter is the pH value of the mobile phase. The pH value should be set in a way that stable retention times are obtained. All buffer additives must be volatile in order to prevent a severe contamination of the ion source. Due to this reason, the addition of 0.1% formic acid to the mobile phase has been proven to be a good compromise.

A relevant criterion for the development of multicomponent methods in LC-MS is the separation of the substances from the column void time. Especially, for very polar compounds, which show no or weak interactions with

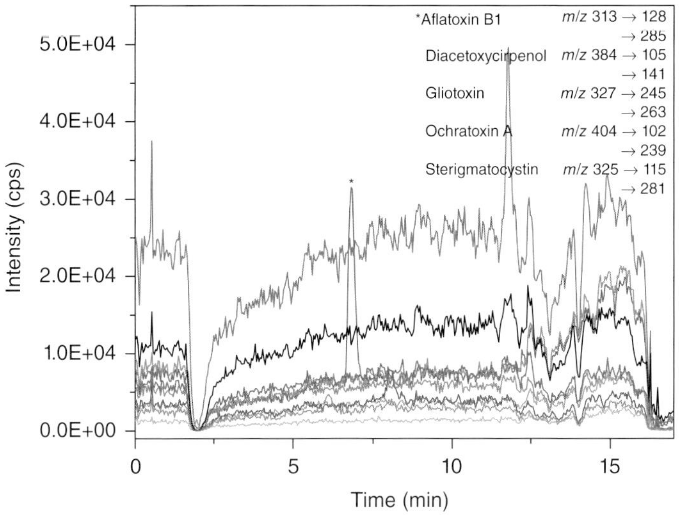

Figure 5.1 Matrix effect chromatogram of a house dust extract. Selected mass transitions of different mycotoxins are shown.

classical reversed-phase (RP) stationary phases, this is problematic because of a severe signal suppression for all compounds at the column void time. This phenomenon is shown in Figure 5.1. The depicted chromatogram is called a *matrix effect chromatogram*. To obtain such a chromatogram, a defined matrix, in this case, house dust extract, is injected into the chromatographic system. After the column, the analytes are added to the eluent using a syringe pump and a mixing tee. During the whole chromatographic run, specific mass transitions of the target analytes are recorded using a triple-quadrupole MS in multiple reaction monitoring (MRM) mode. Assuming optimal conditions, the intensity for each mass transition should be constant. However, at the column void time, the signal intensity of all mass transitions is severely reduced, because at this time, salts and other polar matrix compounds that are contained in the sample elute. Due to this reason, it is an important aim in LC-MS method development to achieve a retention factor (k) higher than 2 for all compounds.

Besides that, another phenomenon is visible in Figure 5.1. The mass transition highlighted by a star shows a distinct signal at a retention time of about 7 min. This signal originates from an isobaric interference from the matrix, which has the same mass transition such as the investigated substance, in that case, aflatoxin B1. If the target analyte would have a similar retention time, a false-positive result could be the consequence. In that particular case, a chromatographic separation is definitely required. Moreover, to increase the MS selectivity for each selected compound, at least two specific mass transitions for each substance should be selected. In addition, it is desirable to obtain a uniform distribution of all components over the chromatogram. Extremely steep gradients can lead to a high peak density at some points in the chromatogram and, depending on the scan speed of the MS, to an insufficient number of data points across each peak. This problem will be discussed in detail in Sections 5.4 and 5.5.

To summarize, for the final method, a linear solvent gradient with a moderate gradient slope should be used whenever possible. The adjustment of the pH value should be orientated on the MS requirements and pK_a values of the investigated compounds.

5.3.4 Choice of the Suitable Column and Column Dimension

At the moment, three different kinds of separation media can be distinguished: fully porous particles with a particle diameter between 1.5 and 5 μm, core-shell particles, which have a thin porous layer on a nonporous core with a particle diameter between 1.3 and 5 μm as well as monolithic columns. In the recent years, sub 2 μm fully porous particles and core-shell particles with a diameter of 2.6 μm have often been used because they exhibit a superior performance over conventional 5 μm particles. Now the question of which particle technology is suitable for LC-MS applications might arise.

The most important aspect is to keep connections as short as possible between the outlet of the HPLC column and the ion source of the mass spectrometer. Otherwise, the separation efficiency will be reduced and the peaks are detected as broad bands. These effects also lead to decreased signal intensity for each compound. Although this may sound very trivial, it is in many cases not possible to connect the HPLC column via short tubing with the MS. For example, prefabricated capillaries with a given inner diameter and length are often installed in the ion source of the mass spectrometer. In addition, the modules of the UHPLC system cannot always be perfectly aligned to the mass spectrometer. Figure 5.2 shows a typical setup as it is found in many laboratories.

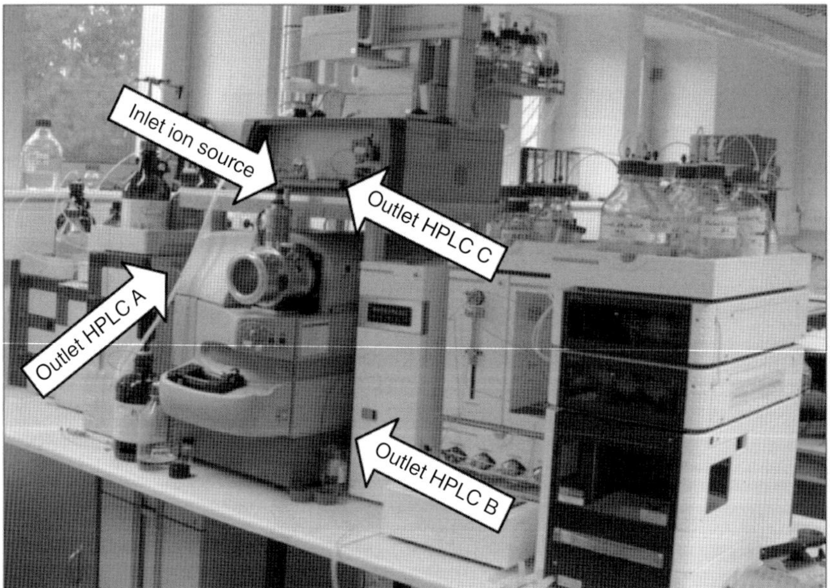

Figure 5.2 Typical system setup of LC-MS hyphenation.

It can be clearly seen that depending on the position of the UHPLC system relative to the position of the mass spectrometer, a more or less distinct extra-column volume occurs. In the worst case, a transfer capillary of up to 1 m must be used to connect the column to the MS. In this particular case, the ion source inlet is located on the right side. If the HPLC system on the left side should be connected to the ion source, a relatively long way from the column outlet (outlet HPLC A) to the MS inlet must be bridged. In contrast, using the HPLC system on the right side (outlet HPLC B), the distance between the column outlet and the ion source inlet can be minimized. Nevertheless, an unfavorable position of the HPLC column in the column oven can still lead to an unnecessarily long transfer capillary.

To minimize the band broadening after the column, the inner diameter of the connecting capillary must be decreased. However, this inevitably leads to a high pressure drop, which is responsible for a significant part of the system's back pressure if long and thin capillaries are used. For a theoretical consideration, the reader is referred to the content of Chapter 3.3.1.2. Depending on the ion source, several system volumes exist, which cannot be further minimized by the user. The influence of these volumes plays a key role if the overall performance of highly efficient UHPLC columns is to be utilized. In many cases, a transfer capillary with an inner diameter of 130 µm or more leads to a significant loss of separation efficiency. The narrow bands generated on the column can partially or even completely merge when leaving the column. Due to this reason, it should be considered whether stationary phases with a particle diameter below 2 µm are really suitable for the available configuration in the laboratory. This also includes the column inner diameter. If a column with a large inner diameter of, for example, 4.6 mm is used, the influence of the system volumes on the chromatographic efficiency and/or resolution is much less noticeable. In contrast, using 2.1 mm ID columns, the existing dispersion volume after the column can lead to decreased chromatographic separation efficiency.

The situation is different if the connection of the column to the mass spectrometer is possible in a direct way. Figure 5.3 shows a system design in which the flexible arrangement of the UHPLC system allows a connection of the column to the inlet of the ion source with nearly no dead volume.

In that case, the adjustment of the column oven to the ion source makes a long transfer capillary unnecessary. Many manufacturers have now responded to this problem and provide "smart" system designs. Using a flexible column oven, the critical system dead volumes after the column can be significantly reduced. The integration of column switching valves is also an important aspect, which might lead to a lower column performance. It must be critically weighted by the operator whether an automated column switching is more important than the minimization of extracolumn volumes. Again, a compromise between the desirable flexibility and optimal chromatographic conditions must be made.

As a rule of thumb, it can be generalized that if a system design as shown in Figure 5.3 and, in addition, a fast scanning mass spectrometer is used, the complete efficiency of fully or partially porous sub 2 µm particles can be utilized. If this is not the case, monolithic columns are a suitable alternative because the

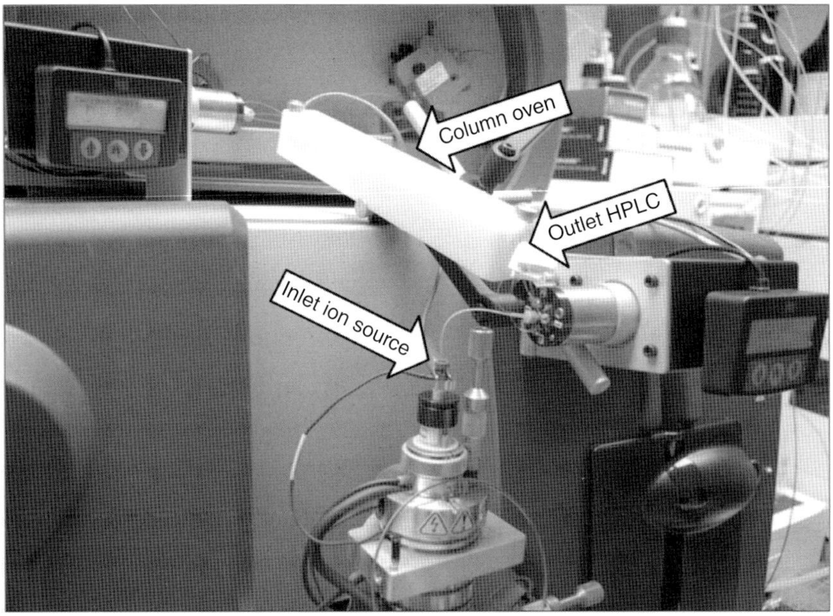

Figure 5.3 Optimized system design for the hyphenation of a flexible UHPLC system with a mass spectrometer.

decreased theoretical efficiency compared to fully porous particles is not noticeable in praxis. In addition, the resulting back pressure compared to fully porous particles is extremely diminished at similar chromatographic conditions because of the high permeability of monolithic stationary phases. In contrast to the classic stainless steel columns, monolithic columns are only available in PEEK (polyether ether ketone) hardware. Therefore, the specified maximum pressure is 200 bar. If the length of the column is below 15 cm, the pressure limit will not be reached at flow rates up to 1 ml min^{-1} even using methanol as organic cosolvent.

5.4 Target Analysis Using Triple-Quadrupole Mass Spectrometry

The instrumental developments during the past years show that in the field of target analysis, not only triple-quadrupole but also high-resolution time-of-flight mass spectrometers or Orbitraps (HRMS) are used because the robustness, linear dynamic range, and sensitivity have been greatly improved. Nevertheless, HRMS are not the preferred choice for target analysis, because usually the sensitivity of triple-quadrupole mass spectrometers (QqQs) will not be reached. Moreover, HRMS systems are usually more expensive than systems with lower mass resolution. For this reason, the general workflow for the development and application of a multicomponent method for target analysis is discussed for QqQ instruments.

Triple-quadrupole mass spectrometers have been established as "workhorses" in the field of residual and trace analysis because selective determination and sensitive quantification of target analytes are possible. It does not mean that compounds that are present in the matrix have no influence on the signal intensity (see Figure 5.1). However, substances that have, for example, an identical mass transition with the target analyte can be masked if the second mass transition is different. This kind of data acquisition is known as *multiple reaction monitoring* mode and means that multiple mass transitions can be recorded to correctly identify a specific compound. Because this acquisition mode is the most important for the quantification with QqQ instruments, it will be discussed in more detail.

After the ionization process and transfer of all formed ions into the high-vacuum part of the mass spectrometer, usually a focusing takes place using a quadrupole, hexapole, or octupole (Q0). The mass spectrometric experiment occurs in three sequentially arranged quadrupoles. Only the first (Q1) and third quadrupole (Q3) are used for the separation of the formed ions by their mass-to-charge ratio (m/z ratio). The second quadrupole (q2) is used for the generation of structure-specific product ions through the collision activation of a neutral gas (nitrogen or argon) in combination with a defined collision energy. In MRM mode, Q1 isolates the precursor ions, q2 is used for fragmentation, and Q3 isolates the formed product ions. This technique allows for a selective determination of the target analytes even in complex matrices, in which several interfering components are expected. In order to confirm the quantification results, at least two mass transitions for each substance should be recorded. The transition with the highest signal-to-noise ratio (s/n) is used for quantification, while the transition with the lower s/n is used for substance verification. The acquisition of different mass transitions needs time. Hence, the number of compounds, which can be determined within one chromatographic run, is limited. In addition to the acquisition time, which is also known as *dwell time*, a pause time between each mass transition must be considered. This is necessary because after each dwell time, the instrument must reset its voltages and, for MRM mode, allow enough time for the collision cell to be emptied. The terms "cycle time," "loop time," and "duty cycle time" all refer to the total amount of time necessary for the instrument to acquire one data point for each mass transition, including any time necessary to switch polarities or pause between mass transitions. In the following, we will use the term cycle time in the rest of this chapter. Usually, most applications are based on ESI working with positive ionization mode (ESI positive), but some analytes can be detected more selective and sensitive using negative ionization mode (ESI negative). Whether a change in polarity during the analysis is useful depends on the number of target analytes as well as on the available time between two peaks. Furthermore, the required switching time between the polarities is a key factor. For the following calculations, a switching of the polarity is not considered, because this will negatively influence the number of data points per peak. In contrast, if the method only includes a limited number of compounds, a polarity switch may be useful when an even more sensitive detection is possible in negative ionization mode for a specific target analyte.

The first step for LC-MS multicomponent method development is the optimization of the mass spectrometric parameters. As described in Section 5.3.2,

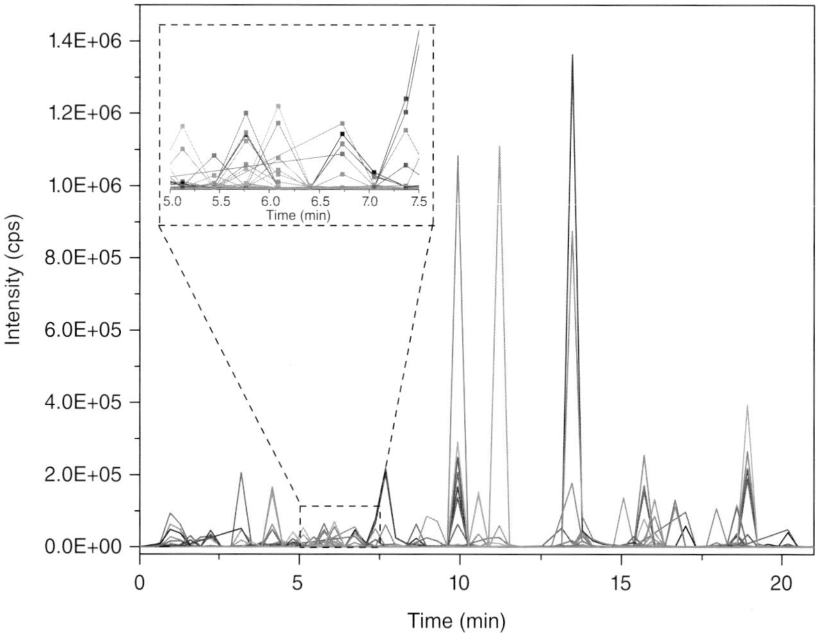

Figure 5.4 LC-MS/MS chromatogram of 182 MRM transitions on a 50 mm × 2.0 mm Chromolith FastGradient RP18 column. Chromatographic parameters: temperature: 40 °C; injection volume: 20 μl; mobile phase: A = water + 0.1% formic acid, B = methanol + 0.1% formic acid; gradient: 5–95% B in 20 min; flow rate: 400 μl min^{-1}; mass spectrometric parameters: pause time: 5 ms; and dwell time: 100 ms.

the optimization of the parameters that influence the ionization efficiency for a specific mass transition can be very complex. In the next step, the retention time of all substances that are included in the multicomponent method have to be determined. This can be done either through an injection of single standards or a standard mixture including all analytes. When all retention times are known, a calibration can be measured to determine the linear working range. At this point, some "bad" surprises may occur. Figure 5.4 shows the resulting chromatogram of a multicomponent standard containing 91 substances.

Looking at the chromatogram, one can see that triangular peak shapes result. Apparently, unfavorable acquisition settings have been chosen. From the magnification of the chromatogram, it becomes clear that only one data point per peak was recorded. Regarding the mass spectrometric acquisition parameters, the values of a "standard method" were selected. What went wrong?

The method shown in Figure 5.4 includes 182 mass transitions, which were continuously recorded over the complete chromatographic run with two MRM transitions for each compound. This retention-time-independent MRM mode has the advantage that no adjustment for the acquisition of selected mass transitions to a defined time or time window has to be made. Small fluctuations in retention time as well as a change of the solvent gradient need no further alignment of the MS method. However, the disadvantage is that the total number of the selected mass transitions is recorded over the complete chromatographic run. On the basis of the following calculation, it can be shown that the number of simultaneously measured MRM transitions is too high. For the example depicted in Figure 5.4, the dwell time was 100 ms and the pause time 5 ms. These settings are standard values, which are recommended by the MS manufacturers. Of course, these are fully justified, because such a large dwell time (100 ms) increases the signal intensity while minimizing the noise. When the MS method includes only a few analytes, which have to be quantified, the risk of an inadequate number of data points per peak is significantly lower. However, for the example chosen here, a cycle time of 19.1 s results. This means that it takes 19.1 s until all 182 MRM transitions are measured to achieve one data point for each substance. This corresponds to a data acquisition rate of 0.05 Hz. Obviously, the standard values of the manufacturer need reviewing. The minimum number of data points required to accurately define a peak may depend on specific guidelines, regulations, or personal preference. In some fields, 12 or even fewer data points per peak may be allowed, while in others, no less than 20 points per peak are acceptable. Some authors even demand 100 data points [1]. So what is the way out of this dilemma?

It could be tried to increase the peak width for every analyte. Such suggestions were really discussed at user meetings at the time when modern UHPLC systems became commercially available. However, such a suggestion is meaningless because in addition to the decreased chromatographic separation efficiency and resulting loss of sensitivity, the run time of the method is increased. This is in clear contradiction to increase the sample throughput. A better alternative is to adjust the acquisition parameters for the dwell time and pause time in the MS method. When the values are set to 10 ms for the dwell time and 5 ms for the pause time, the cycle time is only 2.73 s, which corresponds to an acquisition rate of 0.37 Hz. The resulting chromatogram is depicted in Figure 5.5.

Although the number of data points per peak could be increased, the requirement of 10 or more data points still cannot be achieved. A further decrease of the cycle time is theoretically possible. However, the noise would simultaneously increase in a way that the quantification of analytes in lower concentrations is impossible. An alternative strategy could be the separation of the chromatogram into periods, which are highlighted by the dashed horizontal lines in Figure 5.5. Within these periods, only MRM transitions of those analytes are measured, which elute within those time windows. Even though the number of simultaneously recorded MRM transitions per unit time can be decreased using periods, there are still some pitfalls, which should be considered. In order to subdivide the chromatogram into useful periods, some areas must be identified where no analyte elutes. This is the case for the example shown in Figure 5.5, but small fluctuations of the retention time can cause that a peak is not detected in the selected retention time window. In addition, with increasing complexity of the sample, the possibility to find regions in the chromatogram where no relevant peak elutes decreases. The smaller the chromatographic resolution between two adjacent peaks, the more difficult it is to divide the chromatogram into single periods. The technical progress of the last years makes it possible to detect a high number of substances in one chromatographic run by faster electronics and new measurement algorithms. For this purpose, the MS manufacturers developed the so-called retention-time-dependent MRM mode. This mode is also known

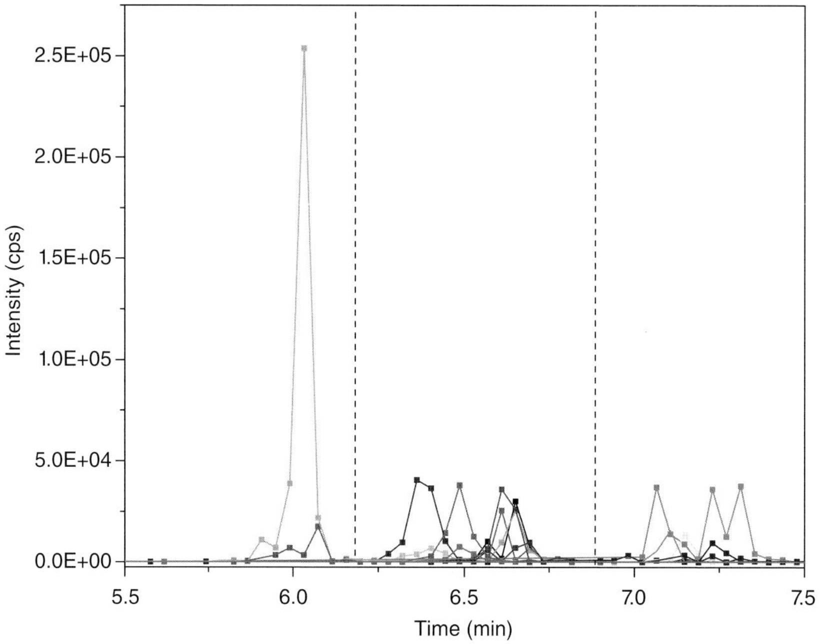

Figure 5.5 Extracted LC-MS/MS chromatogram of 182 MRM transitions on a 50 mm × 2.0 mm Chromolith FastGradient RP18 column. Chromatographic parameters: temperature: 40 °C; injection volume: 20 μl; mobile phase: A = water + 0.1% formic acid, B = methanol + 0.1% formic acid; gradient: 5–95% B in 20 min; flow rate: 400 μl min^{-1}; mass spectrometric parameters: pause time: 5 ms; and dwell time: 10 ms.

as *timed, targeted,* or *scheduled MRM* depending on the nomenclature of the individual manufacturer. This MRM mode provides the possibility to define an individual acquisition time window for every mass transition. The advantage of this acquisition mode is that the MRM transitions are not measured over the entire chromatographic run, but only in that time frame in which the component actually elutes from the column and is transferred to the detector. For a successful application of the retention-time-dependent MRM mode, component-specific retention time shifts should be considered. For example, a small change of the mobile-phase pH affects those substances where the adjusted pH value is located closer to the pK_a. It should be taken into account that small changes of the mobile phase's pH value can lead to considerable shifts of the previously determined retention times. Especially, the early eluting compounds are more prone to retention time variations when a large-volume direct injection is performed. Special attention should be paid to the definition of the detection window for each MRM transition. Very sophisticated algorithms offer the possibility to define an individual acquisition time for each MRM transition. In that way, the resulting number of sequentially measured MRM transitions at a given time can be drastically reduced. The retention-time-dependent MRM mode requires the specification of the so-called target scan time. In contrast to the classic MRM mode, the cycle time is preset. In dependence on the number of

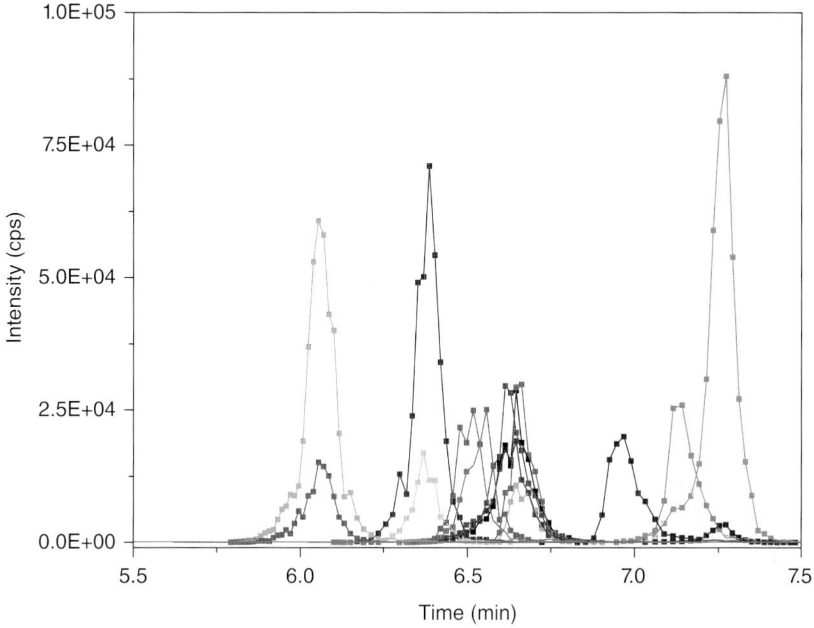

Figure 5.6 Extracted LC-MS/MS chromatogram of 182 MRM transitions on a 50 mm × 2.0 mm Chromolith FastGradient RP18 column. Chromatographic parameters: temperature: 40 °C; injection volume: 20 μl; mobile phase: A = water + 0.1% formic acid, B = methanol + 0.1% formic acid; gradient: 5–95% B in 20 min; flow rate: 400 μl min^{-1}; mass spectrometric parameters: pause time: 5 ms; MRM detection window: 60 ms; and target scan time: 2 s.

sequentially measured mass transitions, the dwell time is automatically adjusted so that in total the defined target scan time is reached including the pause time. Figure 5.6 shows the resulting chromatogram under the optimized conditions highlighting the individual data points.

After the definition of the specific retention time windows, the region with the highest peak density should be investigated because this region corresponds to the largest number of sequentially measured MRM transitions. On the basis of the average peak width, which can be estimated from the chromatogram, the setting of the target scan time can be accomplished to obtain a sufficient number of data points across each peak for a successful quantification.

The following calculation is performed to clarify this issue. Two different scenarios are proposed. The first scenario assumes that a UHPLC system is used and the peak width is expected to be 1 s. The second one considers a typical LC-MS hyphenation where the paths from the outlet of the column to the inlet into the ion source are unfavorable. In that case, the expected peak width is 10 s. The peak width in these examples is defined as the beginning and end of the peak as detected by the automatic integration of the software. Broader peaks are not discussed because the resulting number of data points will be even higher than for the second scenario. In general, it should be noted that broader peaks negatively influence the achievable LOD. Furthermore, it is assumed for both scenarios that despite using the retention-time-dependent MRM mode, there are still regions in the chromatogram where 20 substances – corresponding to 40 MRM transitions – have to be measured in the same retention time window. Using conventional MS devices, the minimum dwell time should not be adjusted below 5 ms. For modern instruments, this value can be reduced to 0.8 ms. However, it is strongly recommended not to use the smallest possible value especially when the method is applied to determine low concentrations. In that case, the noise and the limit of detection will be increased. We chose a dwell time of 20 ms for older devices and 2.5 ms for modern mass spectrometers for the following calculations. Measuring 40 MRM transitions then leads to an acquisition time of 800 and 100 ms, respectively. Additionally, the pause time must be considered. For the older MS device, this value is set to 5 ms, whereas modern mass spectrometers can deal with a pause time of 1 ms. The resulting pause times are 200 and 40 ms, respectively. The resulting cycle time is 1 s for the older device and 140 ms for the modern MS, which corresponds to a data acquisition rate of 1 and 7.14 Hz, respectively. Assuming a peak width of 1 s, only one data point per peak is obtained using the older MS device despite the sophisticated retention-time-dependent MRM mode. In contrast, the modern MS is able to acquire at least seven data points per peak. With a peak width of 10 s, the resulting number of data points is 10 for the first and 71 for the second scenario.

As a general statement, it can be concluded that highly efficient UHPLC separations with peak widths of 1 s in combination with older mass spectrometers

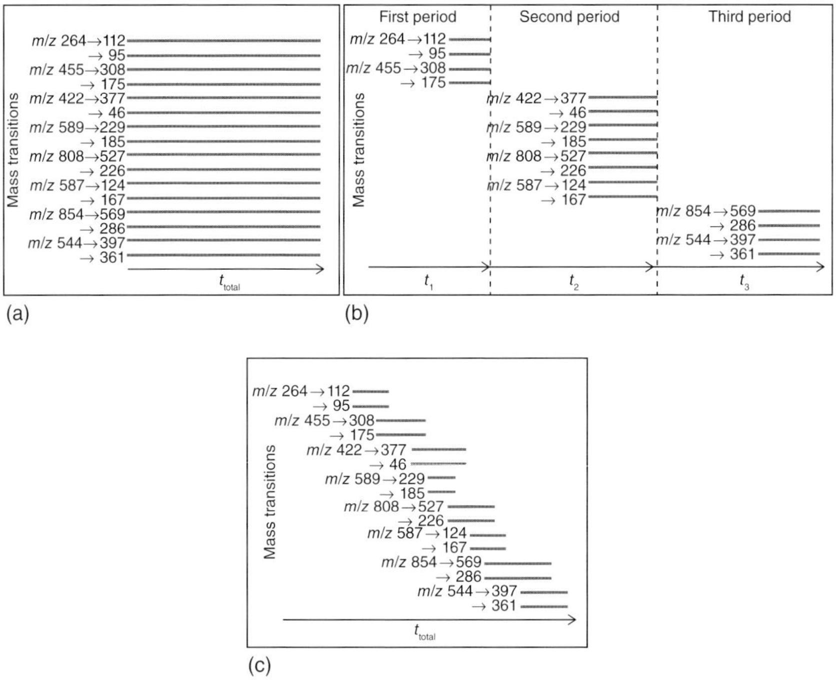

Figure 5.7 Comparative presentation of the multiple reaction monitoring (MRM) principles. (a) Retention-time-independent MRM mode; (b) retention-time-independent MRM mode including a classification of the chromatogram into periods; and (c) retention-time-dependent MRM mode with variable detection windows.

and a large number of simultaneously measured target analytes will lead to an inadequate number of data points per peak. The generated peaks on the column are too narrow. This conclusion does not mean that the hyphenation of older mass spectrometers with a UHPLC system is not useful! One of the biggest advantages of modern UHPLC systems is the consequent reduction of all system volumes. In order to achieve fast analysis cycle times, a low gradient delay volume is a prerequisite. This is especially true when low flow rates between 300 and 500 μl min^{-1} are applied, because the sample throughput is negatively influenced by a high gradient delay volume of, for example, 1 ml using older HPLC systems. One example to illustrate this issue can be found in Sections 2.1.2.2 and 9.3.

The user should now be able to determine the resulting number of data points per peak for her or his own method on the basis of the specification of the available MS device. For this purpose, the average peak width can be used. This assumption is justified because of the band compression during the solvent gradient, which usually leads to a constant peak width across the whole chromatogram. When the peaks elute on an isocratic plateau at the beginning or end of the gradient, a corresponding adjustment of the peak widths should be considered. In Figure 5.7, all MRM modes for the measurement of a specific mass transition are comparatively illustrated.

5.5 Screening Approaches Using LC-MS

In this section, the requirements for suspected- and non-target screening using mass spectrometry are briefly discussed. The common opinion is that only high-resolution devices such as time-of-flight mass spectrometers or Orbitraps can be used. Nevertheless, ion traps, triple-quadrupole mass spectrometers, or QTRAPs (quadrupole linear ion trap, QqLit) can be employed to characterize substances using "data-information depended MS and MS/MS experiments." These combined algorithms are also known as all ion fragmentation (AIF), data-dependent acquisition (DDA), data-independent acquisition (DIA), or information-dependent acquisition (IDA). Examples for suspected-target screening are the pesticide screening for "food safety" or the drug screening in the field of forensic or clinical toxicology [22–25].

In general, all statements of Section 5.3 regarding the LC-MS hyphenation are also valid for the screening approach. This means that the selection of a particular column should be based on the available MS instrumentation. The same is true for the mobile phase as well as the gradient profile. Therefore, the statements outlined in Section 5.3 are not repeated here. Important differences, which should be discussed, concern the choice of suitable settings for the applied ionization technique. In screening analysis, almost no information about the expected analytes is available (see Sections 5.2.2 and 5.2.3 for further explanation). Therefore, generic parameters must be defined to detect the highest possible number of analytes, because identification is only possible for sufficiently ionized compounds. This especially includes the settings for analyte ionization such as ionization voltage, gas flow, and ionization temperature to evaporate the solvent.

Differentiation between full-scan and combined MS and MS/MS experiments for screening approaches must be done for HRMS devices. In general, the maximum sensitivity and mass resolution specified by the manufacturer cannot be achieved at the maximum data acquisition rate. Depending on the analytical problem, a compromise should be made, because it is a crucial difference whether a high number of compounds have to be identified by their sum formula and database comparison in a single run or a limited number of target substances has to be analyzed at a low concentration compared to matrix ingredients. Related to the relevant small-molecule mass range (*m*/*z* 50 to *m*/*z* 1000) adapting the mass resolution of the instruments usually leads to a sufficiently high scan rate (>10 data points per peak) using Orbitraps, time-of-flight (TOF) or quadrupole time-of-flight (Q-TOF) mass spectrometers. Therefore, no restrictions regarding fast chromatography arise.

However, for a proper identification of substances from complex matrices, the mere determination of the sum formula is often not a sufficient criterion. Due to this reason, HRMS with the opportunity to generate product ion spectra from selected precursor ions (Q-TOF systems or Q-Orbitraps) has been established. In this case, the acquisition of additional information using MS/MS spectra needs time. Therefore, the specifications of the manufacturer regarding the scan rate in MS full-scan mode must be critically evaluated or corrected. General rules with respect to the adjustment of these parameters cannot be derived, because choosing "right" screening parameters often depends on the users' experience. Figure 5.8 illustrates the general workflow for screening analysis.

Figure 5.8a shows the total ion current chromatogram (TIC). The TIC is the sum of all ions that are registered at the detector. Usually, the TIC is of low relevance, because small peaks are hidden in the TIC as a needle in a haystack. According to the set scan rate, a complete full-scan spectrum is generated for each data point. An exemplary spectrum is shown in Figure 5.8b. This full-scan spectrum shows the mass spectrum at a given time. On the basis of the *m*/*z* ratio and the isotopic pattern, the sum formula of an unknown compound can be elucidated. However, the determination of the sum formula is not necessarily a sufficient criterion to assign the correct structural formula.

In contrast to the electron impact ionization for GC/MS, the generated mass spectra of LC-MS strongly depend on the selected collision energy. Meanwhile, the MS/MS spectra included into databases are comparable for different manufacturers and mass spectrometers if these are recorded according to a standardized protocol [24]. In addition to an own reference database, commercially or freely available databases are a useful alternative. Especially, open-source databases, such as CheLIST, ChemSpider, DAIOS, DrugBank, HMDB, mzCloud, Norman MassBank, MetFusion, METLIN, PPDB, Stoff-Ident, or TOXNET, become increasingly important. The databases differ in the extent of substances, the number of MS/MS spectra, meta information (e.g., exact mass, sum formula, solubility, REACH-data, literature references) as well as available search options [8–14].

In order to further evaluate the identity of the expected compound, the determination of product ion spectra is useful. Figure 5.8c shows the schematic workflow. For example, in so-called data-dependent experiments, the precursor ion

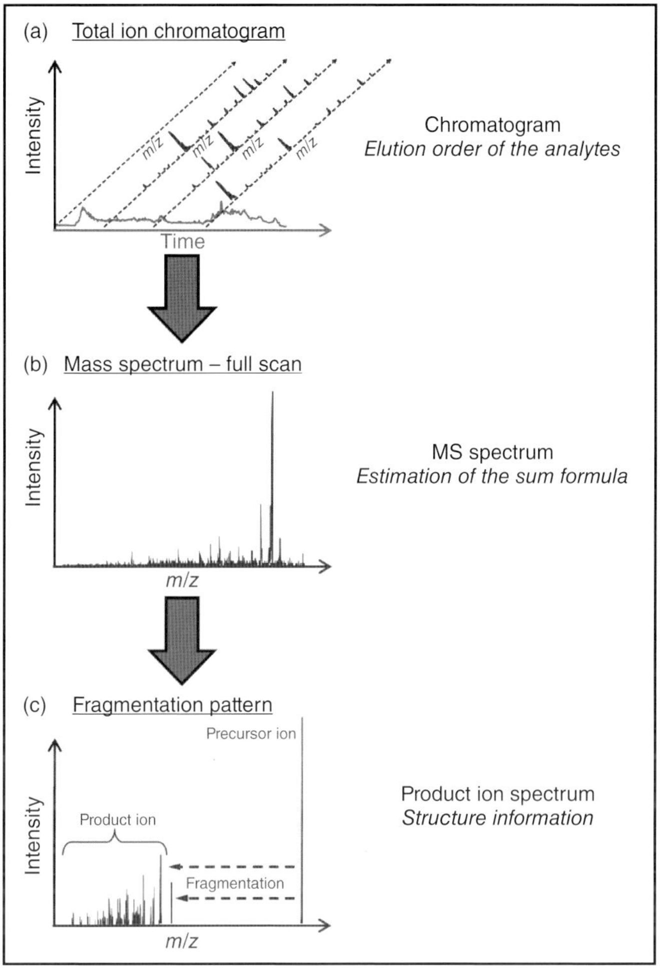

Figure 5.8 General workflow of an LC-MS- and MS/MS-screening analysis.

with the highest intensity can be selected and subsequently fragmented. In this way, the generated product ion spectrum provides additional information, which can be used for structure elucidation. An example can be found in Section 6.8. For this workflow, the number of MS/MS experiments must be defined before a new full-scan mass spectrum is recorded. For example, it is useful to record more than one data-dependent experiment before a new full scan is done. Figure 5.9 illustrates the acquisition cycle for four and eight data-dependent MS and MS/MS experiments, respectively. Again, the number of possible MS/MS experiments depends on the peak width, which is shown in Figure 5.9a for the different system designs.

As depicted in Figure 5.9b, in the first scenario, four "data-dependent" experiments were recorded after the MS full scan. The acquisition time for the full scan was 20 ms, the acquisition time of the product ion spectra was 20 ms, and the

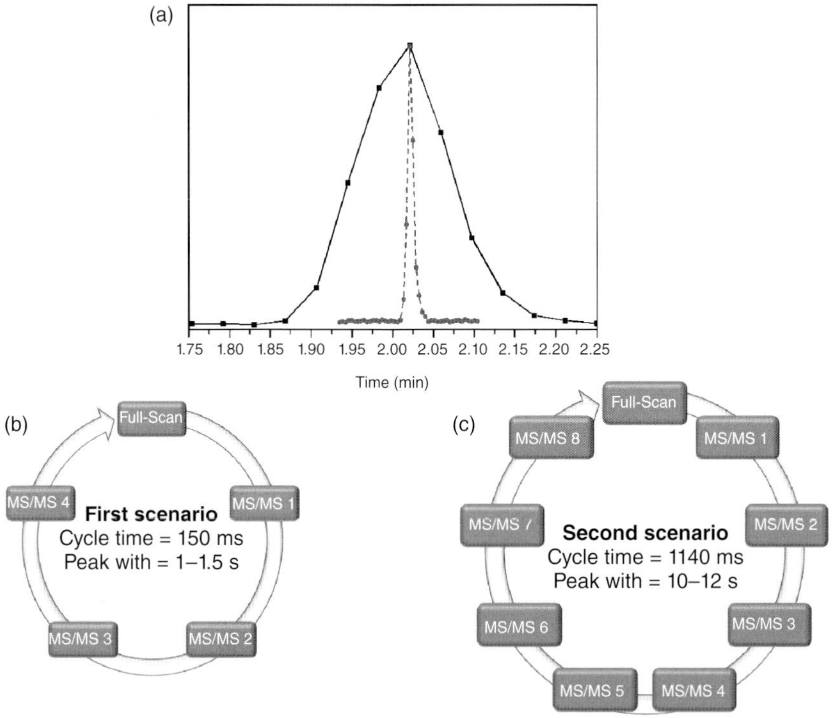

Figure 5.9 Comparison of (a) the obtained peak width using a conventional (■) and highly efficient LC separation (•). In addition, the corresponding cycle times of combined MS full-scan and MS/MS acquisition data are shown for four (b) and eight (c) information-dependent MS/MS experiments.

pause time was 10 ms for each MS/MS experiment. When all times are summarized, the resulting cycle time is 150 ms. For the example shown in Figure 5.9, the system design described in Section 6.6 was used. There, a two-dimensional LC system was connected to a quadrupole time-of-flight mass spectrometer. Because of the minimization of all extracolumn volumes, the peak width was around 1.5 s. As can be seen from Figure 5.9a, approximately seven to eight data points were obtained across a chromatographic peak. This design represents all systems that allow a highly efficient chromatographic separation in combination with mass spectrometry.

In contrast, eight MS/MS experiments were performed after the MS full scan had been finished in the second scenario as shown in Figure 5.9c. This is possible because the peak width was about 10–12 s. The acquisition time for the MS full scan was 250 ms, the time for one product ion scan was set to 100 ms, and the pause time was 10 ms for each MS/MS experiment. Therefore, the resulting cycle time was 1140 ms. Using these settings, 9–10 data points can be obtained. In contrast to target analysis, it is not necessary to generate 10 or more data points per peak for screening analysis. When this method should also be used for quantification, the MS acquisition parameters need to be adjusted.

Summarizing the calculation given earlier, it can be concluded that information-dependent MS and MS/MS experiments require time. Therefore, the number of data points per peak is significantly reduced in contrast to full-scan MS mode. On the basis of the MS system as well as the resulting peak width and specific acquisition parameters, the user should be able to calculate how many data points per peak are obtained for his or her own method using the available instrument.

5.6 Miniaturization – LC-MS Quo Vadis?

This chapter ends with an outlook in terms of miniaturized LC-MS hyphenation. The presentations at scientific conferences over the last years indicate that all manufacturers are dealing with the issue of miniaturization, and this trend will certainly continue. Meanwhile, the predominantly used column inner diameter is 2.1 mm for LC-MS. Regarding the restriction in the flow rate using ESI, it should be mentioned that from theoretical deductions a further reduction of the column inner diameter is desirable. However, the general opinion still maintains that these columns cannot be packed adequately, and, therefore, significantly lower separation efficiency is obtained as expected from the theoretical point of view. Based on the statement of Chapter 3, a key aspect for the minimization of the extracolumn volume is the consequent reduction of the volume of all connection tubings. It is often frustrating to see that the benefits of highly efficient columns cannot be utilized despite of all efforts an experienced HPLC expert invested. That especially occurs if no direct adjustment between two modules such as the column oven and the ion source inlet can be accomplished (remember the system design shown in Figure 5.2). Even when using an optimized system configuration as depicted in Figure 5.3, an inappropriate inner diameter of the emitter tip still leads to an excessive band broadening. As a consequence, the application of columns with an inner diameter of 1.0 mm and less seems visionary.

Therefore, the following experiment should answer the question whether it is possible to use a nano-HPLC column in combination with a micro-LC system and a conventional mass spectrometer. A monolithic column was used with an inner diameter of 100 μm. This column was directly coupled using a filter between the injector of the micro-LC system and the emitter tip of the mass spectrometer. The mass spectrometer was an older device from Sciex (3200 QTRAP). To minimize the band broadening after the column, an emitter tip with an inner diameter of 25 μm was installed instead of the classical 100 μm ID emitter tip. In Figure 5.10, both emitter tip variants are shown. While the classical tip is made of stainless steel, the modified miniaturized emitter tip is based on a PEEKSil capillary.

To ensure the ionization, at the top of the PEEKSil capillary, a stainless-steel tip with the respective inner diameter is installed. Regarding the connection technique, it should be mentioned that the PEEKSil emitter is designed for 1/32″ fittings. High-pressure resistant fittings are screwed to a 1/32″ union. This union additionally offers the advantage of being able to be used as grounding point. The change of the emitter tip can be done easily in a few minutes. As shown

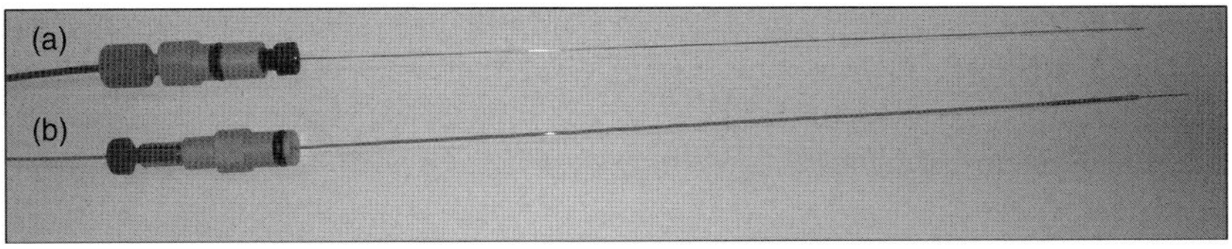

Figure 5.10 Comparison of commercially available emitter tips. (a) Conventional emitter tip with an inner diameter of 100 μm, compatible with 1/16″ fittings and (b) miniaturized emitter tip with an inner diameter of 25 μm, compatible with 1/32″ fittings.

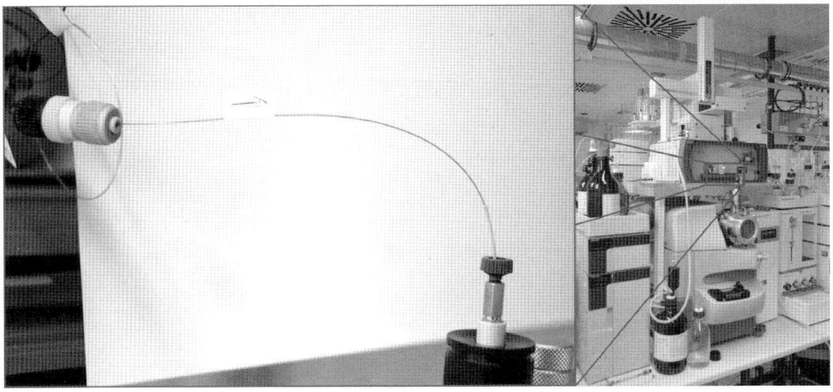

Figure 5.11 System design for the separation of selected pharmaceuticals on a monolithic nano-HPLC column (150 mm × 0.1 mm).

in Figure 5.11, the monolithic column can be adjusted like a transfer capillary between the injector and the ion source of the mass spectrometer.

According to the recommendation of the manufacturer, the back pressure of the monolithic column should not exceed 200 bar. Therefore, the flow rate was adjusted to 5 μl min^{-1}. Figure 5.12a shows the separation of approximately 50 pharmaceuticals for a conventional HPLC system in combination with a 2.0 mm ID monolithic column. In the second scenario (Figure 5.12b), the system design shown in Figure 5.11 was used.

Looking at the peak width, no significant band broadening can be observed using a nano-HPLC column. This result may be surprising at first because even at the operation in a low flow rate range, no modification of the ion source of the mass spectrometer has to be made. It should also be mentioned that for the separation with a nano-HPLC column, no optimized mass spectrometric conditions were chosen. That is why a “fair” comparison regarding the sensitivity between both approaches cannot be made. The comparison shown here should only demonstrate the outstanding potential of the described hyphenation.

An additional benefit of the miniaturized approach concerns the solvent consumption. Compared to the conventional separation, it is just 1%. An even more important aspect concerning the miniaturized LC-MS hyphenation is the lower amount of “dirt,” which reaches the mass spectrometer. This is a central issue because the separation between matrix ingredients and target compounds represents a major challenge in residual analysis. It is observed that very low LOD can be obtained when a reference standard dissolved in high purity solvents

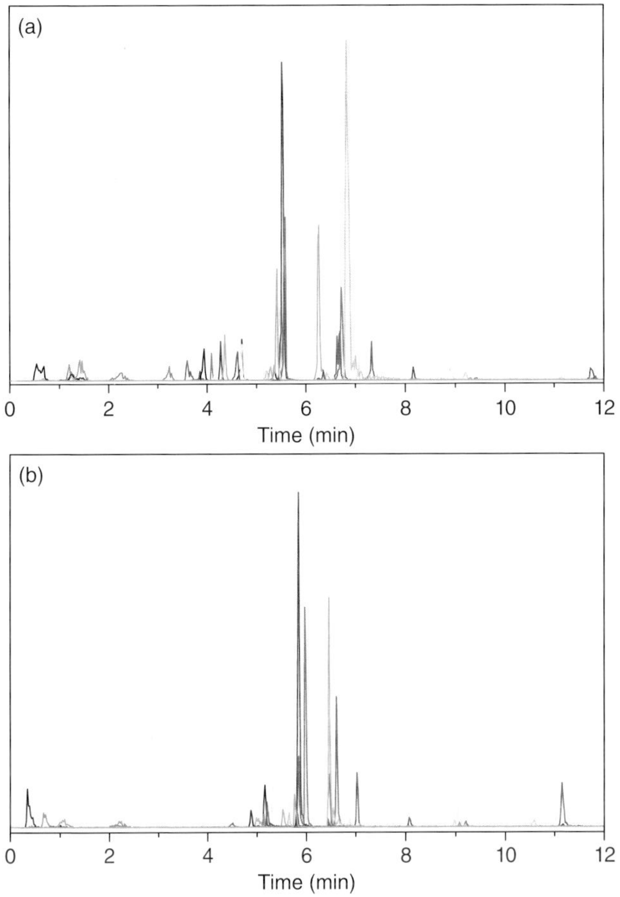

Figure 5.12 LC-MS/MS chromatogram of a separation for 50 pharmaceuticals on (a) 50 mm × 2.0 mm Chromolith FastGradient RP18 column. Chromatographic parameters: temperature: 40 °C; injection volume: 10 μl; mobile phase: A = water + 0.1% formic acid, B = methanol + 0.1% formic acid; flow rate: 500 μl min^{-1}; mass spectrometric parameters: pause time: 5 ms; dwell time: 20 ms and (b) 150 mm × 0.1 mm Chromolith CapROD C18 column. Chromatographic parameters: temperature: room temperature; injection volume: 100 nl; mobile phase: A = water + 0.1% formic acid, B = acetonitrile + 0.1% formic acid; flow rate: 5 μl min^{-1}; mass spectrometric parameters: pause time: 5 ms; and dwell time: 20 ms.

is measured. In contrast, the included matrix components in real samples lead to more or less pronounced matrix effects and therefore to ion suppression. In this respect, the miniaturization is superior for samples that contain a high amount of interfering matrix compounds because the injection volume is reduced. As a consequence, the cleaning intervals of the ion source are extended. The reduction of the column inner diameter to 100 μm is only applicable for specially designed micro- or nano-HPLC systems. These kinds of HPLC systems are commercially available for several years from different manufacturers. In addition, the dispersion after the column needs to be diminished. Without the modification of the emitter tip to 25 μm, the utilization of nano-HPLC columns with the shown system design is not possible.

References

1 Kromidas, S. (2016) *The HPLC Expert: Possibilities and Limitations of Modern High Performance Liquid Chromatography*, Wiley-VCH Verlag GmbH, Saarbrücken.
2 Cole, R.B. (2010) *Electrospray and MALDI Mass Spectrometry: Fundamentals, Instrumentation, Practicalities, and Biological Applications: Fundamentals, Instrumentation, and Applications*, John Wiley & Sons, Inc.
3 Dass, C. (2007) *Fundamentals of Contemporary Mass Spectrometry*, Wiley-Interscience.
4 de Hoffmann, E. and Stroobant, V. (2007) *Mass Spectrometry – Principles and Applications*, John Wiley & Sons, Ltd., Brussels.
5 Krauss, M., Singer, H., and Hollender, J. (2010) *Anal. Bioanal. Chem.*, **397**, 943.
6 Schymanski, E.L., Jeon, J., Gulde, R., Fenner, K., Ruff, M., Singer, H.P., and Hollender, J. (2014) *Environ. Sci. Technol.*, **48**, 2097.
7 Zedda, M. and Zwiener, C. (2012) *Anal. Bioanal. Chem.*, **403**, 2493.
8 BB-X Stoffident. http://bb-x-stoffident.hswt.de/app (accessed 23 June 2016).
9 ChemSpider. http://www.chemspider.com/ (accessed 23 June 2016).
10 Daios Online. http://www.daios-online.de/ (accessed 23 June 2016).
11 MassBank. http://www.massbank.eu/MassBank/ (accessed 23 June 2016).
12 Metlin Scripps. http://metlin.scripps.edu/index.php (accessed 23 June 2016).
13 MSBI. http://msbi.ipb-halle.de/MetFusion/ (accessed 23 June 2016).
14 MzCloud. https://www.mzcloud.org/ (accessed 23 June 2016).
15 DrugBank. http://www.drugbank.ca (accessed 23 June 2016).
16 Human Metabolome Database. http://www.hmdb.ca/ (accessed 23 June 2016).
17 Toxicology Data Network. http://toxnet.nlm.nih.gov/ (accessed 23 June 2016).
18 PPDB:Pesticide Properties Database. http://sitem.herts.ac.uk/aeru/ppdb/en/index.htm (accessed 23 June 2016).
19 Chemical Lists Information System. http://chelist.jrc.ec.europa.eu/index.jsp?id=0 (accessed 23 June 2016).
20 Wolf, S., Schmidt, S., Müller-Hannemann, M., and Neumann, S. (2010) *BMC Bioinf.*, **11**, 148.
21 Gross, J.H. (2011) *Mass Spectrometry: A Textbook*, 2nd edn, Springer Spektrum.
22 Alder, L., Greulich, K., Kempe, G., and Vieth, B. (2006) *Mass Spectrom. Rev.*, **25**, 838.
23 Dresen, S., Gergov, M., Politi, L., Halter, C., and Weinmann, W. (2009) *Anal. Bioanal. Chem.*, **395**, 2521.
24 Oberacher, H., Pitterl, F., Siapi, E., Steele, B.R., Letzel, T., Grosse, S., Poschner, B., Tagliaro, F., Gottardo, R., Chacko, S.A., and Josephs, J.L. (2012) *J. Mass Spectrom.*, **47**, 263.
25 Oberacher, H., Weinmann, W., and Dresen, S. (2011) *Anal. Bioanal. Chem.*, **400**, 2641.

6

2D chromatography – Opportunities and limitations

Thorsten Teutenberg and Juri Leonhardt

6.1 Introduction

Two-dimensional liquid chromatography seems to mature and make it into industrial research and routine laboratories. Due to increasing interest in this technology, a special issue has been devoted in the Journal Analytical and Bioanalytical Chemistry in 2015. Many publications therein are highlighting the steady success of this emerging separation technique [1–10]. Moreover, instrument manufacturers have also significantly improved or introduced systems for two-dimensional liquid chromatography. Although the technical progress, which has been achieved during the past years, is really impressive, the question needs to be asked why one-dimensional liquid chromatography might not be sufficient for addressing the analytical tasks of the future.

6.2 Why Two-Dimensional HPLC?

The ever-rising demands toward chromatographic and mass spectrometric methods make it necessary to analyze samples containing hundreds or thousands of compounds [11]. It is often not intended to identify or quantify every single substance. In fact, the aim is to obtain an overview about the constitution of the sample by screening methods. Generally, it has to be differentiated between target and nontarget screening methods [12]. If a target screening is performed, only known compounds are analyzed where a reference standard is available. In contrast, a nontarget screening is usually focused on acquiring information about all compounds in the sample. Often, reference standards are not available. The reader is referred to Chapter 5 for a detailed overview of the different screening approaches.

Mainly, triple-quadrupole mass spectrometers are used for target analysis. Here, multiple reaction monitoring (MRM) is applied for a sensitive and selective determination of known compounds. In general, two mass transitions that are characteristic for the analyte are chosen. It is therefore possible to

The HPLC Expert II: Find and Optimize the Benefits of your HPLC/UHPLC, First Edition.
Edited by Stavros Kromidas.

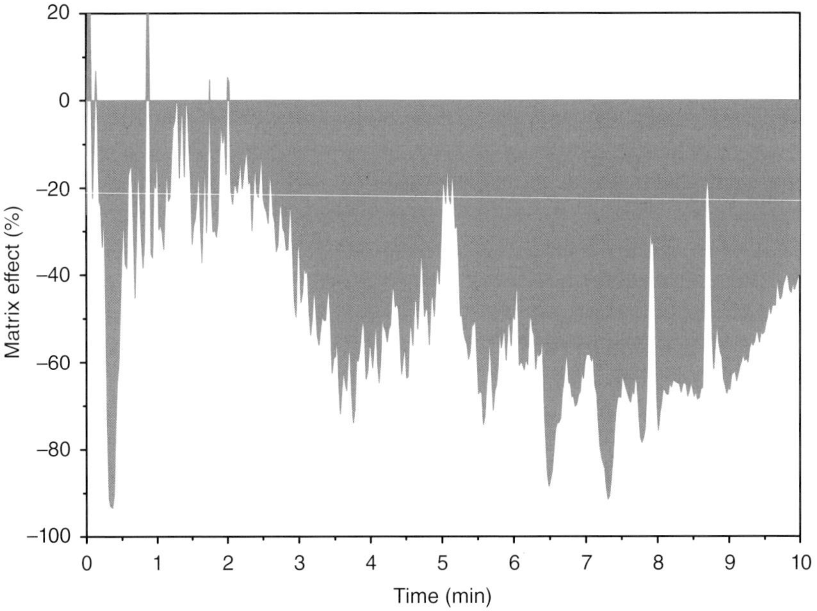

Figure 6.1 Matrix effect chromatogram of a tea extract. For further details, see text.

separate coeluting compounds if they have different MRM transitions. If too many substances are eluting at the same time, however, this might lead to ion suppression. An example of a so-called matrix effect chromatogram is given in Figure 6.1.

Such a chromatogram is obtained if the matrix is injected onto the stationary phase and is eluted with the solvent gradient. Simultaneously, a solution that contains the analytes of interest is infused via a T-piece after the column. The detector acquires the mass transitions of each target analyte during the complete chromatographic run. In an optimal case, a loss of intensity will not occur, and a constant signal is obtained for each mass transition. However, often, a severe loss of the signal is observed. Such effects are mainly caused by interfering compounds of the matrix, leading to ion suppression for the target analyte [13]. In order to guarantee a reliable quantification in difficult matrices, standard addition has been proposed [14].

For screening approaches, highly efficient chromatographic methods should be used to detect as many compounds as possible. This means that the peak capacity, which is the number of separated peaks with unit resolution, has to be increased [15]. The limitation of this approach in one-dimensional liquid chromatography can be seen in Figure 6.2.

There, a chromatogram of the separation of a pesticide standard that was obtained using one-dimensional liquid chromatography hyphenated to time-of-flight mass spectrometry is shown. The injected sample contains hundreds of compounds. Although the deconvolution algorithm is able to make an automated peak recognition, thus distinguishing the compounds by their

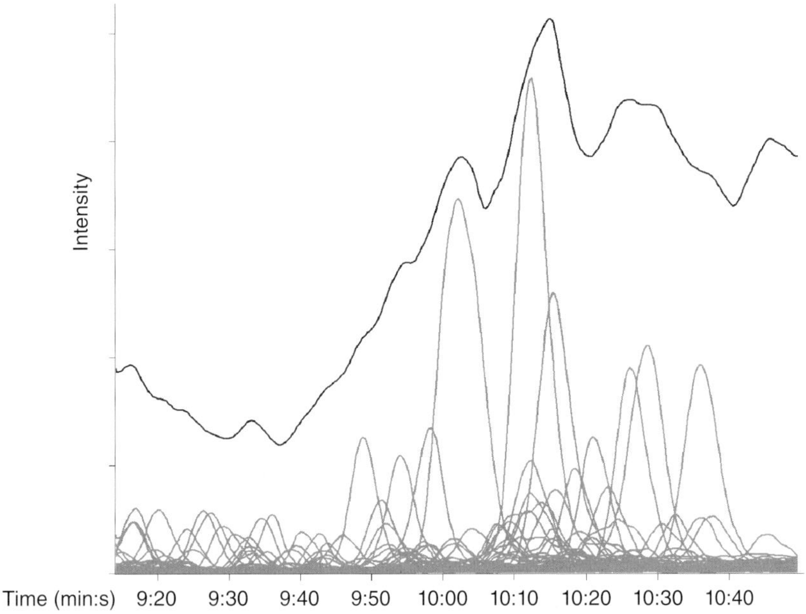

Figure 6.2 Chromatogram of a pesticide standard. The deconvoluted extracted ion chromatograms are marked by vertical gray lines. The bold black line represents the total ion current chromatogram.

different mass-to-charge (m/z) ratios, a coelution of peaks cannot be avoided. Analytes that have a low ionization efficiency and are present in low amounts might therefore not be detected. This might even be the case for compounds that are present in a high concentration when the number of coeluting peaks is very high and signal intensity is suppressed by coeluting species. As should be clear by the chromatogram in Figure 6.2, a complete chromatographic separation will not be possible with one-dimensional liquid chromatography. The only way to increase the number of resolved peaks is to increase the peak capacity on the basis of two-dimensional liquid chromatography.

6.3 Peak Capacity of One- and Two-Dimensional Liquid Chromatography

6.3.1 Peak Capacity of One-Dimensional Liquid Chromatography

The peak capacity is the number of peaks that are placed in an elution window with unit resolution. In a very simplified form, for solvent gradient elution, peak capacity can be calculated as follows:

$$P_c = 1 + \frac{t_g}{w} \tag{6.1}$$

P_c is the theoretical peak capacity, t_g is the gradient time, and $\overline{w}$ is the mean peak width. Although there exist numerous equations in the scientific literature to exactly calculate the peak capacity [16], the simple equation given earlier is well suited to correctly estimate the peak capacity from the chromatogram. Equation 6.1 is based on the assumption that the peak width across the chromatogram is constant for all compounds eluted during the solvent gradient. This assumption is valid because if a linear solvent gradient is applied using a constant gradient slope, the band compression leads to a nearly constant peak width during the complete gradient run [17]. This is in contrast to isocratic elution because later eluting peaks are much broader than early eluting peaks. However, isocratic elution often cannot be used if very complex samples have to be analyzed, because very long analysis times are obtained for hydrophobic compounds. It should be noted that the gradient time is often associated with the duration of the solvent gradient. For a more realistic estimation of the peak capacity in one-dimensional LC, the gradient window should only stretch from the first to the last eluting peak. In this case, the separation space as experienced by the sample is considered more correctly.

6.3.2 Peak Capacity of Two-Dimensional Liquid Chromatography

6.3.2.1 Heart-Cut 2D LC (LC-LC)

When heart-cut 2D LC is applied, usually only one fraction is transferred from the chromatogram of the first dimension (D1) to the second dimension (D2) column. This is schematically presented in Figure 6.3. The advantage of the heart-cut approach is that the transferred fraction can be analyzed on a stationary phase

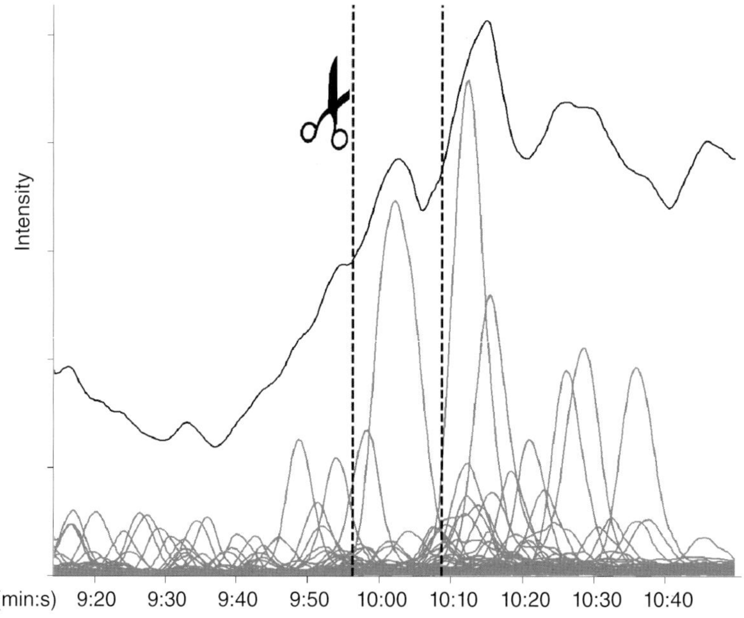

Figure 6.3 Schematic illustration of heart-cut 2D LC. For further details, see text.

with a different selectivity. It might then be possible to completely resolve all coeluting compounds of this fraction.

For a high degree of flexibility, it is essential to use a second detector and a suitable switching valve setup. A comprehensive analysis of the chromatogram of the first dimension is usually not necessary when heart-cut 2D LC is applied.

For LC-LC, the peak capacities of both dimensions can be added. A peak capacity of 200 is obtained when both columns in D1 and D2 have a single peak capacity of 100. The domain of heart-cut 2D LC is mainly a peak purity analysis of a pharmaceutical ingredient. These separation problems are often not very complex. However, the verification of the peak purity is extremely important within the context of pharmaceutical research and the approval of suitable drug candidates by the FDA or other regulative authorities. The interested reader is referred to further literature where an example is given of a general workflow to determine the peak purity of a chiral pharmaceutical compound [18].

6.3.2.2 Comprehensive 2D LC (LC × LC)

The application of comprehensive LC × LC (pronounced: LC times LC) is useful if highly complex samples have to be analyzed. This might be the case for nontarget screening in environmental analysis, where metabolites and transformation by-products have to be characterized [19]. Also, a screening for pesticides in feed and food requires methods with a high resolving power [20]. Moreover, comprehensive 2D LC has been used for a long time in the field of bioanalysis to identify biomarkers from natural/plant extracts [21]. In contrast to heart-cut LC, the whole eluent of the first dimension is transferred in single fractions to the second-dimension column. This procedure is characterized by the adjective "comprehensive" and the sign for multiplication "×" has to be used. The chromatogram given in Figure 6.4 illustrates this issue.

It is important to note that a peak eluting from the first-dimension column should be at least sampled three or four times. This requirement is based on the early work of Murphy, Schure, and Foley and is denoted as M-S-F criterion [22]. If the M-S-F criterion is not fulfilled, the chromatogram of the first dimension cannot be reconstructed from the retention data of the second-dimension column. Moreover, a lower peak capacity is obtained as theoretically predicted. In a comprehensive 2D LC system, the peak capacities of both dimensions will be multiplied. This assumption is only valid if both separation systems are orthogonal and the influence of the sampling on peak capacity can be neglected. If the first column has a peak capacity of 50 and the second column a peak capacity of 20, the overall peak capacity of the system will be 1000. This simple calculation underlines the huge potential of comprehensive two-dimensional liquid chromatography. However, it is extremely difficult to establish a true orthogonal separation system because of practical reasons.

An orthogonal system will be obtained when the retention mechanisms of both dimensions are statistically independent [23]. A separation by hydrophobicity combined with a separation by molecular weight (RP × SEC) might lead to an orthogonal system. Also, the hyphenation of a normal phase with a reversed phase (NP × RP) is a useful combination. In an optimal scenario, all compounds in the mixture are spread evenly across the two-dimensional separation space

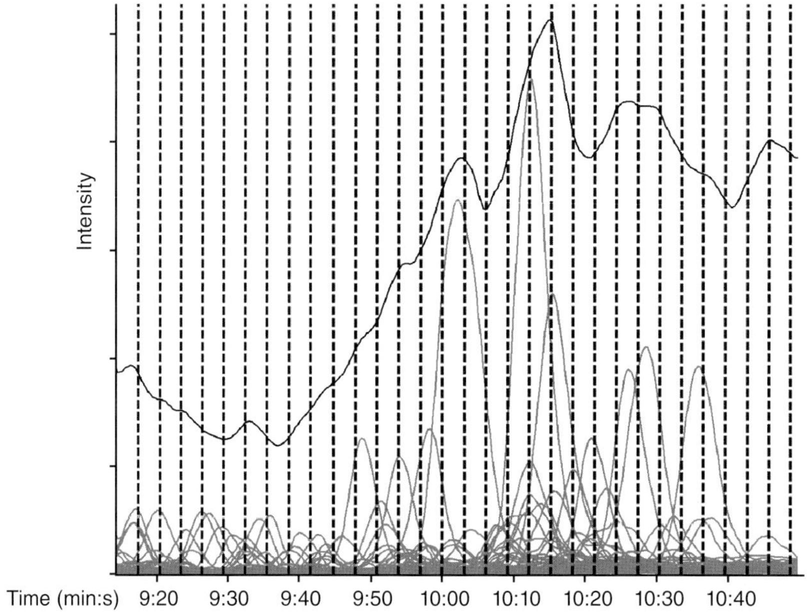

Figure 6.4 Schematic illustration of comprehensive 2D LC. For further details, see text.

as is depicted in Figure 6.5a. The filled circles represent the peak maxima of the compounds, which are separated in the first and second dimension. It can be seen that the peaks are spread across the available separation space. In contrast, the filled circles in Figure 6.5b are aligned along the bisecting line. This means that the retention mechanisms of both dimensions are strongly correlated.

Setting up a truly orthogonal separation system is in many cases not possible. The reason is that the interactions of the analytes with the mobile and stationary phase cannot be reduced to one single mechanism. This is demonstrated by the illustration given in Figure 6.5b. Here, a strong correlation between both dimensions and thus a low orthogonality are obtained because the peaks are not spread evenly across the available separation space.

A strong correlation between the two dimensions also results in a lower overall peak capacity. In order to account for these effects, the general equation for calculating peak capacity in online comprehensive 2D LC has been modified [24]. A broad discussion of these effects is not the primary intention of this chapter. The reader is therefore referred to the respective literature [16]. Moreover, the theoretical discussion in the scientific literature is often not without controversies, which might be highlighted by a citation from a book dealing with two-dimensional chromatography: "[…] why do we so often report the individual peak capacities of the two dimensions and their product as the 2D peak capacity? […] A more realistic answer is that *it is easy to do and appears more impressive than it really is – especially to those who fund our work*. In fact, as a practical metric it is often nonsense [25]". In Section 6.12.3, the discussion about peak capacity and the advantages of 2D LC over 1D LC will be revisited.

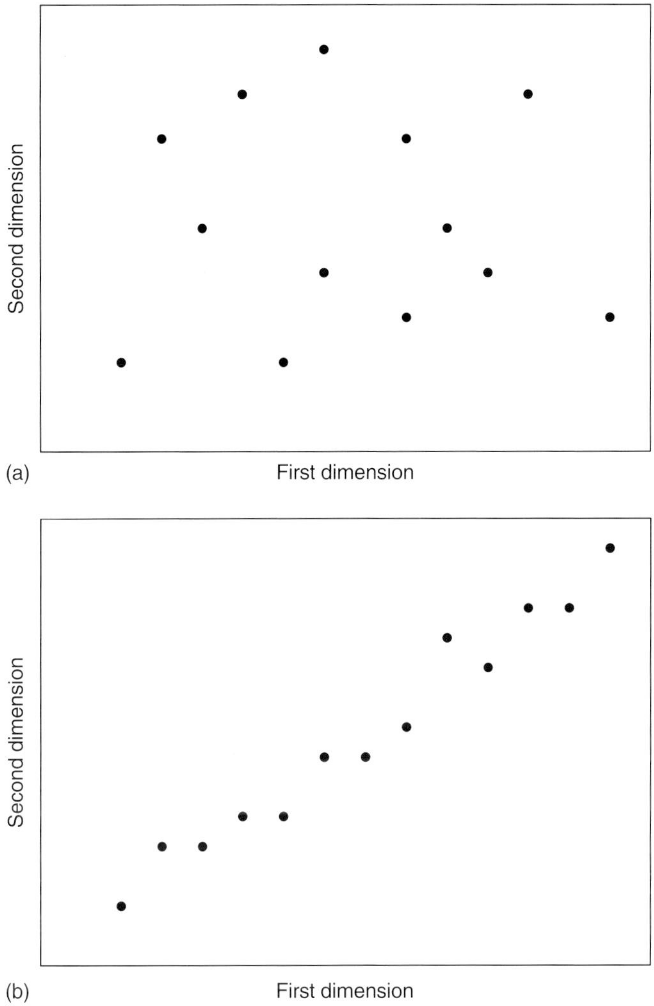

Figure 6.5 Schematic illustration of the correlation of a hypothetical scattering of compounds across a two-dimensional separation space. (a) Low correlation of both retention mechanisms (orthogonal system) and (b) high correlation (no orthogonal system) between first and second separation. For further details, see text.

An interesting approach to estimate the coverage of the available retention space is presented in [26]. This method is based on the simulation of a virtual combination of one-dimensional runs. The authors of that study state that it is necessary to at least include 30 target analytes with a high significance for the respective separation problem. After the acquisition of the one-dimensional runs on different stationary phases, the one-dimensional chromatograms can be rearranged to a virtual 2D plot. This kind of representation then allows classifying the most suitable separation system without the need to run a real two-dimensional separation. The validation experiments can then directly be performed with the best separation system. Such a software-assisted method development can be a real benefit for the user because some effects, which might be related to the modulation, can be neglected during method development. A drawback of this approach is that the selected target analytes have to be representative for the real sample.

Interim Conclusion In a simplified way, Equations 6.2 and 6.3 can be used for calculating the peak capacity for heart-cut 2D LC (Equation 6.2) and comprehensive 2D LC (Equation 6.3):

$$n_{c,2D} = n_{c1} + n_{c2} \tag{6.2}$$

$$n_{c,2D} = n_{c1} \times n_{c2} \tag{6.3}$$

While the peak capacity can be easily calculated for heart-cut 2D LC on the basis of the single chromatograms obtained on each column, Equation 6.3 should only be considered as a rough estimation.

6.4 Modulation

6.4.1 Online Heart-Cut 2D LC

For heart-cut 2D LC, the transfer of a fraction from the first dimension to the second dimension is usually obtained by a valve switching and a sample loop. This process is therefore termed *sampling* or *modulation*. In principle, the complete elution volume of a peak of interest should be transferred to the D2 column. If a peak has a width of 15 s and the flow rate is 1 ml min^{-1}, a volume of 250 μl has to be transferred. This example highlights that the inner diameter of the second-dimension column has to be adapted to the elution volume of the peak and to the flow rate in the first-dimension separation. Otherwise, a volume overload of the second-dimension column will result. Using a trigger signal of a detector that is installed after the first-dimension column, the transfer valve can be switched and the eluate of the first-dimension column will be directed to the sampling loop. The volume contained in the sampling loop is then injected onto the second-dimension column. The workflow described here is focused on online LC-LC. The fraction might also be transferred offline to the second-dimension column. In this case, it might be possible to sample higher volumes because a large part of the transfer solvent can be eliminated by evaporation. Furthermore, a solvent exchange can be accomplished for a better compatibility of the injection solvent with the starting conditions of the D2 mobile phase.

6.4.2 Comprehensive Online 2D LC

For comprehensive online 2D LC, the modulation is more complex and also more difficult to understand then for the transfer of one single fraction as described in Section 6.4.1. In principle, the modulation is achieved by the two-loop technique. Figure 6.6 shows the general setup of an LC × LC system.

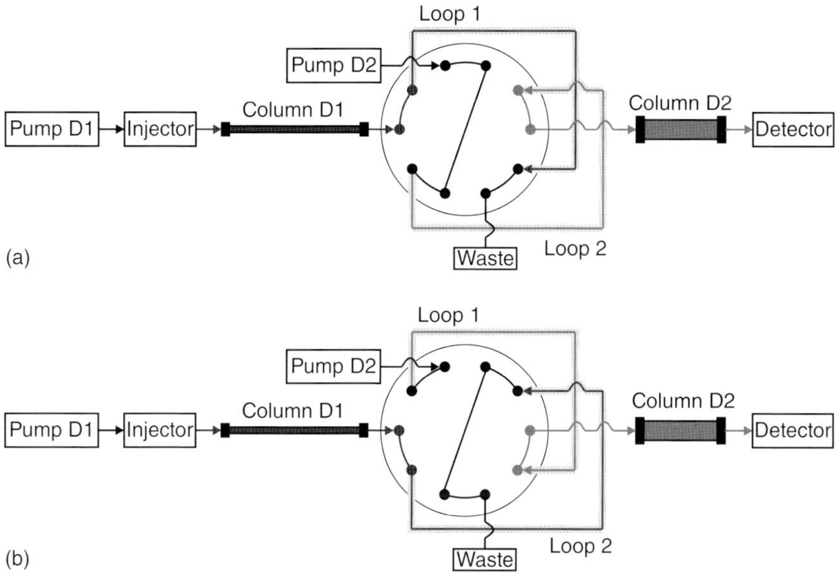

Figure 6.6 Schematic illustration of the principal setup of a comprehensive online LC × LC system. For further details, see text.

After the sample has been injected onto the first-dimension column, the eluate of the first-dimension column will be directed alternatively to the sampling loops. As has already been mentioned in Section 6.3.2.2, a peak eluting from the first-dimension column should be sampled at least three to four times. If a peak has a width of 15 s and should be sampled three times, every 5 s, a complete analysis cycle on the second-dimension column including all re-equilibration steps needs to be accomplished. Such very fast cycle times have not been done for online comprehensive 2D LC. Typical cycle times for the second dimension are between 30 and 60 s. This means that a peak should have a width of about 90 – 180 s on the first-dimension column to fulfill the M-S-F criterion. This might appear "devastating," since in the age of UHPLC, peak widths of 1 s can be easily obtained. In the following, it will be demonstrated that a seemingly terrible separation efficiency on the first-dimension column is not an argument for the alleged inadequacy of comprehensive online 2D LC. In order to guarantee the traceability of the following steps on the basis of simple examples, the cycle time of the second dimension will be set to 1 min. This means that all steps including the re-equilibration of the column after the solvent gradient have to be finished before the next cycle can be started.

If it can be assumed that a peak has a width of 3 min on the first-dimension column and has to be sampled three times, the M-S-F criterion is fulfilled. In principle, the complete chromatogram obtained on the first dimension is now sampled with a sample time of 60 s. The first-dimension column is then connected with either the first or the second sample loop while the valve is switched alternatively. One of the loops is filled with eluate of the first-dimension column while the other loop is connected to the second-dimension column. This process is continuously repeated until the separation is finished on the first-dimension column.

If the peaks would have a width of 15 s or lower when eluting from the first-dimension column, the separation, which has been achieved on this column, would inevitably be "destroyed" and the peaks would merge in the sample loop. According to the definition, which has been established by many authors, the separation could not be called comprehensive.

As can be seen from the setup depicted in Figure 6.6, it becomes clear that the diameter of the second-dimension column should be larger than that of the first-dimension column. The reason is that a higher flow rate has to be used for the second-dimension column to obtain very small cycle times of about 1 min. If the inner diameter of the first-dimension column will be larger than that of the second-dimension column, the transfer volume would be too big, causing a volume overload when transferring the eluate from the first to the second-dimension column.

On the basis of explicit examples, the practical relevance will be explained. If the flow rate of the first dimension will be adjusted to 1 ml min^{-1} and the sampling time is 1 min, a transfer volume of 1 ml will result. In order to avoid a volume overload of the second-dimension column, not more than 10% of the column void volume should be filled with the injection plug. Otherwise, split or severely distorted peaks will be observed. Therefore, a compromise has to be found balancing the need for an optimal ratio between the inner diameter of the column and the flow rate in both dimensions. This is definitely the most difficult step when optimizing an LC × LC system.

Another important aspect of online comprehensive LC × LC is that the whole eluate of the first-dimension column has to be transferred to the second-dimension column. The volume of the sample loops has therefore to be adapted to the flow rate of the first-dimension column. If the capacity of the sample loops is too low, a significant portion of the first column's eluate will be directed to waste. A comprehensive analysis of the sample according to the definition would then be not possible.

6.4.3 Stop-Flow and Offline LC × LC

When performing stop-flow or offline LC × LC, the sampling and analysis of the fractions on the second-dimension column can be uncoupled. Using the stop-flow technique, the flow of the first dimension will be stopped during the analysis on the second-dimension column. Using offline LC × LC, the eluate from the first-dimension column will be sampled and stored in different vials using a fraction collector. The advantage is that fractions containing a higher volume can be sampled. Moreover, a solvent change can be done. This enables the compatibility of phase systems, which are not directly amenable in online LC × LC mode. Because of the fact that establishing online LC × LC presents the biggest challenge and has the potential of a high sample throughput when compared to offline or stop-flow LC × LC, the focus of the following sections will be on practical problems of the online LC × LC approach.

6.5 Practical Problems of Online LC × LC

6.5.1 Compatibility of the Solvent Systems

As has been already discussed in Section 6.3.2.2, the phase systems, which have been selected, should exhibit a different selectivity in order to cover the chromatographic retention space to a high degree. It could therefore be assumed that the combination of a normal phase with a reversed-phase stationary phase should be ideally suited, which has also been demonstrated by some studies [27]. Unfortunately, the solvent systems that are used in both dimensions are often not compatible. The reason is that water is the strongest eluent in normal-phase chromatography but the weakest eluent in reversed-phase chromatography. The transfer and injection of an organic solvent plug onto a reversed-phase column in an NP × RP hyphenation can therefore lead to severely distorted or split peaks [28]. Moreover, it has to be considered that some solvents used for normal-phase chromatography are only partially or even not miscible with water.

6.5.2 Dilution

The sampling requires a peak eluting from the first-dimension column to be sampled three or more times. This and the fact that the inner diameter of the second-dimension column is larger than that of the first-dimension column lead to a strong dilution of the elution band from the first-dimension column. This makes it necessary to think about a suitable strategy to inject a large volume on the first-dimension column to counteract the effect of the sampling process. Otherwise, the sample has to be preconcentrated by either offline or online solid-phase extraction.

6.5.3 High Flow Rate

In order to obtain a low cycle time of 1 min or even less in the second dimension, a high flow rate usually has to be adjusted. Often, the flow rate is between 2 and 5 $ml\,min^{-1}$ to achieve a high linear velocity and to reduce the gradient delay time [29]. This disadvantage originates from the fact that the inner diameter of the second-dimension column needs to be larger than the diameter of the first-dimension column to avoid overloading when sample fractions are transferred. It is not only the high consumption of toxic and costly solvents that represents a drawback, but also the hyphenation with mass spectrometry is limited.

6.5.4 Compatibility with Mass Spectrometry

The application of comprehensive 2D LC makes it necessary to use a mass spectrometer. This is especially the case when complex natural extracts have to be analyzed and information about the composition of the sample needs to be gathered. Time-of-flight mass spectrometers with integrated MS/MS functionality (Q-TOF) have distinct advantages over low-resolution systems. In principle, it is possible to use electrospray ionization (ESI) or atmospheric pressure chemical ionization (APCI), although ESI is used more frequently than APCI. Nonetheless, it should be noted that ESI is performed with lower flow rates below 1 ml min^{-1}, whereas APCI can also be used with higher flow rates up to 2 ml min^{-1}. Therefore, the flow rate in the second-dimension separation should be adjusted to achieve an optimal hyphenation between liquid chromatography and mass spectrometry.

Interim Conclusion In order to solve the problems listed in the previous sections, a column with an inner diameter smaller than 1.0 mm should be used in the first dimension and a low flow rate of less than 50 μl min^{-1} should be adjusted. The inner diameter of the second-dimension column should not be higher than 2.1 mm. Otherwise, high flow rates above 2 ml min^{-1} cannot be avoided to obtain a cycle time of less than 1 min. Miniaturization may therefore play an important role to improve the compatibility of online comprehensive 2D LC with mass spectrometry.

6.6 Development of a Miniaturized LC × LC System

In this section, the concept of a miniaturized LC × LC system is described before different concepts will be discussed. The results that are presented here are mainly based on the publication by Haun *et al.* [30]. The interested reader will find additional information to deepen the understanding of this concept.

6.6.1 Technical Platform

In order to eliminate the drawbacks that are described in Section 6.5, a technical platform was used, which was originally designed for simple cleanup of bioanalytical samples. This system contained two separately controllable pneumatic syringe pumps, which allow for a splitless delivery of the eluent from 50 nl min^{-1} to 200 μl min^{-1}. The heating compartment contained two 10-port-2-position valves. However, only one valve was used for the sampling process as depicted in Figure 6.6. The very compact design of all modules results in a very low extracolumn volume. The temperature could be adjusted from 5 °C above room temperature to 80 °C. However, the stability of the integrated valves needs to be considered when a high temperature is adjusted.

6.6.2 Selection of the Stationary Phase

The selection of the stationary phases was mainly oriented to develop a screening method for water and wastewater samples. As has been noted in Section 6.5.2, the modulation in LC × LC leads to a strong dilution. Therefore, a column combination has been chosen, which allows for a large-volume direct injection (LVDI) of aqueous samples onto the first-dimension column. A stationary phase, which is known under the trade name Hypercarb by ThermoFisher and consists of porous graphitic carbon (PGC), was selected. This phase is already used for the analysis of small polar molecules, for example, acrylamide. In contrast to silica-based reversed-phase stationary phases, PGC is characterized by a strong hydrophobicity because of the fact that it contains only carbon [31]. By applying an LVDI, a high enrichment factor can be achieved. Leonhardt *et al.* demonstrated that for 5-fluorouracil, which usually elutes at the void time of a typical silica-based reversed-phase stationary phase, a retention factor of 146 can be achieved if an aqueous sample is directly injected [32]. A good illustration of this effect can also be found in Chapter 9, where an LVDI has been performed on a 2.1 mm ID PGC column.

A core–shell stationary phase with a particle diameter of 2.6 μm was used in the second dimension to facilitate high linear velocities. The reason is that partially porous particles have a better pressure-to-flow rate ratio than sub 2 μm fully porous particles. A material, which is known under the trade name SunShell by ChromaNik Technologies, was selected. According to the information from the manufacturer, the loadability of this material should be higher when compared to other core–shell phases.

Because the difference of the interaction mechanisms between PGC and silica-based stationary RP phases is well documented in the scientific literature [31, 33, 34], it could be anticipated that a sufficient difference in the selectivity of both stationary phases is given.

6.6.3 Selection of the Mobile Phase and Temperature

For further optimization of selectivity, different mobile phases should be used in the first and second dimension. Therefore, methanol has been selected as protic solvent in D1, and acetonitrile has been selected as aprotic solvent in D2. The reason for the chosen order was the different viscosity maxima of binary solvent systems consisting of water–acetonitrile and water–methanol. A mixture of water–methanol exhibits a much higher pressure maximum when a solvent gradient is applied than a mixture of water–acetonitrile [35] and is therefore more suitable to be used in D1. In D2, a very high linear velocity has to be achieved to minimize the cycle time. In this respect, temperature plays a pivotal role. This is not only to change the selectivity but also to reduce the pressure maximum and thus viscosity maximum of the mobile phase. The temperature was therefore set to 60 °C.

6.6.4 Column Dimension and Modulation

In order to circumvent the problem of the incompatibility between the flow rate of the second dimension and mass spectrometry (see Sections 6.5.3 and 6.5.4), an inner diameter of 0.1 mm has been selected in D1 for the PGC column. The injection volume onto the D1 column was 1.57 μl, which can be considered a large-volume injection if the column void volume is taken into account. Typically, a volume of 50–200 nl is injected onto nano-LC columns. The flow rate was adjusted to 200 nl min^{-1}. This means that a transfer volume of 200 nl results when the sampling time during the modulation is 1 min. In order to obtain a fast cycle time, a flow rate of 40 μl min^{-1} in D2 was chosen. This was close to the pressure limit of the pneumatic pumps. For conventional HPLC applications, such a flow rate appears to be very low, and it might not be evident that fast separation and equilibration are achieved. If the column dimensions of 300 μm ID and 5 cm length are considered, a very high linear velocity results, and the complete column volume is flushed in less than 2 s. In this respect, the gradient delay volume is also an important parameter to reduce the cycle time. This volume has been determined to be 0.9 μl. Therefore, a delay time of about 1.4 s is obtained until the programmed gradient reaches the head of the column. The use of longer columns with a larger inner diameter in D2 will inevitably lead to longer cycle times. Further information in regard to the gradient delay time can be found in Chapter 2.

6.6.5 Gradient Programming and Overall Analysis Time

The gradient programming has been kept as simple as possible for a generic approach. For the D1 separation, a linear gradient was programmed with an isocratic plateau at the end of the gradient. The overall analysis time per injection is about 110 min. In D2, the cycle time was 1 min. In principle, the overall analysis time, which is governed by the runtime of the first-dimension separation, may be easily reduced to 60 or even 30 min if the gradient slope is adjusted at constant flow rate in D1. As a consequence, a sufficient sampling of a peak eluting from the D1 column might be difficult to achieve, unless the cycle time in D2 is concomitantly reduced.

6.6.6 Coupling with Mass Spectrometry

In order to avoid a severe band spreading between the D2 column and the inlet of the mass spectrometer, an emitter tip with an inner diameter of 50 μm was used. The change of the emitter tip can be accomplished within a few minutes. Besides that, a modification of the conventional ESI source is not necessary. The replacement of the tip is mandatory to avoid a significant band spreading after the column. At this point, it should be noted that this setup does not resemble nanoelectrospray, where the position of the tip has to be adjusted with the help of cameras and micromanipulators in a complicated manner.

The hyphenation with a hybrid high-resolution mass spectrometer with an integrated MS/MS functionality is preferred, because not only the accurate mass but also product ion spectra can be obtained. This is a useful feature if the method is applied to a comprehensive screening of the sample that might contain hundreds or even thousands of compounds. Therefore, a Q-TOF instrument was used in order to obtain the sum formula and assign a structural formula to the sum formula. An example highlighting this thesis is given in Section 6.8. Further information regarding the different acquisition modes and also the limitations in terms of acquiring sufficient data points over a chromatographic peak are discussed in Chapter 5.

6.7 Real Applications

6.7.1 Measurement of a Reference Standard

In order to demonstrate the capacity of the miniaturized LC × LC system described in the previous sections, a reference standard containing 99 environmentally relevant compounds was measured. Figure 6.7 shows the total ion current (TIC) chromatogram of this reference standard in a three-dimensional plot.

On the basis of the TIC chromatogram, no further statements regarding the quality of the separation can be made. Because of the fact that special software for automatic data processing was not available, every mass had to be manually extracted and processed. For a general representation of all detected compounds, an analyte map was created, which is given in Figure 6.8.

Every point is related to the peak maximum of an analyte in the first and second dimension. It can be clearly seen that the combination of the PGC and RP-C18 column leads to an orthogonal separation in some regions of this area, because the peaks are spread across a significant part of the available two-dimensional separation space. This is especially evident when the peaks that are located on the dashed vertical line are inspected. These compounds would coelute on the PGC column but can be well separated on the D2 column.

In order to show the separation efficiency of the D2 column, a typical chromatogram that is obtained for the micro-LC dimension is given in Figure 6.9.

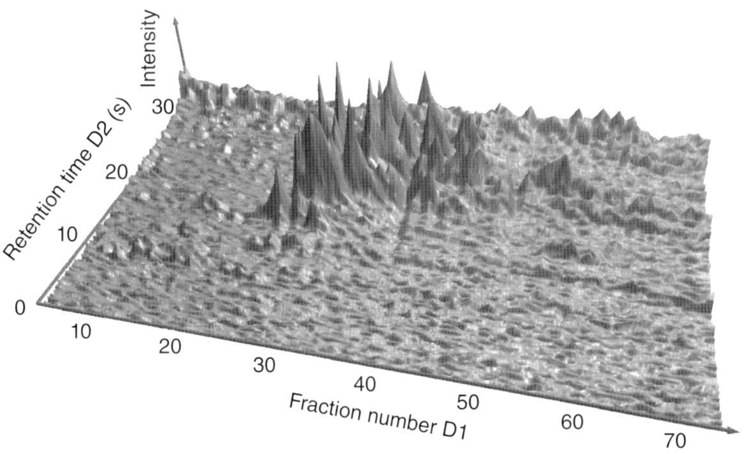

Figure 6.7 Total ion current chromatogram of the two-dimensional separation of the reference standard. For further details, see text.

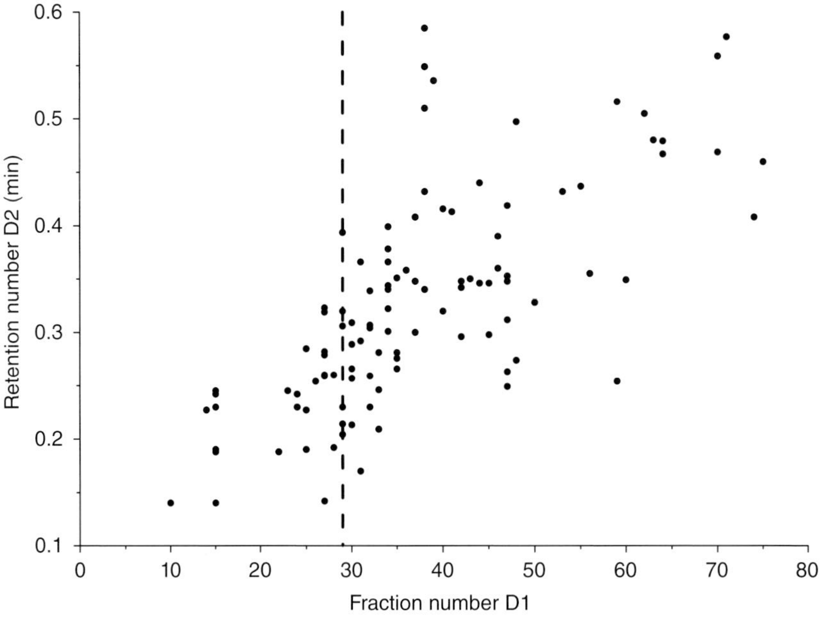

Figure 6.8 Analyte map on the basis of the extracted ion current chromatograms of all compounds detected in the reference standard. For further details, see text.

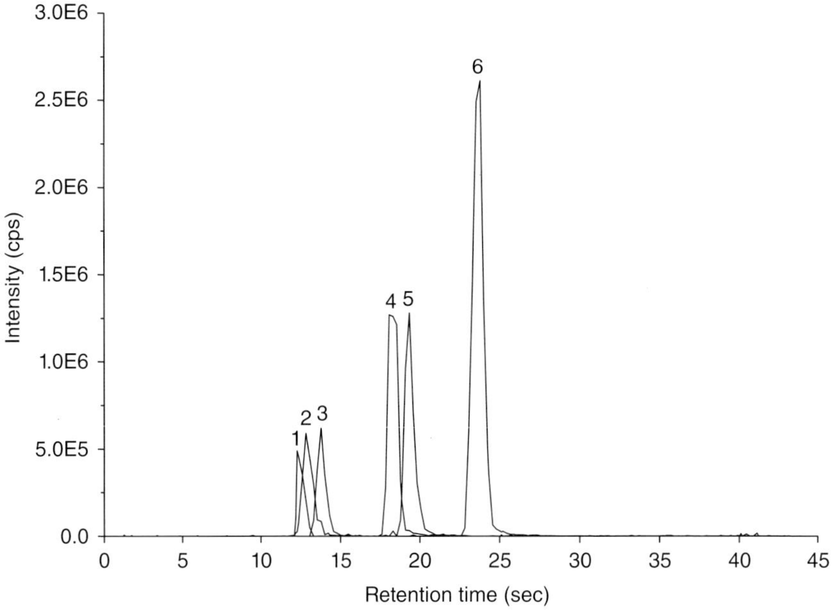

Figure 6.9 Single chromatogram of the micro-LC separation in the second dimension. Different extracted masses originating from the 29th fraction are shown. For further information, see text.

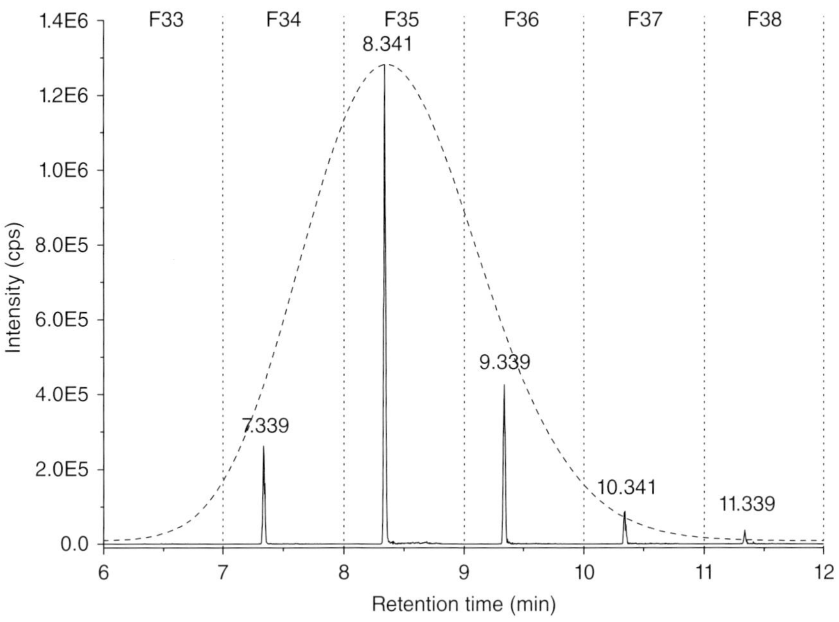

Figure 6.10 Typical extracted ion current chromatogram across five fractions. What can be seen is the nonmodulated signal as registered continuously by the detector. For further details, see text.

This chromatogram correlates to the fraction, which is marked in Figure 6.8 by a vertical line.

The peak widths are between 1 and 2 s. Figure 6.10 illustrates the retention time stability from run to run.

The selected compound was sampled in five fractions. Because of the cycle time of 1 min, the analyte signal should be recorded exactly with a difference of 1 min from fraction to fraction. For a better understanding, a solid line representing the envelope has been integrated. In addition, the fraction numbers are given at the vertical dashed lines. When the retention times from run to run are analyzed, it becomes clear that differences appear at the third decimal point. This also underlines the extremely high reproducibility of the gradient mixing even in the case of micro-LC. It contradicts an often cited myth that micro-LC systems only exhibit low robustness and are not suitable for routine analysis.

6.7.2 Measurement of a Real Sample

In order to show the applicability of the new approach, a wastewater sample after enrichment via solid-phase extraction was measured. It was anticipated that this sample would contain a high number of potentially interfering matrix compounds in a high concentration. In Figure 6.11, the TIC chromatogram of the wastewater sample is shown. In contrast to the corresponding chromatogram of the reference standard, the matrix clearly dominates in some parts of the chromatogram. While for the reference standard, distinct signals can be seen, some

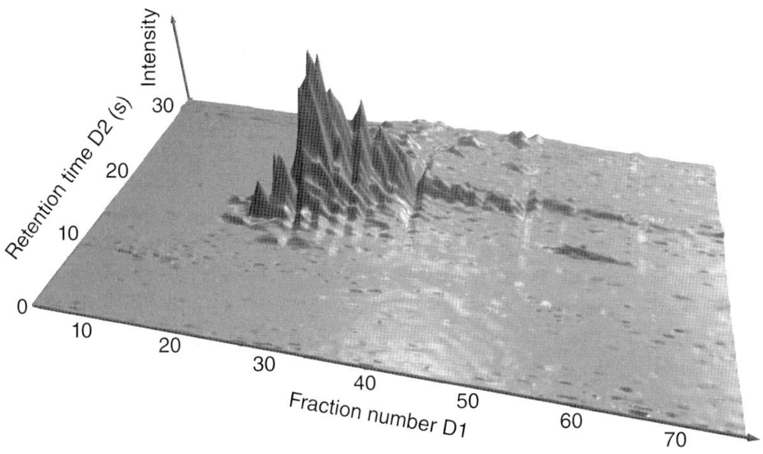

Figure 6.11 Total ion current chromatogram of the two-dimensional separation of the wastewater sample. For further information, see text.

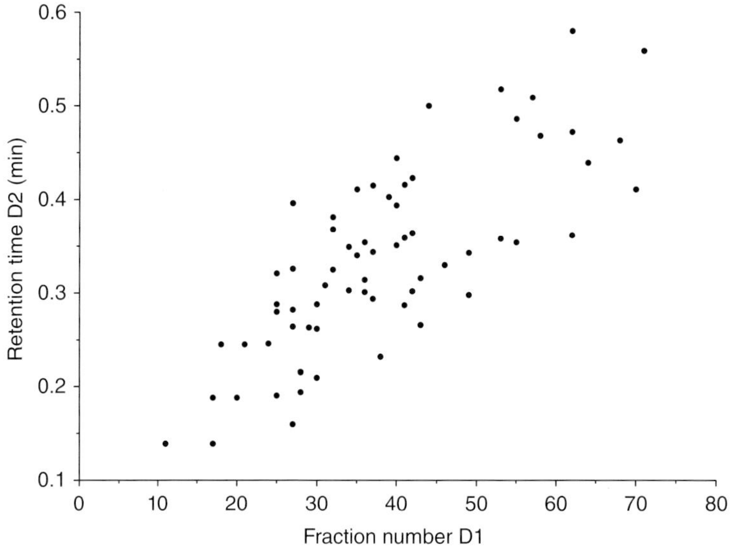

Figure 6.12 Analyte map on the basis of the extracted ions for all compounds that have been identified according to a suspected target screening. For further details, see text.

merged and broadly eluting peaks without any separation are obtained in the 2D LC chromatogram.

On the basis of the manual data processing, a suspected target screening was applied to search for known target compounds. Figure 6.12 shows the corresponding analyte map.

Approximately 60 compounds could be identified. The accurate mass was taken as the major criterion for analyte identification. The deviation should not be higher than 5 ppm. In the context of a suspected target screening approach, such a deviation is usually accepted. However, if a nontarget screening is applied where there is no information about the detected species, a much lower deviation from the accurate mass is required. However, even if the deviation is less than 1 ppm, an identification of a principally unknown compound on the basis of its accurate mass may lead to false-positive hits. A more in-depth discussion is out of the scope of this chapter, and the reader is therefore referred to the respective literature [12].

6.8 Advantages of the MS/MS Functionality

To conclude this section, an example is given that illustrates the advantages of generating product ion spectra with high-resolution hybrid mass spectrometers. Figure 6.13a represents the chromatogram that is obtained for the extracted mass *m*/*z* 261.0323. In contrast to the envelope depicted in Figure 6.10 exhibiting only one peak maximum, the envelope in Figure 6.13 clearly shows two maxima. On the first sight, this might hint at bad chromatographic conditions for the D1 separation. In order to support this hypothesis, the product ion spectra can be consulted. Figure 6.13b,c shows the precursor ions that are correlated to the highlighted peak maxima in Figure 6.13a.

The differences of the product ion spectra at the selected retention times are clearly visible. On the basis of the information acquired during the measurement of the reference standard, the mass spectra can be assigned to the respective compounds. In this case, cyclophosphamide and ifosfamide have been identified. These compounds are structural isomers that cannot be differentiated by their accurate mass. Because a partial separation is possible in D1, pure and characteristic product ion spectra are obtained at the peak maxima.

6.9 General Comments to Specific Aspects of an LC × LC System

6.9.1 Offline LC × LC versus Online LC × LC

A critical variable for optimizing an online LC × LC system is the determination of the modulation parameters. The inherent problem to adjust the inner diameters and flow rates of both dimensions has to be overcome, so that the resulting transfer volume from the first to the second dimension does not cause a significant loss of separation efficiency.

Another approach to circumvent this problem is using offline LC × LC. Here, all eluting fractions from the D1 column will be sampled by means of a fraction collector in different vials. Alternatively, they are directed to a great number of trapping columns. Each fraction can then be analyzed without any time limitation in regard to the cycle time of the second dimension. In this respect, a very high peak capacity is achieved because the full separation efficiency of both columns

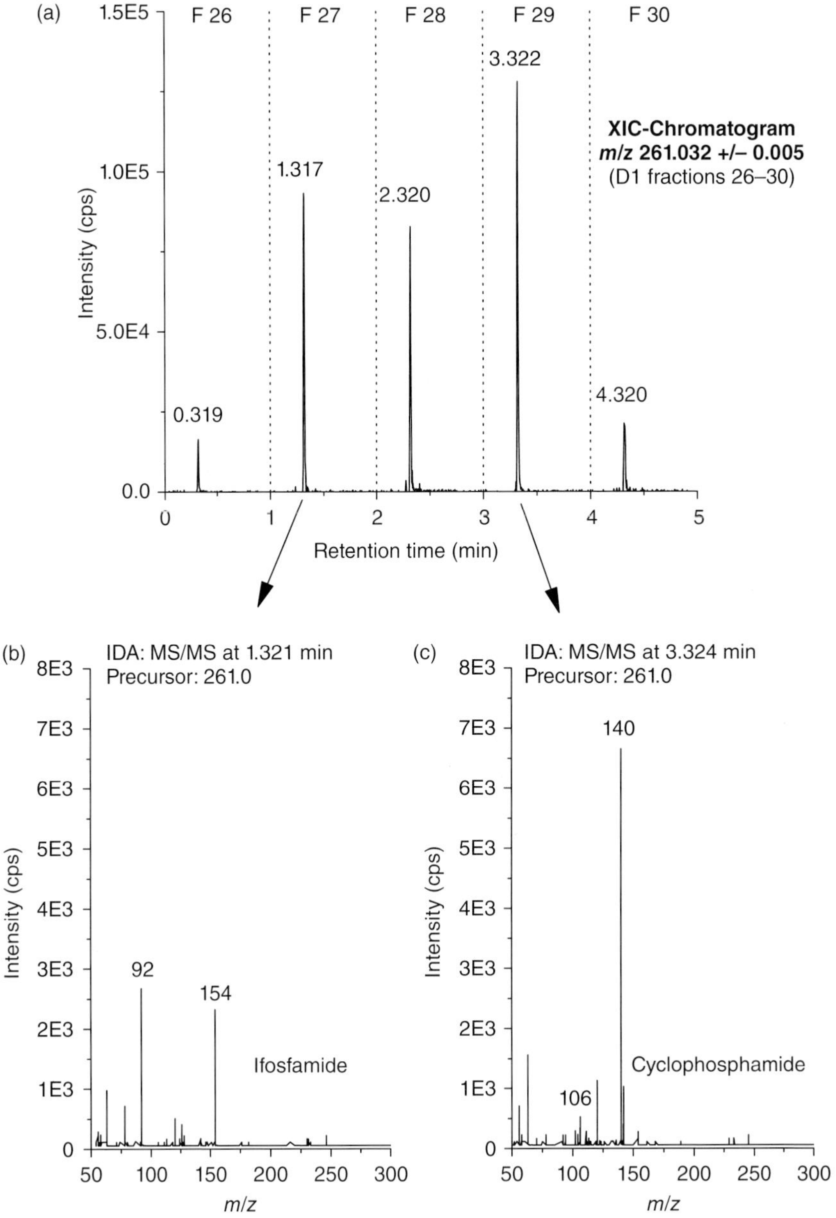

Figure 6.13 (a) Envelope of the extracted ion current chromatogram with *m/z* 261.0323 across six fractions. What can be seen is the nonmodulated signal as registered continuously by the detector. For further details, see text. (b,c) Product ion spectra for the precursor ion with *m/z* 261.0. For further details, see text.

can be used. Unfortunately, the overall analysis time will rapidly increase. Moreover, it has to be guaranteed that the capacity for storing all fractions is sufficient.

On the basis of a recent publication by Schiesel *et al.*, an example is given to understand the necessary steps for establishing an offline LC × LC system [36]. In this study, the application of offline LC × LC is described for the comprehensive impurity profiling of nutritional infusion solutions. The authors use a system on the basis of RP × HILIC hyphenated to an ion trap mass spectrometer and a charged aerosol detector. The procedure consists of three steps, which are temporally independent. At first, the sample is analyzed on a silica-based reversed stationary phase (Gemini C-18). The so-called hydrophobic fraction is analyzed on this column without any further fractionation. The polar fraction eluting between 0 and 8 min will be sampled for further processing. The solvent is evaporated to dryness, and the residue is dissolved using 100 μl water containing 0.1% trifluoroacetic acid. Then, the sample is again separated using two serially coupled RP columns (Gemini C-18 and Synergi Fusion RP), and the eluate is stored in 30 fractions. A solvent change has been employed for the following separation by HILIC. The solvent is evaporated to dryness and is dissolved in a mixture of 50/50 (v/v) water/acetonitrile. In this respect, the compatibility with the HILIC separation is established, because water is the strongest eluent in HILIC mode. Each fraction is then injected onto the HILIC column and is analyzed using ion trap mass spectrometry and charged aerosol detection.

The description of this workflow reveals that the sample is indeed analyzed "comprehensively." The question might be asked whether this complex treatment for a single sample is really justified. The authors of that study claim that the offline LC × LC approach is much easier to implement than using online LC × LC. Furthermore, the polar fraction obtained from the first and second separation step can be enriched. This is an important argument because the inevitable dilution that is caused by the modulation process in online LC × LC is counteracted. In addition, compatibility between RP and HILIC is achieved. A combination of these separation systems usually leads to a higher selectivity and has therefore also been suggested by other authors [37, 38]. It has to be noted, however, that a solvent change bears the risk of a loss of volatile compounds. In the study by Schiesel *et al.*, the analysis of the polar fraction is of utmost importance, because the sample was subjected to a stress test, and the authors were interested to gather information about degradation products. It can be expected that the volatility of these compounds is low and the solvent exchange will not have a negative influence on the result. Nevertheless, if the samples are stored over a long period, unwanted effects might occur as the adsorption of some compounds to glass surfaces.

Although offline LC × LC requires a much higher manual effort, the specific problems of online LC × LC can be avoided. For the characterization of "precious" samples, offline LC × LC is easier to implement from a technical point of view. This also refers to the fact that the separation power of both dimensions is not restricted by a short gradient run in the second dimension, which is usually the case for online LC × LC. However, it is rather doubtful if offline LC × LC will be used on a large scale for environmental screening analyses. Here, the stop-flow approach might be a compromise because it does not require storage of different sample fractions and is not dependent on the modulation time, which governs the cycle time in a true online LC × LC system.

6.9.2 Stop-Flow LC × LC

In stop-flow LC × LC, the flow of the first-dimension column is interrupted after each modulation step. This means that there is in principle no limitation of the runtime of the second-dimension separation. The longitudinal diffusion clearly leads to a broadening of the peaks eluting from the D1 column. However, this effect can often be neglected and is not practically relevant. Using the stop-flow approach, the peak capacities that are obtained correspond to those of an offline LC × LC system with the difference being that a much easier system setup is required. It has to be kept in mind that the overall analysis time is very long.

6.9.3 Multiple Heart-Cut LC-LC and Selected LC × LC (sLC × LC)

A third variant of multidimensional chromatography is the multiple heart-cut or selected LC × LC (sLC × LC) approach. Here, a limited number of fractions are analyzed on a second-dimension column. The sLC × LC approach is useful if samples of lower complexity have to be analyzed and a separation of coeluting species is necessary. Most studies have been carried out by Stoll *et al.* to demonstrate the capabilities of storing a limited number of fractions in a set of sampling loops. The analysis of the stored samples can be done during the gradient of the first dimension. With this concept, it is possible to circumvent the offline sampling of fractions as described in Section 6.9.1 for the offline LC × LC approach. Also, the flow of the first dimension does not need to be stopped, so that the efficiency of the first-dimension separation will not be diminished by the longitudinal diffusion. However, the sLC × LC approach might not be suitable for a nontarget screening of complex samples. The reason is that when using a mass spectrometer, it is hardly possible to identify regions in the chromatogram where no substance elutes. Therefore, the sLC × LC approach can be considered an advanced heart-cut approach that allows for the analysis of more than one fraction during one chromatographic run. The interested reader is referred to the respective literature to gain a deeper understanding of the application areas of this concept [39–41].

Interim Conclusion Offline LC × LC as well as stop-flow LC × LC might be superior in terms of the complexity of the system when compared to online LC × LC or sLC × LC. Especially, for offline LC × LC, a solvent change can be done to increase the compatibility between orthogonal separation systems, for example, NP × RP or HILIC × RP.

However, the benefit of online LC × LC is a higher sample throughput without the need to store a large number of fractions eluting from the first-dimension column.

6.10 Method Development and Gradient Programming

Generally, the same rules apply as for one-dimensional method development. If an unknown sample has to be analyzed or the information about the sample is limited, a generic gradient should be used. This means that for both dimensions, a linear gradient, which covers the full range from, for example, 5–95% B, should be applied. Many studies have described more advanced gradient programming for LC × LC methods including the so-called shift gradients [42]. A shift gradient usually refers to a change in the starting conditions of the second-dimension separation during the linear gradient of the first-dimension separation. If we assume that the gradient in D1 runs from, for example, 5–95% B over 1 h and the cycle time for the D2 separation is 1 min, a shift gradient may change the starting conditions for the B-solvent in steps of 1% for every new cycle. In contrast to a full gradient in the second dimension, where the B-solvent always starts from the same conditions, for example, 5% B, the shift gradient leads to a gradually higher elution strength in the second dimension as the gradient in D1 progresses in time. Many authors regard this technique to be superior to the conventional full-gradient programming. The reason is that it is nearly impossible to reduce the interaction between all compounds in a given sample to one specific retention mechanism on one particular stationary phase to establish an orthogonal separation system. This is indeed a fact that cannot be ignored. With the exception of the analysis of polymers that can be separated by size in one dimension and hydrophobicity in the other dimension, true orthogonal systems for the majority of all different sample classes will not be obtained. Therefore, often both dimensions of an LC × LC system are more or less "correlated." This is especially true for all kinds of RP × RP systems. The potential benefit of shift gradients is that the gradient window for consecutive D2 runs can be adapted so that the resolution of compounds that are eluting during these gradients is higher when compared to full gradients. Unfortunately, this concept does not consider that the approach is no longer generic. Many users would prefer easy-to-use generic methods without changing the gradient parameters. The technical feasibility to apply shift gradients is an advantage if fine-tuning of a separation needs to be done to optimize a two-dimensional separation. For screening methods, the application of full gradients seems to be a better way because it is difficult or impossible to anticipate all theoretical combinations of a composition of a complex sample.

Interim Conclusion For a generic approach without any need for method development, full gradients should be applied to both dimensions. If fine-tuning of an established method is desired, shift gradients may be an interesting option to improve the resolution of critical peak pairs in the second dimension.

6.11 Presentations of the Instrument Manufacturers (in Alphabetical Order)

Although the potential of two-dimensional separations is great, these instruments can be considered expert systems that require a high qualification of the operator. From a personal point of view, the hyphenation of a two-dimensional chromatographic system is most effective when a high-resolution mass spectrometer is used. There is an ongoing trend also in mass spectrometry that the workflows become ever more complex requiring trained personnel. The instrument manufacturers should therefore offer solutions to their customers that are ready-for-use and robust. The following section does not raise a claim of completeness. In a foreseeable future, this list might grow. Moreover, the information that is given is rather general. The interested reader should consult the manufacturer to obtain more specific details.

6.11.1 Commercially Available Solutions for LC × LC

6.11.1.1 Agilent

Agilent offers a two-dimensional HPLC system for comprehensive analysis that is based on the Infinity 1290 series. The system control can be achieved via ChemStation software from Agilent. Data processing can be done by external software delivered from GC Image (GC Image, LLC, Lincoln, NE, USA). The system allows for complex gradient programming, for example, shift gradients (see Section 6.10). Further information can be taken from scientific publications as well as application notes by Agilent [43, 44].

6.11.1.2 Shimadzu

Shimadzu offers a two-dimensional HPLC system for comprehensive analysis that is based on the Nexera series. The system control can be achieved via LabSolutions software from Shimadzu. Data processing can be done by ChromSquare software, which is also directly available from Shimadzu.

6.11.1.3 Thermo/Dionex

Systems capable of comprehensive LC × LC for nano-HPLC were introduced by Dionex with the UltiMate II in 2000. The system control and data processing can be done via the Chromeleon software. ThermoFisher today offers system solutions on the basis of the Dionex UltiMate 3000. The system is able to deliver a broad flow rate range from nano-HPLC to conventional UHPLC. It can be used for online LC × LC as well as offline 2D LC [45].

6.11.2 Further Systems

6.11.2.1 Sciex

Sciex now offers an LC × LC system on the basis of the Eksigent NanoLC 425 platform. This is the only miniaturized system, which is specially designed for nano-HPLC in the first dimension and micro-LC in the second dimension. The system control can be achieved by the Eksigent control software. Data processing is feasible using the GC Image software (GC Image, LLC, Lincoln, NE, USA).

6.11.2.2 Waters

Waters does not offer a system for comprehensive LC × LC, although this would be possible on the basis of the instrumentation. The workflow Waters has focused on is heart-cut LC-LC for purity analysis of an active pharmaceutical ingredient.

6.12 2D LC – Quo Vadis?

In this section, the most important arguments for establishing two-dimensional liquid chromatography in research and routine laboratories are discussed. As has been laid out in the previous sections, HPLC systems for comprehensive analysis are now commercially available. However, the number of users of this technology is still rather small. Besides a more complex method development, the availability of suitable processing software is limited.

6.12.1 Software

A prerequisite for an efficient use of the LC × LC technology is the availability of suitable system control and data processing software. While it can be anticipated that all manufacturers of LC × LC systems possess control software, which comprises all modules of an LC × LC system, there are some gaps regarding the processing software. Many applications, which are described in scientific publications, are based on UV detection, although the analysis of complex samples using LC × LC requires sensitive mass spectrometric detection. Nevertheless, the integration of additional detectors such as evaporative light scattering or fluorescence detection can have some beneficial effects. In most cases, the identification of known and unknown compounds is the main goal of screening methods. Insofar, processing software for LC × LC should contain all features, which are also offered for one-dimensional LC-MS analysis. Regarding a suspected or nontarget screening, a matching of the accurate mass, the isotopic pattern, product ion spectra, or data bases, which are either created in-house or provided by the instrument manufacturer, should be made. Because of the fact that those workflows might still be improved even for one-dimensional LC-MS, there is much room for advancement of two-dimensional processing software. There is the hope that instrument manufacturers will provide innovative software solutions in the near future. With regard to this, companies such as GC Image, a spinoff of the University of Nebraska, as well as Chromaleont, a spinoff of the University of Messina, should be mentioned. They were intensively engaged for many years with the development of processing software for GC × GC, but also for LC × LC. Here, instrument manufacturers are responsible for delivering adequate and mature software solutions. Self-written excel macros for data processing are not helpful for establishing this technology on a broad basis or within the pharmaceutical industry.

6.12.2 System Setup

A central point, which has to be solved, is the preadjustment of an LC × LC system. The possibilities regarding the design of such a system are nearly unlimited. This holds true for the inner diameters and selectivities of the columns but also the resulting flow rates. This all makes it necessary that a profound knowledge of the technical problems is available. Here, research groups from academia as well as instrument manufacturers have the obligation to provide solutions that fit the need for specific problems from industry.

6.12.3 Peak versus Peak Capacity

It is often criticized that although two-dimensional separation systems offer a higher separation power and peak capacity, the real benefit of those methods when compared to one-dimensional separations is not obvious. Two-dimensional liquid chromatography today exhibits its second or third "hype" if oral or poster sessions of scientific meetings and symposia are evaluated. The progress that is achieved in this area is mainly pushed by a limited number of renowned scientific groups with a longstanding expertise. The arguments that are used to demonstrate the superiority of two-dimensional methods over one-dimensional methods and to persuade potential users of this technology are often not very convincing. This might be related to some aspects, which are not taken serious in the ongoing discussion. On the one hand, this refers to the detection techniques, and on the other hand, this is related to the predominantly theoretic point of view of a higher peak capacity of two-dimensional separation systems. The usefulness and importance of mass spectrometric detection have already been discussed in Section 6.8. It is true that a higher peak capacity is often obtained in two-dimensional liquid chromatography resulting in a higher number of compounds that can be chromatographically resolved. The user, however, is not primarily interested whether it is possible to obtain a higher peak capacity. For practical purposes, it seems to be more important whether a two-dimensional system will also lead to a significantly higher number of detected or identified compounds when compared to a one-dimensional separation. Therefore, the sample itself and not the theoretical peak capacity should be the basis for this evaluation. A direct comparison between a one-dimensional and a two-dimensional separation would be useful for an objective judgment of the performance of two-dimensional separation approaches. There are only a very limited number of dedicated studies that prove this hypothesis [46]. Leonhardt *et al.* recently made a comparison between a one-dimensional and a two-dimensional chromatographic approach coupled to hybrid high-resolution mass spectrometry [47]. They used the platform described in Section 6.6. The comparison was based on a screening approach in environmental analysis, where three criteria for compound identification have been employed. First of all, the accurate mass with a deviation of less than 5 ppm was used for identification. If the first criterion was fulfilled, the retention time should not deviate more than 2.5% from the retention time measured in the reference standard. If this criterion was also fulfilled, MS/MS spectra of the reference standard were compared with that of the sample. Table 6.1 summarizes the results of this comparison.

All chosen compounds could be detected with both approaches on the basis of the accurate mass in the reference standard. When a real sample was analyzed, 48 compounds were found with the 1D LC approach, while 65 compounds could be detected with the miniaturized 2D LC approach. Using the second criterion as an additional filter for the elimination of false-positive hits, four compounds had to be removed for the 1D LC approach, while only one compound had to be eliminated for the 2D LC approach. It should be emphasized that retention time is a powerful parameter to ensure the identification of a compound. Of course, a further criterion for an unambiguous identification should be considered. MS/MS

Table 6.1 Overview of the identified analytes by 1D-HPLC-MS and 2DnLC × μLC-MS. Detailed list of detected targets is given in [47].

	1D LC	2D LC
Detected targets in reference standard by < 5 ppm.	99	99
Detected targets in wastewater sample by < 5 ppm.	48	65
Detected targets in wastewater sample by < 5 ppm and retention time < 2.5%	44	64
Detected targets in wastewater sample by < 5 ppm, retention time < 2.5% and MS/MS hit.	16	31

information can help to distinguish even isobaric compounds that have the same *m/z* as was shown in Section 6.8. The limitation of acquiring additional MS/MS information also depends on the peak width. This has been outlined in Section 5.5. As can be seen from Table 6.1, MS/MS information is only obtained for a small number of analytes that have been identified on the basis of the first and second criteria. The reason is that the algorithm for acquiring product ion spectra looks for the most intense peak in the full-scan MS spectrum, which is then fragmented. Peaks with a lower intensity are not selected. Even if 10 MS/MS experiments are carried out during one cycle, the compounds of interest might be missed because they possess an intensity too low compared to highly abundant ions of the matrix. This could be circumvented if the algorithm is programmed such that only ions of known substances are selected for fragmentation experiments. Nevertheless, the number of compounds that can be confirmed by applying all three criteria is significantly higher for the 2D LC approach. This demonstrates that the miniaturized system is far superior for screening purposes than a one-dimensional system, although the absolute injection volume is reduced by a factor of 13, and a further dilution cannot be avoided by the modulation.

References

1 Schmidt, T.C., Schmitz, O.J., and Teutenberg, T. (2015) *Anal. Bioanal. Chem.*, **407**, 117.
2 Jandera, P., Hajek, T., and Stankova, M. (2015) *Anal. Bioanal. Chem.*, **407**, 139.
3 Li, D.X., Jakob, C., and Schmitz, O. (2015) *Anal. Bioanal. Chem.*, **407**, 153.
4 Kilz, P. and Radke, W. (2015) *Anal. Bioanal. Chem.*, **407**, 193.
5 Pretorius, N.O., Rhode, K., Simpson, J.M., and Pasch, H. (2015) *Anal. Bioanal. Chem.*, **407**, 217.
6 Li, D.X. and Schmitz, O.J. (2015) *Anal. Bioanal. Chem.*, **407**, 231.
7 Stoll, D.R., Talus, E.S., Harmes, D.C., and Zhang, K. (2015) *Anal. Bioanal. Chem.*, **407**, 265.
8 Elsner, V., Wulf, V., Wirtz, M., and Schmitz, O.J. (2015) *Anal. Bioanal. Chem.*, **407**, 279.

9 Wang, S.Y., Qiao, L.Z., Shi, X.Z., Hu, C.X., Kong, H.W., and Xu, G.W. (2015) *Anal. Bioanal. Chem.*, **407**, 331.
10 Vanhoenacker, G., Vandenheede, I., David, F., Sandra, P., and Sandra, K. (2015) *Anal. Bioanal. Chem.*, **407**, 355.
11 Di Palma, S., Hennrich, M.L., Heck, A.J.R., and Mohammed, S. (2012) *J. Proteomics*, **75**, 3791.
12 Krauss, M., Singer, H., and Hollender, J. (2010) *Anal. Bioanal. Chem.*, **397**, 943.
13 Stahnke, H., Reemtsma, T., and Alder, L. (2009) *Anal. Chem.*, **81**, 2185.
14 vom Eyser, C., Palmu, K., Otterpohl, R., Schmidt, T.C., and Tuerk, J. (2015) *Anal. Bioanal. Chem.*, **407**, 821.
15 Neue, U.D. (2008) *J. Chromatogr. A*, **1184**, 107.
16 Li, X.P., Stoll, D.R., and Carr, P.W. (2009) *Anal. Chem.*, **81**, 845.
17 Neue, U.D., Marchand, D.H., and Snyder, L.R. (2006) *J. Chromatogr. A*, **1111**, 32.
18 Lee, C., Zang, J., Cuff, J., McGachy, N., Natishan, T., Welch, C., Helmy, R., and Bernardoni, F. (2013) *Chromatographia*, **76**, 5.
19 Kormos, J.L., Schulz, M., Kohler, H.P.E., and Ternes, T.A. (2010) *Environ. Sci. Technol.*, **44**, 4998.
20 Ferrer, I., Garcia-Reyes, J.F., Mezcua, M., Thurman, E.M., and Fernandez-Alba, A.R. (2005) *J. Chromatogr. A*, **1082**, 81.
21 Glauser, G., Guillarme, D., Grata, E., Boccard, J., Thiocone, A., Carrupt, P.A., Veuthey, J.L., Rudaz, S., and Wolfender, J.L. (2008) *J. Chromatogr. A*, **1180**, 90.
22 Murphy, R.E., Schure, M.R., and Foley, J.P. (1998) *Anal. Chem.*, **70**, 1585.
23 Gilar, M., Olivova, P., Daly, A.E., and Gebler, J.C. (2005) *Anal. Chem.*, **77**, 6426.
24 Potts, L.W., Stoll, D.R., Li, X.P., and Carr, P.W. (2010) *J. Chromatogr. A*, **1217**, 5700.
25 Davis, J. (2008) in *Multidimensional Liquid Chromatography: Theory and Applications in Industrial Chemistry and the Life Sciences* (eds S.A. Cohen and M.R. Schure), John Wiley & Sons, Inc., 49.
26 Steiner, F., Grübner, M., Dunkel, A., and Hofmann, T. (2014) *Thermo Scientific*, Application Note, PN71130_HPLC_2014_E_05/14S.
27 Jandera, P. (2006) *J. Sep. Sci.*, **29**, 1763.
28 Haun, J., Teutenberg, T., and Schmidt, T.C. (2012) *J. Sep. Sci.*, **35**, 1723.
29 Stoll, D.R., Li, X.P., Wang, X.O., Carr, P.W., Porter, S.E.G., and Rutan, S.C. (2007) *J. Chromatogr. A*, **1168**, 3.
30 Haun, J., Leonhardt, J., Portner, C., Hetzel, T., Tuerk, J., Teutenberg, T., and Schmidt, T.C. (2013) *Anal. Chem.*, **85**, 10083.
31 West, C., Elfakir, C., and Lafosse, M. (2010) *J. Chromatogr. A*, **1217**, 3201.
32 Leonhardt, J., Hetzel, T., Teutenberg, T., and Schmidt, T.C. (2014) *Chromatographia*, **78**, 31.
33 Griffiths, J.R., Perkins, S., Connolly, Y., Zhang, L., Holland, M., Barattini, V., Pereira, L., Edge, A., Ritchie, H., and Smith, D.L. (2012) *J. Chromatogr. A*, **1232**, 276.

34 Melmer, M., Stangler, T., Premstaller, A., and Lindner, W. (2011) *J. Chromatogr. A*, **1218**, 118.
35 Teutenberg, T., Wiese, S., Wagner, P., and Gmehling, J. (2009) *J. Chromatogr. A*, **1216**, 8470.
36 Schiesel, S., Lammerhofer, M., and Lindner, W. (2012) *J. Chromatogr. A*, **1259**, 100.
37 Gilar, M., Olivova, P., Daly, A.E., and Gebler, J.C. (2005) *J. Sep. Sci.*, **28**, 1694.
38 Montero, L., Herrero, M., Prodanov, M., Ibanez, E., and Cifuentes, A. (2013) *Anal. Bioanal. Chem.*, **405**, 4627.
39 Larson, E.D., Groskreutz, S.R., Harmes, D.C., Gibbs-Hall, I.C., Trudo, S.P., Allen, R.C., Rutan, S.C., and Stoll, D.R. (2013) *Anal. Bioanal. Chem.*, **405**, 4639.
40 Groskreutz, S.R., Swenson, M.M., Secor, L.B., and Stoll, D.R. (2012) *J. Chromatogr. A*, **1228**, 31.
41 Groskreutz, S.R., Swenson, M.M., Secor, L.B., and Stoll, D.R. (2012) *J. Chromatogr. A*, **1228**, 41.
42 Li, D.X. and Schmitz, O.J. (2013) *Anal. Bioanal. Chem.*, **405**, 6511.
43 Naegele, E., Agilent Technologies (2012) Agilent Technologies, Inc., Application Note, 5991-0426EN.
44 Agilent Technologies (2012) Agilent Technologies, Inc., Application Note, 5991-0138EN.
45 Eeltink, S., Dolman, S., Detobel, F., Desmet, G., Swart, R., and Ursem, M. (2009) *J. Sep. Sci.*, **32**, 2504.
46 Wagner, Y., Sickmann, A., Meyer, H.E., and Daum, G. (2003) *J. Am. Soc. Mass Spectrom.*, **14**, 1003.
47 Leonhardt, J., Teutenberg, T., Tuerk, J., Schlusener, M.P., Ternes, T.A., and Schmidt, T.C. (2015) *Anal. Methods*, **7**, 7697.

7

Materials in HPLC and UHPLC – What to Use for Which Purpose

Tobias Fehrenbach and Steffen Wiese

7.1 Introduction

Conventional high-performance liquid chromatography (HPLC) and modern ultrahigh-performance liquid chromatography (UHPLC) systems are versatile and powerful analytical instruments available in variable configurations with a large variety of different components such as in-line solvent filters, pump heads, flow cells, injection valves, sample needles, tubing, and seals. Different materials, for example, alloys, polymers, and composite materials, are used for each component. Each material has its specific characteristics and has to meet different requirements depending on the location of operation within the UHPLC system. Materials within the flow path are in constant contact with the mobile phase, need to withstand high pressure and frictional force, and need to seal the flow path against the atmosphere. A selection of wetted materials that are used in UHPLC systems is given in Table 7.1.

Since method parameters of many chromatographic applications, for example, the column temperature, flow rate, pressure, the mobile phase composition, and the chemical and physical properties of the analyte, can be quite diverse, the materials in a UHPLC system cannot accommodate all applicative requirements and needs in regard to mechanical and chemical resistance.

The general requirements for materials in UHPLC systems is discussed in Section 7.2, while an insight into materials in pumps, autosamplers, and tubing is provided in Sections 7.4 and 7.5. Since the properties of materials used in precolumn heating and postcolumn cooling devices, column switching valves as well as detector flow cells are identical or are comparable to materials in pumps, autosamplers, and tubing, the materials of these components are not further discussed elsewhere.

Besides the general chemical and mechanical resistance of materials in UHPLC systems, the interaction of materials with the analytes and the influence of the material on the detection method are of great importance. In case the UHPLC system is noninert, the analytical result will most likely be affected in some way. In order to meet the requirements of inertness, UHPLC manufacturers offer two

The HPLC Expert II: Find and Optimize the Benefits of your HPLC/UHPLC, First Edition.
Edited by Stavros Kromidas.

Table 7.1 Wetted materials (selection) typically used in UHPLC systems.

UHPLC module	Wetted materials
Degasser	PTFE, FEP, PEEK™, ECTFE, amorphous fluoropolymer, stainless steel, titanium
Pump	Stainless steel, titanium, zirconium oxide, sapphire, aluminum oxide, PEEK, PTFE, ECTFE, FEP, tantalum, gold, UHMW-PE composite, ruby, Nitronic® 60, PPS
Autosampler	Stainless steel, gold-coated stainless steel, Vespel® SCP, PEEK composite, DLC, PCTFE, ECTFE, PTFE
Column Oven	Stainless steel, titanium, PEEK, Vespel®, Tefzel®
Detector	Fused silica, quartz, stainless steel, PEEK
Tubing	Stainless steel, titanium, MP35N®, PEEK, Fused Silica

PPS, poly(*p*-phenylene sulfide) and PCTFE polychlorotrifluoroethene.

types of UHPLC systems. While UHPLC systems with stainless-steel components can be used for a wide range of applications, "biocompatible" or "bio-inert" UHPLC systems are specifically designed and optimized for biomolecule analyses. In the field of biochromatography, the interactions of biomolecules with surfaces of different alloys, for example, stainless steel, or with dissolved iron- or metal ions released from the wetted surfaces of the UHPLC system are of central concern. These interactions can be excluded to a great extent by using a biocompatible UHPLC system. In addition, biocompatible systems are highly resistant to (chloride-containing) buffers, which are characterized by extreme pH values and salt concentrations. Such buffers are usually applied in biochromatography. Nevertheless, "inertness" is quite complex, and for each application, the user needs to be aware of possible interactions between material, mobile phase, and analyte.

Important Note The information given in the following chapters does not claim to be exhaustive and complete. Information related to the corrosion resistance of alloys has partly been extracted from lists of general chemical compatibility of the respective material. These lists are partly accessible in the Internet. Presented data – especially data related to the nature of the passive layer on alloy surfaces and biocompatibility – are in most cases taken from scientific studies with experimental setups, which do not reflect real operating conditions in UHPLC systems. Therefore, the results described in the following chapters may only be conditionally applied to chromatographic applications and UHPLC systems. Additional physical and chemical effects, which are not described in the following chapters, are possibly present in UHPLC systems. Such effects can influence the mechanical and chemical resistance and analyte compatibility in a different way than described here. Specifications, instructions, and guidelines given by the instrument manufacturer, which are available in operating instructions or which can directly be requested from the manufacturer, must be considered and obeyed. The given information and literature cited in the following chapters are thought to be a reference for the interested reader and provide the opportunity for a profound reflection on materials and their influence in chromatography.

7.2 Requirements for Materials in UHPLC

Because of higher system back pressures, column temperatures, and linear velocities of the mobile phase in UHPLC, the materials that are used in UHPLC systems have to meet higher requirements compared to the materials used in conventional HPLC. The users' expectations are similar lifetimes and highly reliable analytical results – no matter if the analysis is running on a UHPLC or an HPLC system. Essentially, the materials must meet three key requirements, which are the mechanical and chemical robustness of the system and its compatibility with the analyte. These three aspects are discussed in the following sections.

7.2.1 Mechanical Stability

The mechanical stability is defined as the resistance of a material to the effects of physical force, which are, for example, high pressures, high temperatures, and frictional forces in liquid chromatography (LC). In respect thereof and depending on the location of the material in a UHPLC system, the material has to meet various mechanical demands for, for example, its hardness, structural robustness, ductility (ability to deform plastically under stress), and its wear resistance to different degrees. The wear resistance of materials is measured in the field of tribology, which is the science of interacting surfaces in relative motion including the principles of friction, wear, and lubrication. The mobile phase and the sample can also have an effect on the wetted materials in a UHPLC system at a given application-specific pressure and temperature potentially leading to a change of the mechanical properties over time. Therefore, information about, both the chemical resistance of the materials and application parameters should be included in considerations of mechanical stability. Mechanical stability of materials in a UHPLC system is of high value for the user, because it provides high system utilization, extended overall operational system lifetime, and reduced costs of ownership.

7.2.2 Chemical Stability

The chemical stability of a material is defined as its resistance to the effects of chemicals. Ideally, when in contact with a certain medium, the material remains unchanged in its mechanical, physical, and chemical properties over an extended period of time. Since in liquid chromatography, the interior surfaces of the low-pressure and high-pressure fluidics are wetted by the mobile phase, the user needs to consider the resistance of UHPLC system components to the impact of the mobile phase and the sample. Components of the high-pressure flow path of a UHPLC system are generally made of alloys such as stainless steel and titanium. These alloys form a stable corrosion-resistant oxide layer (passive layer). The properties of passive layers of different alloys are discussed in Section 7.6.2 in more detail. Depending on the mobile-phase composition, the passive layer and the bulk metal phase underneath can be prone to several kinds of corrosion types, which are, for example, general corrosion, pitting, and crevice corrosion as well as erosion corrosion. In general corrosion, the metal uniformly corrodes over the entire surface almost at the same rate (surface corrosion). In many cases, general corrosion is not relevant for metals with a passive layer, but as soon as defective sites are present in the passive layer, these alloys get significantly damaged by highly localized corrosion types. These types of corrosion are pitting corrosion, crevice corrosion, and erosion corrosion, which can be of concern in liquid chromatography [1–4].

In pitting corrosion, localized pits of different sizes and depths are formed in the passive layer. Not only the passive layer is damaged, but also corrosion penetrates the bulk metal very rapidly. Minor discontinuities in the protective oxide layer, or inclusions and defects in the material, or dirt and contamination on the surface are some of the factors that cause pitting corrosion. In addition, certain chemical agents such as chlorides and other halides initiate and significantly accelerate pitting, especially of stainless steel even in diluted concentrations.

Crevice corrosion is quite similar to pitting corrosion and is initiated in crevices, which are mostly formed due to metal–metal protrusion, joints with gaskets, or due to deposition. Erosion corrosion is a combination of corrosion and erosion of the material. The corrosion process is accelerated due to an abrasive stress on the passive layer. The passive layer is damaged and removed by the flow of a turbulent fluid on the surface. Since tubings with smaller inner diameter are more frequently used and the column temperatures are higher in UHPLC applications, it can be expected that erosion corrosion is more significant in UHPLC systems than in HPLC systems [1].

7.2.3 Analyte Compatibility/Biocompatibility

Another requirement for materials in liquid chromatography is their compatibility with the analyte (analyte compatibility). As soon as physical or chemical interactions of the instrument and its materials with the analyte are present, the analyses will be influenced to some extent, and the analytical result will be biased. Therefore, any form of system/analyte interaction needs to be avoided and/or appropriate countermeasures, for example, the addition of an internal standard, need to be considered.

The nature of the protective oxide layer of alloy-based materials and its resistance to corrosion are critical factors influencing the analyte compatibility of the material. The analyte can interact either directly with the passive layer or with metal ions, which are dissolved from the passive layer or the bulk metal phase. In this context, corrosion cannot only significantly damage the instrument and its components over time, but metal ions can subtly leach out of the material into the mobile phase due to slow corrosion processes (metal leaching).

A special case of analyte compatibility of materials is their compatibility with biomolecules (biocompatibility). Biocompatibility is of special concern in many disciplines, for example, in biomedical engineering in which biomaterials are used in the form of implants in the human body. The biomaterial needs to be physiologically and toxicologically safe and, therefore, needs to have a high biological tolerance or inertness, which correlates with chemical inertness against physiological media in the human body. The chemical resistance of the material has also an impact directly on the physical and mechanical integrity of the implant and therefore influences the function and the lifetime of the material. The material type is chosen according to its location and its residence time in the human body in a specific biomedical application. The important topic of inertness in the context of analyte and biocompatibility in UHPLC is discussed in Section 7.6.

7.3 Flow Paths in UHPLC Systems

7.3.1 Low-Pressure and High-Pressure Flow Path

For the discussion of the requirements for materials in UHPLC, it is helpful to classify the fluidics of a UHPLC system according to different aspects. Usually, there is a distinction being made according to the predominant pressure in the flow path. Fluidic system components that are exposed to a much lower pressure than system pressure are considered parts of the low-pressure flow path, while fluidic components that are exposed to higher pressures such as the system pressure are parts of the high-pressure flow path (Figure 7.1). The low-pressure flow path includes all fluidic components downstream between the eluent bottles with in-line solvent filters and the pump heads and also the waste tubing behind the optical detector. Within the low-pressure flow path, which is located before the pump, the mobile phase is delivered by means of hydrostatic pressure generated by the liquid in the solvent lines assisted by a vacuum generated by the pump and the vacuum degasser, which also removes dissolved gases from the mobile phase. In the pump heads, the mobile phase is compressed to the pressure needed to deliver a constant flow of mobile phase to pass the stationary phase within the column. In respect thereof, all fluidic components located downstream between the pump heads and the column are parts of the high-pressure flow path. Furthermore, all tubings between the column and the interface of the detector, for

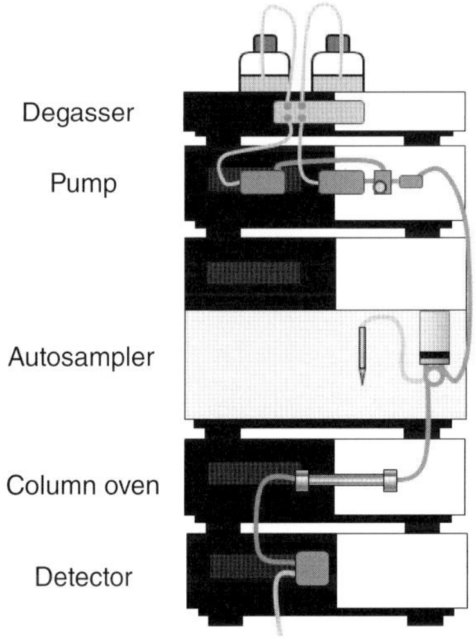

Figure 7.1 Low-pressure and high-pressure flow path in a UHPLC system with fixed-loop (pulled-loop) autosampler. The low-pressure flow path is shown in light gray; the high pressure flow path is shown in dark gray. Reproduced with permission of Thermo Fisher Scientific.

example, mass spectrometer, and – in case of an optical detector, for example, UV detector – the detection flow cell also belong to the high-pressure flow path.

Since in the low-pressure flow path, a lower pressure prevails, other materials are used than in the high-pressure flow path, even though both flow paths are in contact with the mobile phase. In addition to the low-pressure flow path, the fluidic components of two other independent flow paths, which are the wash liquid system of the autosampler for needle wash procedures and the (optional) rear seal wash system of the pump (both are not shown in Figure 7.1), are also prone to lower pressure.

7.3.2 Mobile-Phase and Sample Flow Path

Besides the pressure that prevails in a respective flow path, the fluidics can be classified according to a contact that exists between the system components and the mobile phase (mobile-phase flow path) or the sample (sample flow path). A schematic is shown in Figure 7.2. The flow path of the mobile phase comprises all the components of the low-pressure flow and high-pressure flow path as was discussed in Section 5.1. In contrast, the sample flow path comprises all the components, which are wetted by the sample. The sample flow path starts at the autosampler with the sample needle, sample loop, injection valve, and needle seat (if applicable), includes all the tubing downstream and the column, and finally ends with the detector outlet.

At this point, it should be noted that in a UHPLC system equipped with a flow-through autosampler, the complete sample flow path including the sample

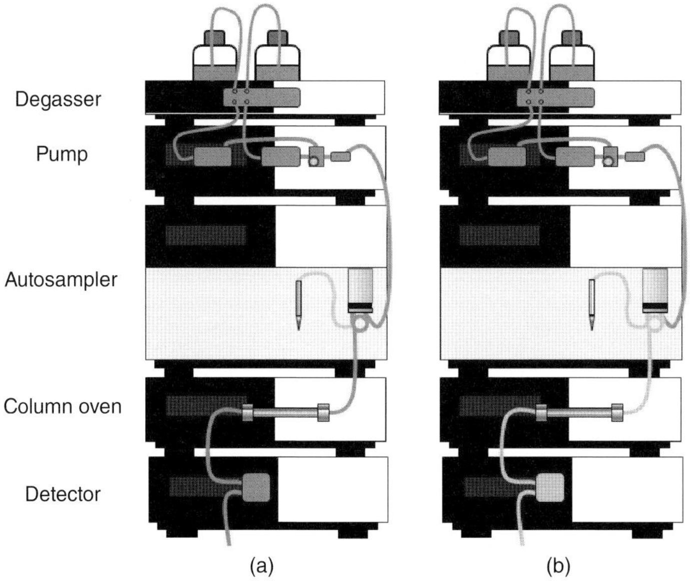

Figure 7.2 Comparison of the mobile-phase flow path (a, dark-gray) and the sample flow path (b, light-gray) in a UHPLC system with fixed-loop (pulled-loop) autosampler. Reproduced with permission of Thermo Fisher Scientific.

needle and needle seat is part of the high-pressure and also of the mobile-phase flow path (with the injection valve in "inject" position). However, in a UHPLC system with fixed-loop autosampler (Chapter 2.1.3.1), the sample loop is the only part that belongs to the high-pressure and mobile-phase flow path. Sample needle, needle seat (if applicable), and tubing connections from the needle to the injection valve are exposed to lower pressure only.

7.4 Low-Pressure Flow Path

The low-pressure flow path starts with the solvent bottles located in the solvent rack, includes the in-line solvent filters and solvent lines from the solvent bottles to the vacuum degasser, and ends with the connection at the pump, either at the gradient proportioning valve (GPV) of low-pressure gradient pumps, directly in the optional solvent selectors of high-pressure gradient pumps, or directly in the pump heads. A selection of materials that are part of the low-pressure flow path of a UHPLC system is given in Table 7.2.

The material of choice for solvent bottles is borosilicate glass, because it is composed of a higher proportion of silicon dioxide and a lower proportion of alkaline metal and alkaline earth metal ions compared to soda-lime glass [5]. In addition, the proportion of boric oxide makes the glass much more resistant to water, acids, bases, and salt solutions. Borosilicate glass is more stable against hydrolysis as common soda-lime glass. If conventional glass types with low chemical resistance are being used, the surface removal is very likely increased with alkaline solvents. The removal can cause higher wear on pump seals and pistons as well as blockages in the system. In respect thereof, it should be mentioned that borosilicate glass is also soluble in strong bases at elevated temperatures.

The function of in-line solvent filters is to prevent intrusion of coarse particles into the system. The maximum flow through the subsequent solvent lines is determined by the pore size of the solvent filters. If the pore size is not adapted to the flow rate, a low pressure might be generated in the solvent tubes at high flow rates, so that outgassing effects occur and gas or air bubbles are generated. Then, depending on the size of the bubbles, the vacuum degasser will not be able to completely remove the trapped gas or air any more. Therefore, in semipreparative applications, it is essential to check the pore sizes, whether they meet the requirements of the applied semipreparative flow rate or not. Common materials

Table 7.2 Wetted materials (selection) typically used in the low-pressure flow path of UHPLC systems.

Component	Wetted materials
Solvent bottles	Soda-lime glass, borosilicate glass
Inline solvent filters	Stainless steel, titanium, PEEK, PTFE, UHMW-PE
Solvent lines	PTFE, FEP, PEEK, ECTFE
Vacuum degasser	Amorphous fluoropolymer, PTFE

for in-line solvent filters are stainless steel or biocompatible materials such as titanium, polyether ether ketone (PEEK), polytetrafluoroethylene (PTFE), or ultrahigh-molecular-weight polyethylene (UHMW-PE). In case of corrosive mobile phases, metal ions dissolve over time from the metal-based in-line solvent filters into the system right at the beginning of the low-pressure flow path. Metals and their corrosion properties are discussed in Sections 7.5.1.2, 7.5.3.3, and 7.6.2 in more detail. Biocompatible solvent filters might be required for certain applications.

Polymer-based low-pressure tubing is commonly used for solvent delivery between the solvent bottles and the pump. The types of material are usually polymers, for example, fluoropolymer-based materials, such as PTFE, fluorinated ethylene propylene (FEP), and ethylene chloro-trifluoroethylene copolymer (ECTFE) with a very wide chemical compatibility. Fittings, including appropriate ferrules, are made of PEEK.

In vacuum degasser units, usually fine porous membranes made of PTFE or amorphous fluoropolymer are used. To ensure that the mobile phase is efficiently degassed before entering the high-pressure flow path, the specification of the maximum flow, which is defined by the degasser cartridges, must be considered. The maximum flow, which is tolerated for efficient degassing, is in the range up to 5 ml min^{-1} for analytical applications. Usually, the maximum flow can be found in the operating instructions or can directly be requested from system manufacturers.

The advantage of polymer-based components in the low-pressure flow path and in the vacuum degasser is the wide solvent compatibility. But often, the materials contain plasticizers such as phthalates and other contaminants, which are released into the mobile phase. This might disturb liquid chromatography–mass spectrometry (LC-MS) analyses as shown in Figures 7.3 and 7.4.

Other common contaminants are lubricants, such as erucamide, which originate from the packing material [6].

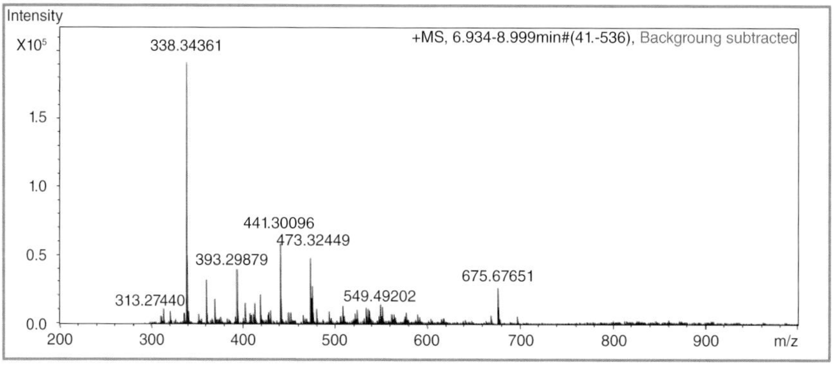

Figure 7.3 LC-MS detection of diisononyl phthalate and erucamide in FEP solvent tubing extract. The FEP solvent tubing had been filled with H_2O/ACN 20/80 (v/v) and incubated at 40 °C for 24 h. Erucamide: $m/z = 338.34$ $[M+H]^+$ and 675.67 $[2M+H]^+$ and diisononyl phthalate: $m/z = 441.30$ $[M+Na]^+$.Reproduced with permission of Thermo Fisher Scientific.

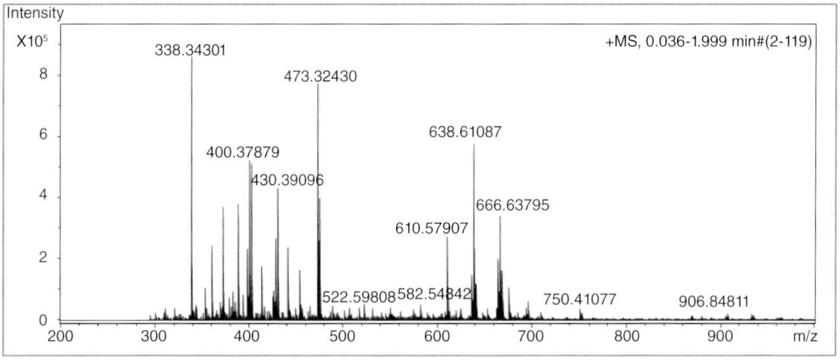

Figure 7.4 LC-MS detection of erucamide in PEEK-frit extract. The PEEK frit was incubated in H_2O/ACN 20/80 (v/v) at 40 °C for 24 h. Erucamide: *m/z* = 338.34 $[M+H]^+$. Besides erucamide, other unidentifiable substances had been detected.Reproduced with permission of Thermo Fisher Scientific.

7.5 High-Pressure Flow Path

7.5.1 Pumps

Once the mobile phase has passed the degasser unit, it enters the UHPLC pump. In the pump head, the mobile phase will be compressed by the piston from atmospheric pressure to system pressure (Figure 7.5). The pump generates the defined flow of the mobile phase in the high-pressure flow path, that is, tubing between the pump and the autosampler, the autosampler, tubing between the autosampler and the column, the column, tubing between the column and the detector, and the detector cell. Assuming a flow rate of 1.2 ml min^{-1} and a piston stroke volume of 60 μl, a low-pressure gradient pump performs 20 piston strokes per minute or a total of 28 800 piston strokes within 24 h of operation time. This extreme workload requires a high wear resistance of the pump pistons and pump seals and robust inlet and outlet check valves. In contrast to conventional HPLC pumps in UHPLC pumps, a higher compression heat is generated during the compression phase. If the additional heat is not properly dissipated, materials in the pump will be affected.

Important pump components are the GPV in low-pressure gradient pumps, the pump heads, and wear parts such as the inlet and outlet valves, pistons and piston seals, and the tubing, which connects these parts (Figure 7.5). Wetted materials frequently incorporated in UHPLC pumps are given in Table 7.3.

7.5.1.1 Inlet and Outlet Valves

Inlet and outlet valves mounted to the pump heads are most commonly ball check valves consisting of a ball and a seat [7]. The valves need to open and close very rapidly, so that the mobile phase, which exits in the low-pressure flow path, is delivered ideally pulsation-free into the high-pressure flow path and onto the column. The valves need to seal the high-pressure flow path over the complete

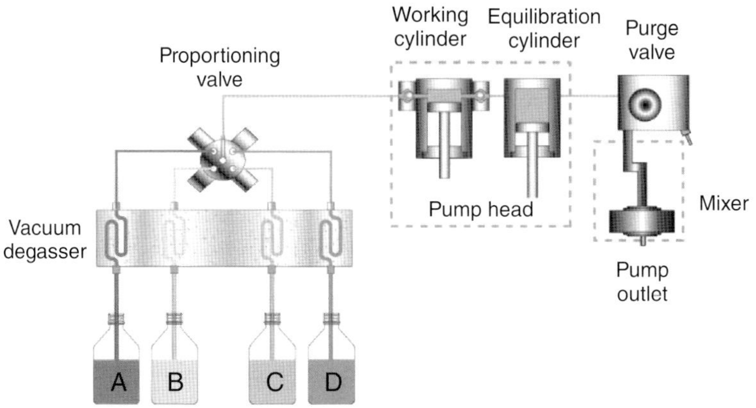

Figure 7.5 Components of a low-pressure gradient pump. Shown are degasser unit, gradient proportioning valve, pump head with working and equilibration cylinder, purge valve, and mixer. Reproduced with permission of Thermo Fisher Scientific.

Table 7.3 Wetted materials (selection) typically used in UHPLC pumps.

Components	Wetted materials
Gradient proportioning valve (GPV)	PEEK, fluoropolymers
Pump head (body)	Stainless steel, titanium
Inlet and outlet valve	Stainless steel, titanium, ruby, sapphire, zirconium oxide, aluminum oxide
Piston seal	PTFE, UHMW-PE, stainless steel, titanium
Piston	Sapphire, ceramics, SiC
Capillaries	Stainless steel, titanium, MP35N

pressure range, so that there is no backflow of mobile phase into the low-pressure flow path. The ball and seat have significant impact on the performance of a UHPLC system and are exposed to high mechanical stress. The shape and surface structure of the ball and seat need to be perfectly aligned, so that a backflow of mobile phase through the valve is prevented. Therefore, the tribological characteristics of the surfaces of the ball and seat are of high importance.

Stainless steel and other alloys are mechanically and chemically not robust enough and do not meet the requirements of UHPLC. Materials with high corrosion and wear resistance are needed. In inlet and outlet valves, therefore, the materials are ruby, sapphire, and ceramics, for example, aluminum oxide (Al_2O_3) or zirconium oxide (ZrO_2) [8, 9]. Ruby and sapphire are synthetic aluminum oxide crystals, while sintered ceramics have amorphous-like structures. Classical valves have a ball made of ruby and the seat made of sapphire, but these types are prone to pressure fluctuations when gradients are formed with water and acetonitrile. The reason is the polymerization of acetonitrile and the deposition of the polymer film on the sapphire seat. The ball sticks to the seat so that the open/close mechanism of the valve is compromised. Ceramics are generally not prone to adhesion of such polymer films, but the drawback of check valves made of ceramics is that they obviously show a high leakage rate at low operating pressures [8]. Besides ball check valves, some manufacturers rely on active inlet valves. In active inlet valves, sealing is mechanically forced via an electromagnet [7]. These valves, however, are more expensive than simple ball check valves.

7.5.1.2 Pump Head

The two most important alloys employed in UHPLC pumps, that is, in pump heads and pump tubing, are stainless steel and titanium. A schematic of a pump head is shown in Figure 7.5. In the following sections, the chemical composition and properties of stainless steel and titanium are described in more detail.

7.5.1.2.1 Stainless Steel Corrosion- and acid-resistant steel alloys of type 316 (AISI notation) and its variants are used in UHPLC pumps on a standard base [1, 10]. A selection of various types and their chemical composition is given in Table 7.4. These alloys belong to a series of chromium–nickel–molybdenum steels with an increased corrosion resistance, which is attributed to the high chromium content of 17–20% and the formation of an oxide layer.

Nickel stabilizes the austenitic structure of these alloys, improves the repassivation in reducing media, and increases the stability against corrosion by mineral salts. The addition of molybdenum increases the resistance of the oxide layer to chloride-containing media. Chromium–nickel–molybdenum steels are therefore more stable against pitting and crevice corrosion than chromium–nickel steels, which do not contain molybdenum, for example, the most frequently used stainless steel alloy type 304 [2, 3].

Influence of Salt-Containing and Acidic Solutions Even though molybdenum generally increases the corrosion resistance of stainless steel, all chromium–nickel–molybdenum steels are prone to corrosion upon contact with chloride-containing media [2, 4]. Therefore, mobile phases that contain halides, for example, sodium chloride, or their corresponding acids, for example, hydrochloric acid, should not be used in UHPLC systems with components made of stainless steel. In general, the chemical resistance of austenitic stainless steel in acidic salt solutions is lower than in neutral salt solutions. Many system manufacturers recommend avoiding mobile phases with a $pH \leq 2$ because

Table 7.4 Chemical composition of stainless steel type 304, 316, 316 L, and 316 Ti [2].

Material/ AISI-No.	Fe (%)	Cr (%)	Ni (%)	Mo (%)	C (%)	Mn (%)	Si (%)	P (%)	S (%)	Ti (%)
1.4301/304	Balance	18–20	8.0–10.5	—	0.08	2.0	1.0	0.045	0.03	—
1.4401/316	Balance	16–18	10–14	2–3	0.08	2.0	1.0	0.045	0.03	—
1.4404/316L	Balance	16–18	10–14	2–3	0.03	2.0	1.0	0.045	0.03	—
1.4571/316Ti	Balance	16–18	10–14	2–3	0.08	2.0	1.0	0.045	0.03	0.7

they are corrosive [11–13]. Sulfuric acid can significantly attack stainless steel even at low concentrations of ≥5% and low temperatures of ≥20 °C (surface corrosion). Special caution is also required with salts of sulfuric acid, especially the hydrogen sulfates. Nitric acid can also lead to surface corrosion in very high concentrations, especially at higher temperatures, for example, 66% nitric acid (boiling) and 99% nitric acid (20 °C), even though nitric acid is used in lower concentrations to passivate stainless steel. Passivation is discussed in more detail in Section 7.6.4. In any case, hydrochloric acid should not be used due to the chloride-induced pitting corrosion, which was discussed previously. However, stainless steel is stable against phosphoric acid over a wide concentration range at room temperature. This also applies to solutions of weak organic acids. In aqueous solutions of formic acid, acetic acid, and citric acid, stainless steel generally shows a good corrosion resistance at room temperature [2, 4, 14].

Influence of Basic Solutions In contrast to acidic solutions, stainless steel is quite stable against a wide range of dilute halide-free inorganic salt solutions with neutral and alkaline pH values, for example, ammonium nitrate, ammonium bicarbonate, sodium hydrogen carbonate, or potassium carbonate. Weak bases such as ammonium hydroxide and organic aliphatic amines do not attack austenitic Cr–Ni steels over a wide concentration and temperature range. But in strong bases, for example, NaOH, at high concentrations and temperatures, corrosion effects are being observed. Therefore, mobile phases with high alkaline pH values of ≥12 should definitely be avoided.

Influence of Organic Solvents Acetonitrile and methanol are the most common organic solvents in reversed-phase (RP) chromatography. They have frequently been used as mobile phase on a daily base in many laboratories, which is a convincing indicator that stainless steel in LC systems is chemically inactive to these solvents. However, in experimental studies of Mowery on the chemical resistance of a pressure regulator valve seat made of stainless steel type 316, different types of local corrosion, for example, erosion corrosion, were observed, when acetonitrile and methanol were passed through the valve seat [15].

For liquid chromatography, the pertinence of these findings needs to be critically questioned. Acetonitrile and methanol have been successfully used for years in HPLC, and both solvents are not directly associated with corrosion at high rates leading to obvious damage of any of the system components. One possible explanation is that acetonitrile and methanol are rarely used in a concentration of 100% in the high-pressure flow path in RP applications – except for a short time at the end of a gradient. However, these findings show that local corrosion effects can occur with undiluted acetonitrile or methanol, if a turbulent flow, high flow rates, and/or rapid flow direction changes prevail in fluidic components made of stainless steel [1]. In respect thereof, the study conducted by Mowery is also cited in scientific literature, which describes interactions of analytes and materials in HPLC systems (Section 7.6.3.3). Therefore, in case metal leaching is observed in an application with acetonitrile or methanol, it could be useful to check the UHPLC system on the facts reported by Mowery. For evaluation, water (≥1%) or phosphate buffer (pH 2.5) can be used to dilute acetonitrile or methanol. According to Mowery, once one of the respective aqueous phases was added to the organic solvent, corrosion was decreased and the lifetime of the valve increased. Furthermore, passivation of the valve with 6M nitric acid minimized pitting corrosion, but erosion corrosion effects and the premature failure of the valve could not be prevented. Also, the addition of lithium nitrate as a corrosion inhibitor resulted in a significant reduction of pitting corrosion, and the addition of hydrazine showed positive effects.

Finally, it should be noted that halogenated organic solvents such as chloroform tend to form hydrochloric acid, which in turn also attacks the stainless steel.

7.5.1.2.2 Titanium Alloys Due to the high specific strength, high corrosion resistance, and biocompatibility, titanium alloys have proven their worth in numerous applications in aerospace engineering, chemical industry, and offshore technology, and as metal-based biomaterial in biomedical engineering [16]. Titanium alloys vary in their chemical composition and crystal structure. The crystal structure influences the mechanical properties of the material. Important titanium alloys are the pure titanium grades ("commercially pure titanium," cp-Ti) in different degrees of purity such as grade 1 and 4, and numerous other titanium alloys, which contain other components such as aluminum, vanadium, niobium, and tin in different weight percentages. A selection of important titanium alloys and their chemical compositions is given in Table 7.5. The pure titanium grades and the material Ti–6Al–4V (titanium grade 5) are considered commercially important. Ti–6Al–4V is the best characterized and tested titanium material and features high strength with good ductility and high corrosion resistance [16].

Titanium is, therefore, often used in applications in which austenitic stainless steel is not sufficiently corrosion-resistant. It is used in biocompatible UHPLC pumps as a construction material of various components. It should be noted that titanium contains a small amount of iron, in pure titanium up to 0.5%. Each biocompatible UHPLC system that relies on titanium is therefore not entirely iron-free.

In contrast to the chromium–nickel steels, which are described in Section 7.5.1.2.1, titanium is significantly more corrosion-resistant to numerous neutral and chloride-containing media [20, 21]. Pure titanium generally also has a high resistance to other inorganic(nonreducing) salt solutions. However, the resistance decreases with acidic pH and higher temperatures. Hydrochloric acid, sulfuric acid, and phosphoric acid can cause corrosion effects at certain concentrations and temperatures, but oxidizing substances such as nitric acid

Table 7.5 Chemical composition of titanium alloys [17–19].

Material-No./grade	Ti (%)	Al (%)	V (%)	Fe (%)	O (%)	N (%)	C (%)	H (%)
3.7025/Grade I	Balance	—	—	0.20	0.18	0.03	0.10	0.015
3.7065/Grade IV	Balance	—	—	0.50	0.40	0.05	0.08	0.013
3.7165/Grade V	Balance	5.5–6.75	3.5–4.5	0.40	0.20	0.05	0.08	0.015

can be added as corrosion inhibitors. Pure titanium is highly corrosion resistant in ≤7% hydrochloric acid, ≤5% sulfuric acid, and ≤30% phosphoric acid at room temperature (corrosion rate ≤0.13 mm per year) [21]. Moreover, at room temperature, it proves to be resistant to organic acids, for example, formic acid, citric acid, and acetic acid, to alkaline media, for example, solutions of sodium hydroxide and ammonium hydroxide, and to oxidizing media, for example, nitric acid, perchloric acid, and hypochlorous acid. When using methanol, however, stress corrosion cracking can be induced depending on the titanium quality. Therefore, when using methanol in conjunction with commercially pure titanium, the addition of about 3% of water is recommended, and the passivity of titanium against stress corrosion cracking is increased [16, 21].

7.5.1.3 Pump Pistons and Piston Seals

7.5.1.3.1 Pump Pistons Besides inlet and outlet valves, the pump pistons and piston seals are considered performance-critical components of a UHPLC pump. The mobile phase in the pump head is compressed by the piston up to system pressure and passed into the high-pressure flow path. The piston chamber is sealed off by the sealing ring from the pump drive, which is located behind the piston chamber. The piston should have a high hardness and a low roughness. These features ensure a low wear of the piston and the seal material.

Sapphire and oxide ceramics such as ZrO_2 are traditionally used as piston materials, because they are characterized by an excellent abrasion and corrosion resistance. As part of further developments of UHPLC pumps, also new materials such as silicon carbide or sintered silicon carbide are used as piston material [22]. Silicon carbide has a low friction coefficient and a high thermal conductivity, whereby the frictional heat is effectively dissipated.

7.5.1.3.2 Piston Seals The piston seal rings used in UHPLC piston pumps are end-face mechanical seals, made of the sealing material and a sealing ring spring (Figure 7.6). The spring preloads the sealing material, so that it is pressed smoothly onto the piston with a constant force.

The piston seal lifetime depends not only on the tribological properties such as the roughness of the material but also on the chemical composition of the mobile

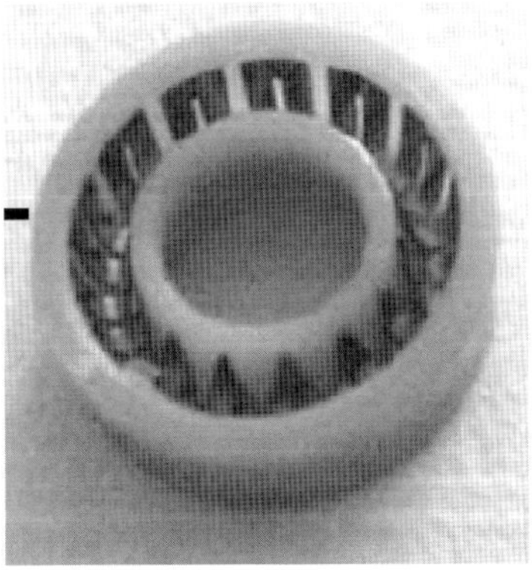

Figure 7.6 End face mechanical seal with seal and spring. Reproduced with permission of Thermo Fisher Scientific.

phase, the system pressure, and the flow rate (the number of piston strokes per time unit). Polymers and polymer-based composite materials are used as sealing materials. Composite materials are composed of a continuous phase, that is, the matrix or base material, and a discontinuous phase, that is, the filler material. Common base materials include UHMW-PE and PTFE or polyethylene (PE). In composite materials, the added fillers such as graphite or carbon fiber improve properties such as the creep resistance, wear resistance, hardness, thermal conductivity, and chemical compatibility of the base material. In contrast to other polymers, PTFE is characterized by a higher chemical compatibility with organic normal-phase solvents, but the wear resistance is generally low. Therefore, fillers such as graphite are added to improve the tribological properties while the chemical compatibility stays comparable.

UHMW-PE-based piston seals are resistant to organic solvents, which are commonly used in RP chromatography. However, the compatibility with solvents that are used in the NP chromatography is not given, so operating the system with UHMW-PE seals with such solvents is generally not recommended. Therefore, UHPLC manufacturers usually offer seals for RP as well as NP operation.

7.5.1.4 Practical Aspects

Since a worn pump piston can damage a sealing ring, the piston should always be checked for abrasive wear during a routine sealing ring exchange. If the piston shows signs of wear, a simultaneous change of seal and piston will be required; otherwise, the lifetime of the sealing ring will be significantly reduced.

In addition to RP- and NP-sealing rings, dedicated biocompatible sealing rings for applications requiring extreme pH values and high salt concentrations are available. In these cases, the spring of the piston seal should be made of a biocompatible, corrosion-resistant metal alloy such as titanium or a nickel alloy, for example, MP35N®, Inconel®, or Hastelloy®.

It should be noted that some manufacturers specify a limited back pressure range for some of the piston seals, so that the pressure range of a UHPLC system cannot entirely be exploited [23, 24].

In order to keep the friction between the piston and the piston seal as low as possible, the user needs to ensure that the surfaces of both components are wetted during operation by using the rear seal wash system, if available. In case the rear seal wash system is not activated or is not used and/or the UHPLC system runs dry of mobile phase, the lifetime of the pump seals will be reduced significantly. The rear seal wash liquid for RP operation, for example, water/isopropanol (90/10) or water/methanol (90/10), should not be recirculated but should rather be pumped unidirectionally. This avoids accumulation of buffer salts and particles in the wash liquid over time. If a unidirectional flow of the wash solution is not supported by the design of the seal wash system, it is recommended to replace the old washing solution with a new one on a daily base.

Highly concentrated buffers as mobile phase can also significantly reduce the lifetime of the sealing ring, because precipitated salt might damage the sealing ring. In this case, the use of the rear seal wash is mandatory. In addition, the system needs to be flushed sufficiently with water to remove buffer salts once the application has been finished.

7.5.2 Autosamplers

Autosamplers are fluidically connected directly behind the pump. Important fluidic components, which belong to the sample flow path, are the injection valve, sample loop, and sample needle. The wash liquid flow path with low-pressure characteristics is also integrated in the autosampler. The wash port is usually filled with wash liquid by syringes or peristaltic pumps to flush the sample needle from the outside. In some autosampler types, the wash liquid has a double function. It is used not only as a needle wash liquid but also as a sample push solvent. In this case, the liquid also fills the volume of the tubing that connects the syringe to the sample needle and hence guarantees aspiration and dispensation of the sample. Some types of flow-through UHPLC autosamplers have a metering device, which replaces the syringe, included in the high-pressure flow path just behind the sample loop.

7.5.2.1 Materials

Alloys with and without coating, composite materials, and polymers are commonly used as wetted parts in autosamplers. A selection of these materials is given in Table 7.6.

Since the materials of the sample flow path get in contact with the sample and the analyte, the compatibility of these materials to the analyte is crucial. The analyte compatibility of materials in UHPLC is discussed in more detail in Section 7.6, so that this topic is not part of this section. Since sample loops have a similar design as system tubing, only the materials and functionality of the sample needle and the injection valve are discussed in the following sections. Additional information on the materials of sample loops is given in Section 7.5.3.

7.5.2.2 Sample Needles, Sample Vials, and Closures

Besides the sample vial, the sample needle is the first component in an LC system, which comes into contact with the sample. It represents the beginning of the sample flow path, pierces the septum of the sample vial, and transfers the sample into the UHPLC system.

Sample needles are available in different materials and are often made of stainless steel, MP35N, or ceramics, but also *fused-silica* and PEEK are being used. The sample needle is part of the high-pressure flow path in flow-through

Table 7.6 Wetted materials (selection) typically used in UHPLC autosamplers.

Flow path	Materials (selection)
Sample flow path (high-pressure flow path)	Stainless steel, MP35N, Vespel SCP, PEEK, PEEK composite, gold-coated stainless steel, DLC, ceramics, stainless–steel-cladded PEEK, FFKM
Wash liquid flow path (low-pressure flow path)	PTFE, FEP, PEEK, ECTFE

FFKM, perfluoroelastomer.

autosamplers if the injection valve is in "inject" position. Since in this design, the sample needle and the injection port are exposed to system pressure (≥1000 bar), the connection of the sample needle in the injection port needs to be reliably leak-tight. Furthermore, the material of the sample needle needs to be resistant against extreme pressures and wear. Only alloys and ceramics can meet these requirements. To seal the sample needle in the injection port, polymers or polymer-based composite materials, for example, PEEK or Vespel SCP, are usually used. The seal is established by the contact of the outer wall of the needle with the appropriately shaped interior wall of the needle seat and by locking the sample needle and needle seat by high mechanical pressure [25]. Therefore, neither PEEK nor *fused-silica* sample needles can be used in a flow-through autosamplers. Sample needles of materials such as *fused-silica* or PEEK are usually applied only in fixed-loop autosamplers (pulled-loop type).

In addition to the mechanical requirement of pressure stability, samples needles must be able to pierce septa and sealing mats of microtiter plates without cutting out the septum material; otherwise, particles might get transferred into the vial or injected into the system. Since a large variety of closures is available on the market, it is obvious that not every autosampler needle can be compatible with all types of septa. Under certain circumstances, issues with closures of a certain material, which were previously applied reliably in an autosampler, can arise when the material has aged, has been purchased from another manufacturer, or came from a different batch.

Furthermore, the more often a sample is injected from the same vial and the septum is pierced multiple times, the more likely it is, depending on the septum material, that particles contaminate the sample and are injected into the system. The issues that are usually observed during an HPLC analysis due to cut-out septa material include retention time and peak are fluctuations. In case a sample needle is blocked by the septa material, it is even possible that the sample is not injected into the system and the sequence is not automatically stopped. Furthermore, the cut-out material of the septum, which potentially contains plasticizers or other extractables (Section 7.4), can contaminate the sample.

Instrument manufacturers usually recommend using certain vials and septa with their autosamplers. These recommendations are given either in operating manuals, brochures, or data sheets or can be found with interactive vial selection tools in the Internet. But in case vials and closures that do not comply with the recommendations are being used, it is advisable to select them according to compatibility lists from manufacturers of vials and closures [26–28]. In some cases, it can be helpful to perform compatibility tests by oneself.

7.5.2.3 Injection Valves

The main requirements for injection valves in modern UHPLC autosamplers are a high chemical resistance, low friction, and band dispersion as well as fast switching times. Two-position, six-port rotary shear valves, which consist of a valve actuator and a valve body with a stator and a rotor, are applied in UHPLC autosamplers (Figure 7.7).

The stator consists of six bores including a thread and a flow passage, which allows liquid to pass the stator to the opposite side to the rotor or vice versa.

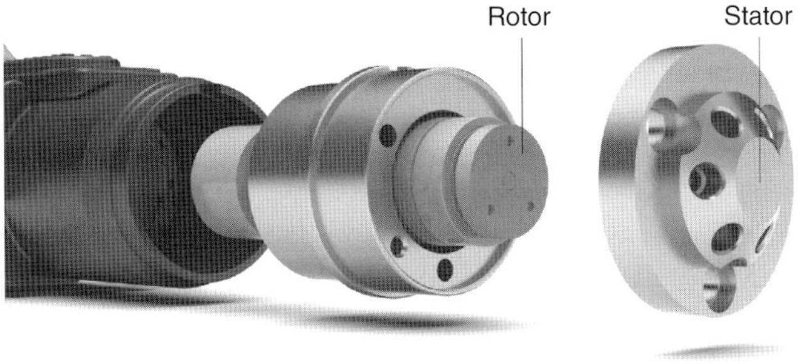

Figure 7.7 Two-position-six-port rotary shear valve including rotor and stator. Reproduced with permission of IDEX Heath & Science GmbH.

The distribution of the six bores on the stator is characterized by its rotationally symmetrical arrangement. The tubing and fitting, for example, for the connection of the pump to the autosampler or the autosampler to the column, are screwed to the bores of the stator. The rotor has three arched flow passages (grooves) to connect two adjacent stator ports. When the valve is switched, the rotor slides on the stator between two positions A and B or "Inject" and "Load," respectively. Stator and rotor are wetted by the mobile phase and are subjected to significant mechanical forces, because both components are held together by a spring in the valve body. The force of the spring is slightly higher than the maximum pressure specification of the UHPLC system [29]. Important properties of rotary shear valves are the material of the stator and rotor, the maximum tolerable pressure, and the pH stability of the individual components.

7.5.2.3.1 Materials for Injection Valves The stator typically consists of a hard material, while the rotor is composed of a softer material. Stainless steel or titanium is usually used as base material for stators, while the materials for rotors are mostly polymers or polymer-based composite materials. Continuous phases for the rotors are, for example, polyimide, polyaryl ether ketone (PAEK), PTFE, PEEK, or ethylene-tetrafluoroethylene (ETFE). The filler materials are fluoropolymers, for example, PTFE, but also graphite, glass or carbon fiber, titanium dioxide, and molybdenum sulfide. The rotor materials are frequently proprietary and are often known by their trade names such as Vespel or Valcon H [25, 30–32]. In order to receive a robust and tight connection with good sliding quality between the rotor and stator, at least one of the contact partners (stator or rotor) must be coated. Most often, the proprietary coating is applied to the stator. These coatings include, for example, diamond-like carbon (DLC) or a tungsten carbide/carbon composite material. Due to the coating, the lifetime of the valve is significantly increased compared to uncoated valves [29, 33, 34]. In addition to stators made of stainless steel, PEEK- or PAEK-stators are being used in conventional HPLC systems with lower pressure capabilities. The advantage of PAEK/PEEK valves is that they are metal-free and biocompatible. PAEK/PEEK valves, however, have a pressure specification of about 345 bars, and therefore, they are not suitable for UHPLC. Thus, biocompatible UHPLC valves consist of either a titanium stator or a stator with a steel body and a biocompatible stator insert (stator face), for example, made of ceramics. The advantage of the biocompatible valve with stator face is that the liquid-wetted surfaces are metal-free.

7.5.2.3.2 Pressure Stability and pH Compatibility The pressure limitation of a rotary shear valve with hard metallic stator and polymer-based rotor is dominated by the low hardness and low wear resistance as well as the compressive strength of the rotor material. PEEK-based rotors usually withstand a lower pressure compared to polyimide-based rotors. Besides the maximum pressure, the pH range, which is tolerated by the valve, is of importance. PEEK-based rotors usually cover a broad pH range of 0–14, while polyimide-based rotors are only compatible with a pH value of 0–10. Information on the chemical resistance of rotor materials can be found in the system-operating manuals or are accessible via the Internet and the web pages of the respective valve manufacturer. Due to the common design of rotary shear valves, which is the combination of a hard metallic stator with a polymer-based rotor, the maximum pressure load that can be applied is currently not more than 1250 bar. For higher pressures, both the stator and the rotor need to be made of a hard material such as ceramic or metal. In this way and in combination with a protective coating, a much higher pressure as 1250 bar can be realized while maintaining the lifetime of the valve [35, 36].

7.5.2.4 Practical Aspects

The lifetime of valves is determined by the number of switching cycles until a failure occurs. In the case of rotary shear valves with two positions, a switching cycle is defined as switching from position A to B and then back to the starting position. The criterion that defines a failure is a certain leakage rate of the mobile phase, which is measured at a given pressure. A leak flow can either form from the flow passages of the rotor radially outwards or develop from port to port (cross-port leakage) concentrically to the rotational axis of the rotor. (Figure 7.8) [29].

Defect on the rotor leading to a leakage is material erosion at certain locations on the rotor (Figure 7.9).

To increase the lifetime of the rotors in UHPLC injection valves, sometimes, the rotor has modified grooves and the injection valves are operated with an optimized switching cycles [37].

There are many important factors that influence the lifetime of an injection valve. The stator and rotor material and the contact pressure of both components,

Figure 7.8 Schematics of a rotor with two types of leakages: radial leakage (shown in dark-gray) and cross-port leakage (shown in light-gray). Reproduced with permission of Thermo Fisher Scientific.

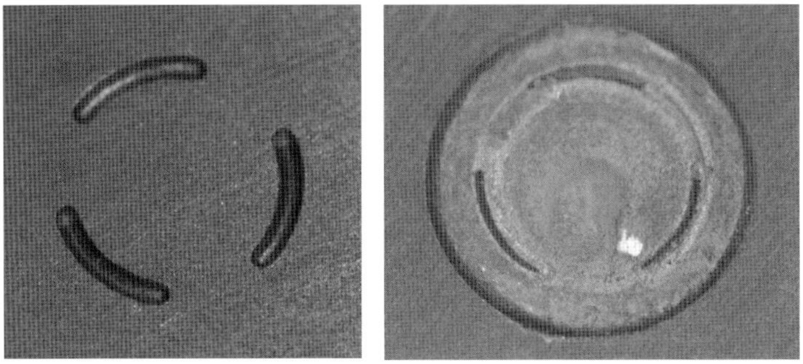

(a) (b)

Figure 7.9 Image of a new, unused rotor (a) and a used rotor (b) with defects, which cause cross-port leakage.

but also the flow rate, system back pressure as well as the composition of the mobile phase and of the sample, that is, the sample solvent, buffer salts, pH value, and the particle content, determine the number of possible switching cycles until the valve fails. The filtration of buffers with 0.1 or 0.2 μm filters and the use of commercial and prefiltered solvents with a very high degree of purity with low particle load (e.g., LC-MS grade) are part of a good UHPLC practice and will increase the lifetime of the valve.

However, a general statement regarding the lifetime of valve components cannot be given due to the multitude of influencing factors. Some system manufacturers offer software-integrated pressure tests for individual UHPLC modules, for example, the pump or the autosampler, which enable the user to determine leakage rates based on an observed pressure drop. Therefore, it is recommended to perform such tests in regular intervals for the injection valve (if available) and to document the leak rate. In this way, a failure can be detected at an early stage and appropriate countermeasures, such as an exchange of the rotor or the entire valve, can be initiated.

7.5.3 Tubing and Fitting Systems

7.5.3.1 Outline

As components of the sample flow path in the high-pressure segment of a UHPLC system, capillary tubing and fittings do have contact to the analyte and the mobile phase. Ideally, the fluidic connections are chemically inert to the mobile phase and inert to the analyte. The analyte compatibility of materials is discussed in more detail in Section 7.6.3.3.

Tubing and fittings in the sample flow path contribute to the extracolumn volume of a UHPLC system and therefore have significant influence on the peak dispersion and the overall separation performance of the instrument. The internal diameter and the length of the tubing should be selected according to the requirements of the separation column and its location in the UHPLC system (Section 2.1.6 and Chapter 3). Fittings are used for a proper connection of the tubing with the counterpart, for example, the injection valve, the column, or the inlet of the detector. Ideally, the fitting fixes the tubing and its counterpart in such a way that both are reliably connected free of any leakage and without dead volume. In order to minimize the extracolumn volume for exploitation of the full separation power of short columns (≤50 mm) with small inner diameter (≤2.1 mm), tubings with smaller dimensions have been used in recent years with increasing frequency. Thus, system manufacturers have also included such tubings as standard accessories in the shipment of UHPLC systems. While in conventional HPLC systems, tubings with an inner diameter of 0.25 mm were used for *normal-bore* columns with an ID (inner diameter) of ≥4 and ≤5 mm, nowadays for connecting the modules in UHPLC systems, tubings with an ID of 0.18 are employed. UHPLC systems that are specifically designed for applications with *narrow* and *micro-bore* columns with an ID of ≥1 and <4 mm are operated with tubing with an ID of 0.13 mm or even ≤0.1 mm. In addition, flexible capillary tubings with 1/32″ outer diameter have prevailed against rigid 1/16″ capillaries, because they allow a flexible and effortless connection of the system components.

7.5.3.2 Materials

Materials for tubing and fittings in UHPLC are stainless steel, titanium, MP35N, PEEK, stainless-steel-cladded PEEK, *fused-silica*, and PEEK-cladded *fused-silica*. PEEK- and *fused-silica* capillaries with small inner diameters are usually used behind the detector or are used as transfer tubing to the mass spectrometer. An overview of materials used for UHPLC capillaries is given in Table 7.7

Table 7.7 Wetted materials (selection) typically used in tubing and fittings.

Material	UHPLC-compatible	Notes on biocompatibility
Stainless steel	Yes	Not biocompatible
Titanium	Limited UHPLC-compatible due to its brittleness; only in larger inner diameters available	Biocompatible, because virtually iron-free, but not metal-free
MP35N	Yes	Biocompatible, because virtually iron-free, but not metal-free
PEEK	No	Metal-free
Stainless-steel-cladded PEEK	Yes	Metal-free
Fused-silica	Limited UHPLC-compatible due to its brittleness when subjected to mechanical stress from the outside	Metal-free, but interactions with free silanol groups possible, limited pH range
PEEK-cladded *fused-silica*	Yes	Metal-free, but interactions with free silanol groups possible, limited pH range

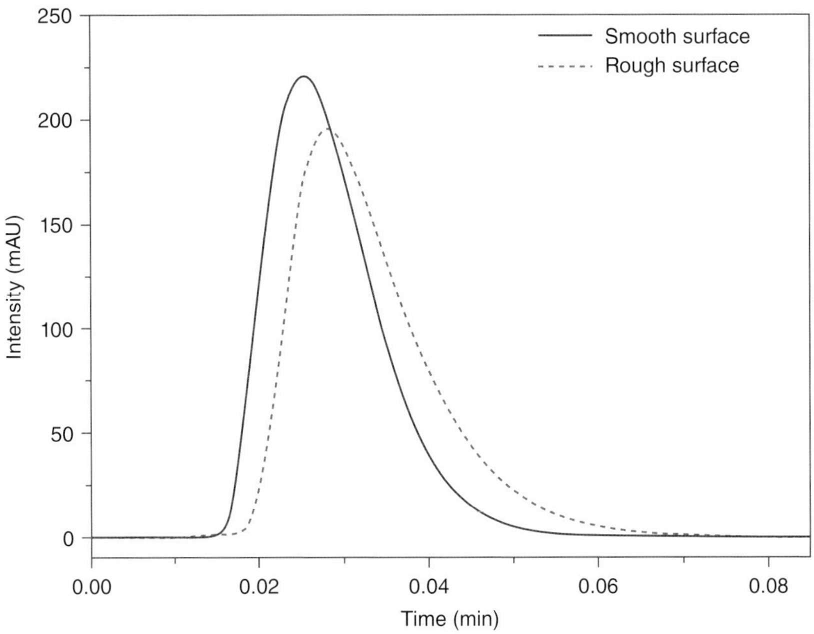

Figure 7.10 Results of a flow injection analysis of caffeine (concentration 100 mg l^{-1} in water, flow 0.5 ml min^{-1}) with two different 75 mm × 0.18 mm stainless-steel capillaries with a smooth and a rough interior surface. The peak volume increases by about 13% when the capillary tubing with rough interior surface is used. Reproduced with permission of Thermo Fisher Scientific.

7.5.3.3 Tubing

For tubing with small inner diameter, the quality of the interior surface is more crucial than for tubing with large inner diameter. The material itself and also the manufacturing and finishing processes are important factors influencing the texture and roughness of the interior surface of the tubing. In this context, PEEK tubing has a lower roughness as, for example, stainless-steel tubing. A higher roughness of the interior surface as well as defects and grooves directly affect band dispersion. Increased band dispersion results in an overall loss in separation performance of the system. The influence of roughness of the interior stainless-steel surface on band dispersion is shown in Figure 7.10.

7.5.3.3.1 Metal-Based Tubing UHPLC capillary tubing is most commonly made of stainless steel. The pressure stability of stainless-steel tubing is found to be well above 1000 bar (≤0.25 mm ID). The wall thickness, which is the difference in the outer and inner diameters, is essential for the pressure resistance. Hence, tubing made of stainless steel with an inner diameter of 0.13 mm and an outer diameter of 1/16″ is more pressure-stable than a flexible 1/32″ capillary tubing with the same ID. As of 2015, the pressure stability of a 1/32″ tubing was significantly higher than the pressure capability of commercially available UHPLC systems (≤1500 bar). Capillary tubing is available in a welded or seamless format. Both types are stable up to ≥1000 bar pressure. The handling of 1/16″-capillaries for

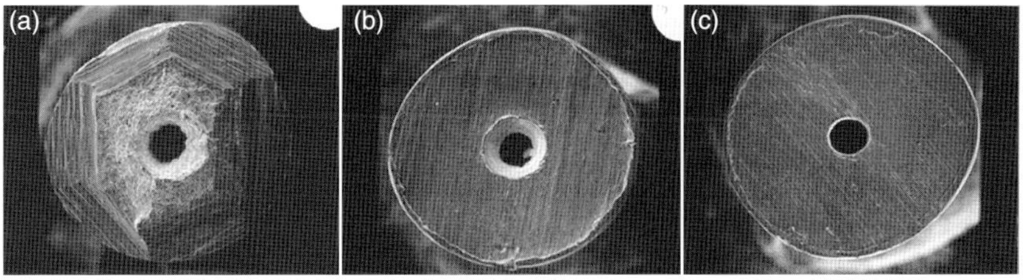

Figure 7.11 End faces of different stainless-steel tubings: (a) file-cut tubing, (b) tubing cut by a commercially available tube cutter, and (c) factory precut tubing. Reproduced with permission of IDEX Heath & Science GmbH.

connecting the components of a UHPLC system is more challenging than the use of 1/32″ OD capillaries. In addition, capillary tubing should not be cut with a file or a capillary tube cutter, because an uneven nonplanar surface on the end face of the tubing and a partial damage of the oxide layer will be the result (Figure 7.11).

Due to abrasion during the cutting process, particles are likely to be released into the system, and an additional dead volume can arise leading to additional band broadening. Therefore, electrolytically precut and electropolished tubing with defined lengths is recommended instead of self-cut tubing.

However, if stainless steel gets into contact with mobile phases with a high content of salts and/or chloride ions as well as extreme pH values, it is prone to significant corrosion as was discussed at the beginning of Section 7.5.1.2. Titanium is, therefore, the preferred biocompatible material in pump and autosampler components, but if the material is used to manufacture tubing, it becomes really brittle [38]. It can release particles leading to further issues down the flow path if it is strongly bent. Therefore, titanium tubing is only available in large inner diameters, for example, in a size of 0.18 mm.

In addition to titanium, the cobalt-based alloy Co–35Ni–20Cr–10Mo (MP35N) has gained acceptance as an additional biocompatible alloy in liquid chromatography within recent years. MP35N is a so-called super alloy containing nickel, cobalt, chromium, and molybdenum. In Table 7.8, the chemical composition of MP35N is given.

Originally, this material was developed for aerospace fasteners, but due to its high mechanical strength and biocompatibility, it is also used in biomedical engineering, for example, in orthodontic applications and for hip replacements [40, 41]. MP35N is characterized by an increased corrosion resistance, which is being attributed to the formation of an oxide layer (Section 7.6.2.1.2). The oxide layer formed on MP35N shows a high resistance to salt water, other chloride-containing solutions, and inorganic acids (nitric, hydrochloric, and

Table 7.8 Chemical composition of MP35N [39].

Co (%)	Ni (%)	Cr (%)	Mo (%)	C (%)	Ti (%)	Fe (%)	Mn (%)	Si (%)	P (%)	S (%)
Balance	33–37	19–21	9–10.5	0.025 (max)	1 (max)	1 (max)	0.15 (max)	0.15 (max)	0.015 (max)	0.01 (max)

sulfuric acid). Moreover, it is stable against pitting and crevice corrosion and stress corrosion cracking [40].

7.5.3.3.2 Polymer-Based Tubing In contrast to stainless steel, PEEK has a high chemical compatibility with aqueous mobile phases with extreme pH values, high salt, and chloride ion concentration and is metal-free. This is the reason why it is used as bioinert material in conventional bio- and ion chromatography. The pressure stability of PEEK tubing depends on its dimension and the manufacturer. PEEK tubing, for example, with 1/16″ outer diameter and an inner diameter of 0.13 mm can cope with pressures up to 480 bar, while the pressure resistance of PEEK tubing with 1/32″ outer diameter is reduced to 345 bar only [30, 42]. The pressure specifications usually relate to conditions such as room temperature and water as mobile phase. Using acetonitrile as cosolvent in the mobile phase and with temperatures higher than room temperature, the pressure resistance of PEEK tubing drops significantly down to about 200 bar [43]. But to provide access to the benefits of PEEK, that is, it is chemical inertness and biocompatibility, to UHPLC users, tubing manufacturers have developed stainless-steel- and nickel-cladded PEEK tubing. According to the specifications, stainless-steel-cladded PEEK tubing is stable up to 1200 bar while nickel-cladded PEEK is pressure-stable up to even 2000 bar [30, 44]. In respect thereof, it is important that appropriate fittings are also available in a biocompatible material and that they support the pressure specification of the respective cladded tubing.

7.5.3.3.3 Fused-Silica Tubing Depending on the manufacturer, *fused-silica* tubing provides pressure resistance up to ≤690 bar. Hence, they are less pressure-stable than stainless-steel tubing, but more pressure stable than PEEK tubing. One major drawback is their handling compared to other tubing, because for making a connection, a PEEK sleeve and a conventional two-piece fitting are usually required. *Fused-silica* tubing with polyimide-coated outer surface is quite fragile and tends to break easily upon impacts of external mechanical force. Thus, they are suitable for UHPLC only to a limited extent. *Fused-silica* tubing is also available in PEEK- or nickel-cladded formats, which are sold under (trade) names such as *PEEK-sheathed, PEEK-coated,* or *Nickel-clad fused-silica, nanoViper™*- or *PEEKSil™*-tubing. The cladding provides a higher resistance to mechanical stress, hence, the pressure stability increases up to ≤1390 bar (50 μm ID), and the flexibility and the lifetime of the tubing improve significantly [30]. In this context, the additional pressure resistance of up to 1400 bar, however, can only be exploited, if a fitting that properly supports this pressure range is available. Due to its bare interior silica surface, *fused-silica* tubing is only compatible with aqueous solvents in a pH range of 0–10.

7.5.3.4 Fitting Systems

Besides the structure of the interior surface of the tubing, the quality of the connection of the tubing and the receiving port, for example, of the injection valve, the column, or the detector inlet, is a major concern with respect to dead volume and pressure stability. Since the aim is to avoid any additional dead volume, the tubing passage needs to be perfectly aligned with the bore of the receiving

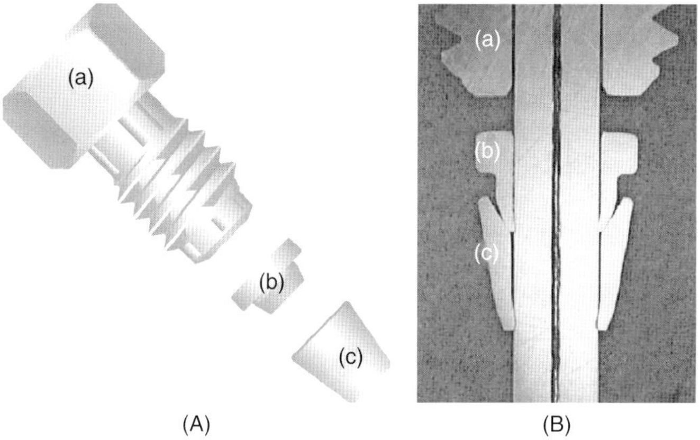

Figure 7.12 Schematic representation of a three-piece fitting system (A) consisting of a nut (a), a collet (b), and a ferrule (c) and a cross section of a tubing, which is fixed by the three-piece fitting system VHP-200 from IDEX (B). Reproduced with permission of IDEX Heath & Science GmbH.

port, while the end face of the tubing needs to directly meet the surface of the receiving port.

The pressure stability of a fluidic connection in a UHPLC system depends on the material of the tubing as well as the operating principle and the material of the fitting. For example, tubing that is made of stainless steel is pressure-stable well above 1000 bar. But the pressure range of the tubing can only be exploited to its full extent, if the fitting also ensures a robust connection within the same pressure range. Therefore, it is crucial that fitting and tubing are considered as a single unit.

Conventional two-piece fitting systems comprise a threaded nut and a ferrule. The ferrule compresses against the tubing at a single compression point and thus holds the tubing in place. In Chapter 2.1, the operation principle of two-piece, stainless-steel-based fittings is discussed in more detail.

In three-piece fitting systems (Figure 7.12), which are using a collet in addition to the nut and the ferrule, the tubing is fixed with two compression points. In this way, these fitting systems are more stable against higher pressure at the same torque compared to two-piece fitting systems [45]. Two-piece fittings have the tendency that the tubing slips off the ferrule or that the inner tubing passage is getting deformed as soon as a higher torque is used [30]. Hence, these issues can lead to additional dead volume and loss of separation performance.

Due to the low-pressure stability of ≤400 bar, polymer-based one-piece fittings can only be used behind the UHPLC column in a meaningful manner. Besides two- or three-piece fitting systems, there are fitting systems on the market that are directly connected to the tubing. The advantage of those tubings (Viper™ and nanoViper™) is that they are much easier to handle and that they generate virtually dead-volume-free connection to the receiving port due to the tip sealing principle [46, 47].

7.6 When and Why Can an Inert UHPLC System Be Required?

Inertness in liquid chromatography is always a question of the application and application-related parameters. There are a considerable number of variables, such as the chemical composition of the mobile phase and the sample solution, the chemical structure of the analyte, the matrix components of the sample, the expected back pressure and the flow rate, the column type, column temperature, and the detection method. Once these variables and the materials in UHPLC systems have been reviewed in detail, only then a discussion about inertness and the requirement of an inert UHPLC system is reasonable. This complexity actually reveals that a straightforward answer to the question of inertness does not exist.

7.6.1 Concept of Inertness

What is inertness? Inertness can be described as a state of absence of interactions in, for example, a physical system. In this context, a physical system is spatially well differentiated from its surroundings and characterized by a system-inherent process. In an ideal world, the system does not affect the process and the process does not affect the system. But in reality, the system and its process are always related to some degree and are mutually dependent on each other. Thus, we can never achieve the ideal state of inertness, but by the right decisions and means, we can approximate to that state.

In the case of liquid chromatography, inertness is achieved once the application (process) runs without being affected by the UHPLC system and if the UHPLC system is not affected by the application (process) in any way. So if inert conditions are predominant, the chromatogram will only contain unbiased information of the injected sample. Artifacts due to possible interactions between the UHPLC system and the application are not present in the chromatogram.

In the following discussion, we will differentiate between *general inertness* and *analyte-specific inertness* of a UHPLC system.

7.6.1.1 General Inertness

The term *general inertness* of a UHPLC system is defined as the chemical resistance of the system to the mobile phase, the sample solution, and the wash solution at defined application-related parameters such as the column temperature and back pressure. In order to assess whether a UHPLC system is generally inert to aqueous solutions and organic solvents used in an application, the following questions should be considered:

- Which materials are used in the high-pressure and low-pressure flow path? Are those chemically inert to the mobile phase, sample solution, and wash solvent at the given temperature and pressure?
- Can corrosion of metal alloys, swelling of polymers, or increased wear be expected?
- Under certain circumstances, is the physical and mechanical integrity of the UHPLC system or of certain components being compromised?

- Are slow corrosion processes, which might occur in a UHPLC system due to the incompatibility of alloys to the mobile phase, possible? If yes, does metal leaching influence detection?

In case of *general inertness*, we always review the influence which the process, that is, mobile phase, temperature, and back pressure, might have on the instrument, its components, and materials. The requirements with respect to *general inertness* are discussed in Sections 7.6.3.1 and 7.6.3.2.

7.6.1.2 Analyte-Specific Inertness

The *analyte-specific inertness* of a UHPLC system is defined as the absence of any influence that the UHPLC system might have on the analyte (analyte compatibility). For assessment of *analyte-specific inertness*, the following considerations should be kept in mind:

- Which materials are part of the sample flow path and are, thereof, in direct contact with the analyte?
- Are there any physical and chemical interactions of the analyte with the wetted surfaces of the UHPLC system to be expected?
- Is it possible that due to the presence of slow corrosion processes, metal ions are being released from the wetted surfaces of the high-pressure and low-pressure flow path into the UHPLC system? Do these metal ions possibly interact with the analyte?
- Do dissolved metal ions possibly bind onto the stationary phase and can the enrichment lead to secondary interactions with the analyte?
- How do possible interactions of analyte and system affect the separation and hence the chromatogram?

Interactions that can occur between the analyte and the components and materials in a UHPLC system are discussed in Section 7.6.3.3.

7.6.2 Nature of the Passive Layer

Important aspects of *general* and *analyte-specific inertness* are the corrosion resistance and biocompatibility of alloys. Both features depend on various properties of the passive (oxide) layer, which are the morphology, chemical composition, and the depth of the passive layer. Since the passive layer is formed on the surface, the surface-finishing processes of alloys are of great importance.

Different methods, for example, thermal oxidation, electro polishing, or chemical passivation with a chemical agent such as nitric acid, are being employed for passivation of materials. The aim of passivation is to replace the air-formed, nonuniform passive layer by a smooth, uniform, and dense passive layer. To achieve a corrosion resistance, which is specifically optimized for a particular application, impurities from the manufacturing process need to be removed from the metal surface [2, 48]. The integrity and quality of the passive layer seem to have a larger impact on corrosion resistance than the depth of the passive layer [48].

The chemical composition and the depth of the passive layer depend on the passivation method, its passivating agent (species and concentration), the duration and temperature of the passivation, and the history of the material. Within the passive layer, often two layers with different oxygen content and resistances are being formed.

Not only the oxide layer changes its composition during passivation, but also the pure metal phase located just below the passive layer is exposed to modifications in its chemical composition. Hence, the metal phase close to the surface is different from the metal alloy in the core of the material. The passive layer is, however, aging over time. During aging, changes in the chemical composition and the depth of the layer and a release of metal ions take place depending on the medium [49–51]. Relevant factors include the oxidizing and reducing properties, the concentration of individual contents of the medium, and its pH value. The passive layer is formed and dissolved in a continuous process. In very corrosive media, it is possible that the protective oxide layer dissolves completely [49]. Aging processes of alloys are also known in biomedical engineering. Various properties of stainless steel and titanium in medical implants change depending on the location in the human body and the duration of their deployment [52]. Interactions with proteins and phosphate in solution are well known. For example, once adsorbed to the surface of metallic biomaterials in physiological solutions, proteins show, depending on their structural characteristics, either corrosion-inhibiting or corrosion-accelerating effects [53, 54].

The adsorption of bovine serum albumin (BSA) accelerates, for example, the corrosion of stainless steel and releases metal ions into the surrounding medium [55]. Under certain circumstances, serum proteins can increase the corrosion of stainless steel and cp-Ti (Tables 7.4 and 7.5), but in contrast, they do not affect the titanium alloy Ti–6Al–4V [56]. The formation of organometallic complexes is discussed as a possible mechanism for the corrosion-accelerating effects. According to this theory, proteins are able to bind to oxidized metal ions located in the passive layer and remove them, so that the passive layer gradually dissolves and the metal is prone to further corrosion [53–55].

In addition to proteins, phosphate tends to adsorb on the passive layer of alloys. In soil science, the binding of inorganic phosphate to metal hydroxides is well known, while in orthopedic applications, phosphate is incorporated into the oxide layer of cp-Ti implants *in vivo* [52, 57].

Furthermore, experimental studies with phosphoric acid and in simulated physiological solutions show that phosphate intercalates into the oxide layer of cp-Ti and Co–Cr–Mo alloys [58, 59]. Phosphates can also influence the adsorption of proteins on the passive layer. Furthermore, it is known that phosphate reduces protein adsorption on to the surface of titanium [60, 61]. But in comparison, phosphates do not affect the protein adsorption on Co–Cr–Mo alloys in the tested concentrations [62]. The following chapters describe and discuss the passive layers of different metal alloys in more detail.

7.6.2.1 Passive Layers of Chromium Alloys

The high corrosion resistance of metal alloys with high chromium content, such as stainless steel and MP35N, can generally be attributed to an enrichment of chromium in the passive layer compared to the metal phase below the passive layer.

7.6.2.1.1 Stainless Steel A stable chromium oxide/hydroxide layer is formed on steel alloys with a chromium content of more than about 10% [63]. In addition, chromium, nickel, and molybdenum are enriched directly below the passive layer in the metal phase, while iron is only present in a smaller proportion. It is assumed that iron dissolves selectively from the metal during passivation [53, 63].

Depending on the passivation method and passivation medium, the properties of the passive layer change. For example, the passive layer, which is formed during passivation of stainless steel type 316L by electropolishing in an acidic medium at high potential, consists of, for example, Fe_2O_3 while at a low potential and under neutral-to-basic conditions, the passive layer is mainly composed of chromium oxide [48]. Nickel is also present in the passive layer, but in a lower proportion [51].

In case stainless steel is passivated with nitric acid, the chromium oxide concentration in the passive layer and the layer depth clearly depend on the acid concentration, temperature, and duration of passivation. These parameters have a greater impact on the depth of the passive layer rather than the chromium content [64].

The chromium enrichment in the passive layer of stainless steel type 316 and the pitting corrosion resistance in 1 mol l^{-1} NaCl solution reach their highest value when passivated with HNO_3 in a concentration of 25% for 1 h at room temperature. Increasing the concentration to levels between 25% and 50% causes a reduction of the chromium-to-iron ratio in the passive film and a reduction of the corrosion resistance. Due to the passivation, a smooth metal surface is achieved and inclusions of manganese sulfide are being removed from the surface, which are one of the main sources for pitting corrosion [65].

Besides nitric acid, citric acid is used for passivating stainless steel. A study that compared nitric acid, citric acid, and Citrisurf®, a commercially available citrate-based passivation agent, demonstrated that the content of iron oxide rapidly decreases in the passive layer within the first 1–2 h of passivation, while the content of chromium oxide increases. When using nitric acid, the iron oxide content reaches a minimum value after about 1 h but then rises again continuously until the end of the measurement period (6 h). In contrast to the results with citric acid and Citrisurf, the nitric acid passivation was not completed, but with both citrate-based solutions, the chemical reactions on the surface reached an equilibrium state after approximately 4 or 2 h, respectively [66].

In acidic and neutral solutions containing chloride, for example, 0.9%(w/v) isotonic sodium chloride, Ringers or Hanks solution, and in phosphate-buffered saline (PBS), a passive layer is also formed on orthopedic steel. These solutions are used for infusion or for cell culture applications, which demand a relatively high concentration of chloride ions in the range of about 140–160 mmol l^{-1}. The studies have shown that the passive film mainly consists of chromium oxide and iron oxide. The addition of complexing agents such as citrate also influences the composition of the metal surface in the oxide layer and the underlying metal phase [53]. The presence of citrate leads to a reduction of the depth of the passive layer and an enrichment of chromium oxide due to the predominant formation of iron(II)-, iron(III)-, and nickel(II) complexes. At the same time, nickel accumulates in the metal phase below the passive layer.

Studies suggest that the passive layer of stainless steel, which forms in physiological solutions, is composed of two layers. The inner layer consists mainly of Cr_2O_3, while iron and nickel oxides build up the outer layer. The stoichiometry of the components in the outer layer depends on the applied potential and the duration of the passivation [53].

7.6.2.1.2 MP35N The high corrosion resistance of MP35N is attributed to a nickel–chromium–cobalt oxide layer in the passive layer. The depth of the passive layer, which forms in a buffer system of acetate/acetic acid (1 mol l^{-1}, pH 4.8, 15 ppm As_2O_3), is about 1.5 nm. Furthermore, chromium and molybdenum are enriched in the passive layer compared to the metal phase [67]. In studies on the corrosion resistance of 35N LT, which is a Co–36Ni–20Cr–10Mo alloy with a lower titanium content compared to MP35N, it was shown that the corrosion resistance in PBS buffer increased by a factor of 4 and the depth of the oxide layer increased by a factor of 2–3 by means of passivation with 10.5% nitric acid. The proportion of chromium oxide in the passive layer increased significantly, while neither nickel nor molybdenum oxide could be detected. The content of cobalt oxide in the passive layer decreased in comparison before passivation. Cobalt and nickel dissolved during passivation of the alloy. After passivation, 35N LT showed a lower solubility of metal ions from the metal in PBS buffer/H_2O_2 [68].

7.6.2.2 Passive Layers of Titanium Alloys

The high corrosion resistance of titanium is attributed to the formation of a highly resistant, adherent oxide layer. The oxide layer increases solely by contact with air within 70 days to a depth of up to 50 Å (5 nm) [49]. Studies indicate that the oxide layer is a system of two layers similar to the passive layer on stainless steel [49, 50]. The passive layer that builds up on cp-Ti by anodization in air-free 0.5 mol l^{-1} sulfuric acid at 25 °C has an estimated depth of 10–40 nm and consists of an outer and an intermediate layer. The outer layer is made of amorphous TiO_2 and has a depth of 10–20 nm. The intermediate, dense TiO_x layer is located between the TiO_2 layer and the metal surface and contains titanium and oxygen in nonstoi-chiometric ratio. The outer layer is sensitive to its environment, and its depth decreases in corrosive solutions due to aging processes [50]. In a different study on passivation of cp-Ti in PBS buffer, a system of two layers also formed on the surface, but with a different composition compared to the previous study. The inner, dense layer consists of TiO_2, while the outer layer is composed predominantly of hydroxides. It was shown that the outer layer grows when exposed to H_2O_2 in PBS solution [69]. The passive layer that forms on Ti–6Al–4V in physiological solutions is composed of titanium oxides of different stoichiometry as well as Al_2O_3. Surprisingly, vanadium could not be detected in the passive layer, even though it is a major component of the alloy [61].

7.6.3 Requirements and Interactions

7.6.3.1 Mechanical and Physical Integrity of the UHPLC System

Conventional HPLC and UHPLC systems are usually made of stainless steel and are designed specifically for applications of reversed-phase (RP) chromatography. The materials are inert to a wide variety of aqueous and organic solvents. Common RP eluents such as aqueous solutions of 0.1% trifluoroacetic acid, 0.1% formic acid or triethylamine, phosphate, or ammonium acetate buffer in concentrations, for example, 100 mmol l^{-1}, and organic solvents such as acetonitrile or methanol are not critical in terms of chemical stability of stainless-steel UHPLC systems. For applications of normal-phase chromatography, however, rather nonpolar eluents such as hexane are used. Therefore, the UHPLC system needs to be modified; otherwise, the lifetime of individual components is significantly reduced and the reliable operation of the UHPLC system is impaired.

In applications that make use of aqueous solutions with extreme pH values and high salt and chloride ion concentrations, corrosion will emerge in stainless-steel UHPLC systems. Thus, the mechanical and physical integrity of the UHPLC system will significantly be affected. Corrosive eluents are primarily used in the chromatography of biomolecules, for example, the separation of monoclonal antibodies and oligonucleotides. A variety of common aqueous mobile phases for the analyses of biomolecules is given in Table 7.9.

Besides, for reconditioning of ion chromatography columns 1 mol l^{-1} NaCl solutions and for cleaning of monolithic or polymer-based anion-exchange columns, aqueous solutions consisting of 1 mmol l^{-1} HCl with 50 mmol l^{-1} NaCl ($pH \geq 3$) or 1 mol l^{-1} NaOH with 1.25 mol l^{-1} NaCl ($pH = 14$) or 375 mmol l^{-1}

Table 7.9 Selection of aqueous buffer systems used in biochromatography.

Method	Mobile phase
Reversed–phase chromatography	Water/acetonitrile or water/methanol with 0.1% TFA or 0.1% FA
Ion-exchange chromatography	5–150 mmol l^{-1} phosphate- or MES-buffer, pH value 4–12, with addition of 20–500 mmol l^{-1} or 1 mol l^{-1} NaCl or 20 mmol l^{-1} Tris–HCl buffer + 1 mol l^{-1} NaCl
Hydrophobic interaction chromatography	0.1 mol l^{-1} NaH_2PO_4, 2 mol l^{-1} $(NH_4)_2SO_4$, 7% 2-propanol, pH 7.0
Size-exclusion chromatography	50 mmol l^{-1} KH_2PO_4, 30 mmol l^{-1} NaCl, pH 6.6

MES, 2-(*N*-morpholino)ethanesulfonic acid.

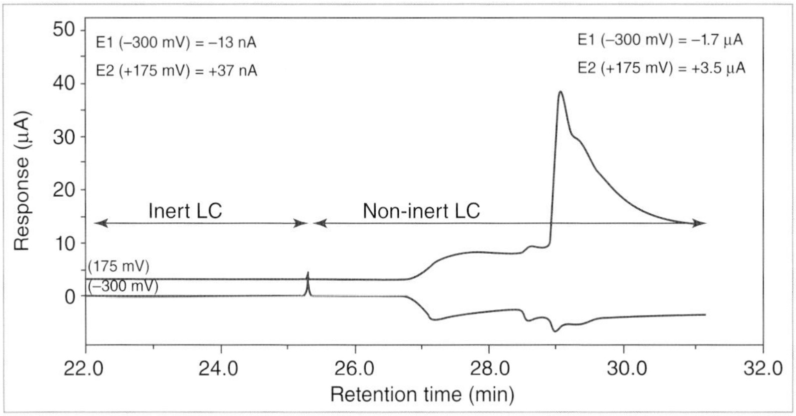

Figure 7.13 Influence of a noninert LC system on the baseline and noise of an electrochemical detector. Reproduced with permission of Thermo Fisher Scientific.

$NaClO_4$ solutions are frequently used. The purchase of a biocompatible UHPLC system is advisable, in case similar eluents are employed.

7.6.3.2 Requirement of the Detection Method

In addition to the aspects of the general corrosion resistance of a UHPLC system, the materials need to be compatible with the detection method. This will now be explained in more detail using the example of electrochemical detection. As described in Section 7.6.2.1, the passive layer of stainless steel contains different metal oxides, which can slowly dissolve in the mobile phase due to slow corrosion processes. The dissolved metal oxides and the underlying unprotected metal, which can also dissolve in the mobile phase, are electrochemically active and can cause an increased noise and a baseline drift in the electrochemical detector. In Figure 7.13, the influence on the baseline and background noise of an electrochemical detector is demonstrated. At the beginning of the run, an inert LC system is fluidically connected to the electrochemical detector. By switching a valve at about 25 min, a noninert LC system is switched into the flow path and the inert LC system is disconnected. After a short time, baseline fluctuations and increased noise are being observed.

The dissolved metal oxides shorten the lifetime of the electrochemical detector cells, because the metal oxides deposit onto the electrodes. Because each stainless-steel component in a stainless-steel-based UHPLC system is a source of corrosion, these parts should be replaced by metal-free components.

In case a stainless-steel-based UHPLC system is used for electrochemical detection, the instrument including the tubing should be repassivated in regular intervals. Nevertheless, for electrochemical detection, a biocompatible UHPLC system is recommended.

7.6.3.3 Interaction of Analyte and UHPLC System

Another important aspect beyond the chemical resistance of the materials in a UHPLC system to the mobile phase is the analyte compatibility or biocompatibility.

7.6.3.3.1 Adsorption of Proteins on Metal A possible interaction of proteins is the adsorption to the interior liquid-wetted surfaces of the UHPLC system. The phenomenon of protein adsorption is of high interest especially in the field of biology, biomedical engineering, biotechnology, and the food industry [70].

The so-called structurally hard proteins such as α-chymotrypsin, which have a high internal structural stability, adsorb mainly on hydrophobic surfaces. These proteins adsorb only with low affinity on hydrophilic surfaces unless attractive electrostatic interactions are present. Structurally soft proteins, such as BSA and immunoglobulins, generally tend to adsorb on all types of surfaces – even in the presence of repulsive electrostatic forces between the surface and the protein.

The adsorption of proteins on metal surfaces is dependent on charges on the metal as well as the protein. Since the hydroxyl group can get protonated or deprotonated, metal surfaces with an oxide layer are amphoteric in nature. They are negatively charged at high pH values or positively charged at low pH values [70]. Thus, the adsorption property of BSA on stainless steel depends on the pH value of the solution and the temperature. BSA adsorbs preferably at a pH value of 5, which is close to its isoelectrical point (pH 4.7). However, the ionic strength of the solution has also a significant influence on the adsorption. With increasing ionic strength, BSA adsorbs to a greater extent on the surface [71].

Proteins also adsorb on titanium. In a study with calf serum, it was demonstrated that serum proteins have adsorptive properties in aqueous solutions toward titanium. However, in phosphate buffer, these effects are significantly reduced [60].

Serum proteins have a low affinity for binding on "bio-inert" materials such as aluminum oxide and zirconium oxide ceramics. In experiments carried out in phosphate/citrate buffer, actually only a small proportion of protein coats the surfaces of these materials. The investigated samples of both materials adsorbed approximately 0.2 mg m^{-2} plasma protein (in PBC buffer (phosphate-buffered citrate solution), pH 7.5 at room temperature). An adsorption of human serum albumin (HSA) could not be detected [72]. In contrast to these values, BSA adsorbs with an amount of 2.1 mg m^{-2} on stainless steel (in 0.06 mol l^{-1} cacodylate buffer, pH 6.8 at 4 °C) [70].

Since the protein adsorption on steel and other surfaces can be irreversible, harsh conditions such as rinsing with NaOH solution are required for desorption of proteins [70].

At this point, it should be noted that the cited studies on protein adsorption are from the field of surface and biomaterial science. The tests were carried out under conditions that do not correspond to those in liquid chromatography. Hence, these results cannot directly be applied to chromatography, but they demonstrate that the materials of various system components, for example, the sample needle, the sample loop, and tubing capillaries need to be optimized in respect to applicative parameters, for example, pH value and ionic strength of the mobile phase or the sample solution, in order to minimize and avoid interactions between the material and the analyte.

7.6.3.3.2 Adsorption of Phosphorylated Analytes Phosphorylated organic analytes such as oligonucleotides, phosphopeptides, phospholipids, and phosphorylated sugars adsorb under acidic conditions (0.1% acetic acid) to electrospray steel capillaries and injectors with Nitronic 60 valve and a PEEK sample loop [57]. Nitronic 60 is an austenitic chromium steel with high nitrogen content and a high proportion of silicon and manganese. It has a higher resistance to chloride-induced pitting corrosion than stainless steel type 316 [73]. The reason for the observed adsorption effects are corrosion processes, which are initiated and accelerated by the electrospray process. The phosphorylated compounds have a higher affinity to corroded surfaces with an increasing number of well-accessible phosphate groups. This observation is consistent with the binding mechanism of immobilized metal affinity chromatography (IMAC) with iron(III) ions [57]. In addition to iron(III) ions, metals such as nickel, copper, cobalt, and zinc are also used for the enrichment of low-abundance phosphopeptides in IMAC. Hence, similar interactions of phosphorlyated compounds with corroded metal surfaces of metal-based, iron-free UHPLC systems are likely to occur.

Stainless-steel components that are part of the sample flow path can cause peak tailing of phosphorylated molecules. In a flow injection analysis of adenine, adenosine, and adenosine-5′-monophosphate (AMP) in a methanol/water mixture (50 : 50, v/v), it was shown that the interaction of AMP with the stainless-steel surface of the tubing is induced by the phosphate group of the analyte. In contrast, when using PEEK tubing, such adsorption effects were not observed. The peak tailing could be suppressed by passivation of the stainless-steel tubing in 400 $mmol\,l^{-1}$ phosphoric acid in methanol (50 : 50, v/v) for 2 h at a flow of 0.2 $ml\,min^{-1}$ or by using a mobile phase with the same composition for the flow injection analysis. Passivation of the stainless-steel tubing with a mixture of 400 $mmol\,l^{-1}$ acetic acid or nitric acid in methanol (50 : 50, v/v) could not improve the peak shape of AMP [74].

7.6.3.3.3 Interaction with Metal Ions Corrosion processes in a UHPLC system and contaminants from the solvent promote the formation of stable metal–analyte complexes. Metal ions can be released from the system into the mobile phase or are introduced into the system as impurities of buffer salts and organic solvents. Phosphorylated peptides do bind, for example, to Fe(III) or Al(III) ions and form metal complexes. As result, the MS signal of metal-free protonated phosphopeptides in liquid chromatography–electrospray ionization–mass spectrometry (LC-ESI-MS) spectra is being reduced. Highly phosphorylated peptides can interact with Fe(III) ions bound on the stationary phase. As a result of this interaction, broader peaks are observed. At low peptide concentrations, the interaction can even lead to a complete adsorption of the peptide, so that no detect or signal of the analyte is being observed. However, the formation of phosphorylated peptide/iron complexes initially depends on the primary sequence of the peptide. Apparently, the vicinity of basic or acidic amino acids near the phosphate group seems to be crucial. The adsorption of phosphorylated peptides can be prevented, for example, by the addition of 25 $mmol\,l^{-1}$ ethylenediaminetetraacetic acid (EDTA) to the sample [75].

Besides phosphorylated bioanalytes, proteins, for example, monoclonal antibodies, can also interact with iron(III) ions, which are enriched on stationary phases. The result is a significantly reduced peak height and loss of resolution [76].

7.6.3.3.4 Interactions with Fused-Silica and PAEK/PEEK In addition to the loss of phosphorylated molecules in chromium-steel-based LC-MS systems, adsorption of these chemical species can also occur in metal-free LC-MS systems. In an experimental study with a mobile phase of 0.1% (v/v) acetic acid in water/methanol (50 : 50), interactions with *fused-silica* capillaries in the sample flow path and in the injector equipped with a PAEK/PEEK injection valve and a PEEK sample loop were observed. Interestingly, *fused-silica* capillaries of different manufacturers showed different adsorption properties [77]. Upon injection of the analytes under neutral conditions and after silanization of *fused-silica* capillaries with dimethyldichlorosilane, adsorption was significantly reduced. However, even in the system with metal-free components, aluminum(III)– and iron(III)–analyte adducts were detected. Most likely, the cations came from sources, for example, from the eluent or the sample vial, other than the LC-MS system.

Phosphopeptides form strong hydrogen bonds with active silanol groups of *fused-silica* capillaries and silica-based stationary phases at low pH values. As a result, these molecules bind to the capillary surfaces or adsorb onto the column. By adding phosphoric acid to the sample, the adsorption of phosphopeptides on the separation column can be significantly reduced, but apparently not if the analyte is present in very low concentrations [78].

In a study on the separation and determination of arsenic, selenium, and chromium species by micro liquid chromatography–inductively coupled plasma–mass spectrometry (LC-ICP-MS) on an anion-exchange column, secondary interactions have been observed. These interactions were attributed to the partial hydrolysis of *fused-silica* capillaries [79]. As a hypothesis, the authors assume that due to a pH value of $\geq$8.5 (ammonium nitrate buffer), siloxane entities are being hydrolyzed and degradation products with free silanol groups are being formed. These molecules get accumulated on the anion-exchange column and interact with analytes, which contain hydroxyl groups, and thereof strong peak tailing is being caused. In further experiments with ammonium buffer with a pH value of $\leq$8.5 or with phosphate buffer, these interactions were not observed.

7.6.3.3.5 Chemical Modification of the Analyte Besides adsorption effects, the chemical modification of the analyte is another possible type of interaction in liquid chromatography. Analytes can get chemically modified while in contact with materials in an LC system as demonstrated in the following example.

When secondary amines are being analyzed in a basic NH_4OH buffer with acetonitrile, nitroso-analyte derivatives are formed due to the corrosion of column frits. Dissolved NH_3 reacts in the presence of iron ions and oxygen to form NO and N_2O_3. Hereupon, these molecules oxidize the secondary amine to the

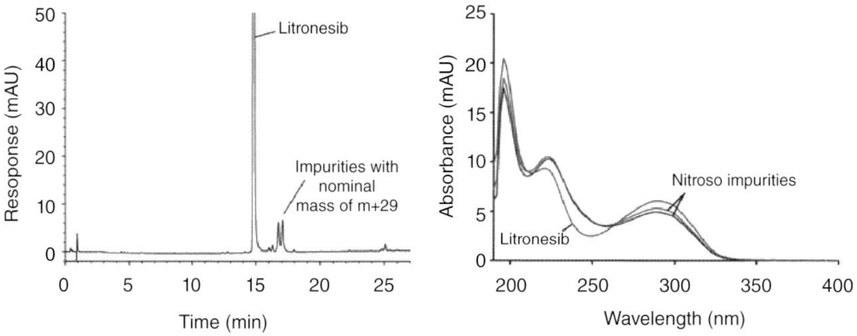

Figure 7.14 Chromatography of litronesib and detection of nitroso derivates. Chromatographic conditions: mobile phase: (a) 0.05% ammonium hydroxide in water, (b) acetonitrile; stationary phase: waters X-bridge C18, 4.6 mm × 75 mm, 2.5 μm; flow rate: 1 ml min^{-1}; gradient: 0–1 min 75% A, 1–23 min 75–20% A, 23–25 min 20% A, 25–28 min 75% A; injection volume: 10 μl; temperature: 30 °C; and UV-detection: 290 nm. (a) LC-UV chromatogram (290 nm) and (b) UV/VIS spectra of litronesib und nitroso derivates. Myers 2013 [80], Reproduced with permission of Elsevier.

nitroso derivative (Figure 7.14). If methanol is used as organic modifier in the mobile phase, this reaction does not occur. The addition of 50 μmol l^{-1} EDTA to the buffer system reduces the formation of nitroso derivatives significantly [80].

In this case, the analyte is chemically modified by a reactive species, which is formed by a chemical reaction of substances of the mobile phase with steel components of the instrument. Besides that, organic molecules can get chemically modified directly in case metal is present. Chinones, for example, are reduced to their corresponding catechol amines due to the reaction with steel components. The chemical reaction can occur, for example, at the outlet frit of the column, so that coelution of the chinone and the corresponding catecholamine is observed. The amount of catecholamine being formed is dependent on the flow rate and the injection volume [81].

7.6.4 Passivation Strategies and Methods

Metal alloys and their corrosion resistance and biocompatibility are the main focus in any discussion on inertness in liquid chromatography. However, the instruments are made exclusively not only of metal alloys but also of various other materials, such as polymers, composite materials, and *fused-silica*. All components of a UHPLC system influence the mechanical and physical integrity of a UHPLC system to some degree. They even might affect the detection or interact with the analytes as the discussions in the previous chapters have demonstrated. The studies indicate that interactions either between the system and the analyte directly with the surfaces of the oxide layer of alloys, *fused-silica* capillaries, or PAEK/PEEK injection valves, or indirectly with dissolved species coming from system components, such as metal ions or degradation products of *fused-silica* capillaries are generally possible. Indirect interactions usually occur mostly downstream of the source in solution (mobile phase) or on the separation column. The result is usually a loss of analyte due to adsorption or chemical reaction, or a change of the retention mechanism of the stationary phase. The chemical modification of analytes either takes place directly on the surfaces of the system or is indirectly mediated through reactive chemical species formed by catalytic effects in the mobile phase. The effects on chromatography are finally band broadening, the occurrence of nonidentifiable or interfering peaks, a low sensitivity, and poor reproducibility.

Appropriate strategies and measures ensuring general and analyte-specific inertness are diverse. In order to renew the passive layer of stainless steel and to restore the general corrosion resistance of a UHPLC system, nitric acid or citric acid can be used for repassivation. But it should be noted that stainless-steel-based UHPLC instruments are already shipped in a passivated state. Therefore, there is no general need for repassivation. In cases in which highly corrosive media such as chloride-containing buffers are used, it is better to use a biocompatible UHPLC system.

Essentially, corrosion should be avoided *a priori*. This implies that the UHPLC system is operated with noncorrosive mobile phases. The instrument should be cleaned at regular intervals to keep the interior wetted surfaces free from residues of the mobile phase (buffer salts) or the sample. For this purpose, cleaning solutions such as mixtures of acetonitrile, methanol, or 2-propanol with water, pure 2-propanol, or depending on the resistance of the materials, sodium hydroxide solutions in concentrations of $\leq 1\,\mathrm{mol\,l^{-1}}$ (4% (w/w)) and $\leq 30\%$ phosphoric acid can be used [6, 12, 14]. Please note that in any case, the compatibility of the different components in a UHPLC system to these cleaning solutions needs to be ensured. The column and the detector and all other system components that are sensitive to these cleaning solutions should be removed, and the instructions given in the operation manuals need to be followed.

Another possibility is to deploy corrosion inhibitors such as hydrazine and nitrates as additives in the mobile phase (Sections 7.5.1.2.1), provided that all components in the system are compatible with these chemical species and the chromatography (resolution and detection) is not affected.

However, corrosion processes, such as erosion corrosion and pitting corrosion, are complex processes, which are dependent on many factors and thus cannot be completely controlled. Because corrosion is a gradual process, an influence of, for example, dissolved metal ions on chromatography can already occur, even though more obvious corrosion effects, such as rust, are not being observed.

In cases of direct interactions of analytes with the oxides of an intact passive layer or with silanol groups of *fused-silica* capillaries, the strategies for maintaining general inertness of the UHPLC system might not be successful. According to the studies presented in Section 7.6.3.3, eventually phosphate-based passivation might help to increase the compatibility of stainless steel toward sensitive analytes such as phosphopeptides and proteins. Other strategies are, for example, adjusting the pH value and the ionic strength of the sample solution or the mobile phase, the addition of phosphoric acid to the sample solution,

Table 7.10 Passivation methods with citric acid, phosphoric acid, and nitric acid for stainless-steel-based tubing or UHPLC systems.

Passivation agent	Passivation procedure	References
Citric acid (30% v/v)	For self-cut tubing • Soaking in citric acid over night • Soaking in water for 30–60 min • Rinsing with water and with methanol	[82]
400 mmol l^{-1} H_3PO_4/MeOH (1 : 1)	At a flow rate of 0.2 ml min^{-1} for 2 h	[74]
5 mmol l^{-1} NaH_2PO_4 in water	• 20/80 (v/v) ACN/water with 1% formic acid at 0.4 ml min^{-1}, ≥3 h, heated with precolumn heater at 80 °C • Phosphate solution at 0.4 ml min^{-1} at 40 °C, 12–60 h	[83]
30% nitric acid	• Rinsing with water to remove organic solvents from the system • 30% nitric acid, 1 ml min^{-1}, ≥250 ml • Rinsing with water at 1 ml min^{-1}, 1 l, over night • Activation of the rear seal wash system after passivation	[84]
50% nitric acid	Not specified	[38]

Before passivation, instructions of the instrument manufacturer regarding chemical compatibility of the instrument toward the passivating agents need to be followed.

or using phosphate buffer as the mobile phase. The addition of EDTA might help to prevent that the analyte binds to dissolved or column-bound metal ions. But note that EDTA in higher concentrations can lead to corrosion of stainless steel [11, 12].

Even though strategies such as the aforementioned are being followed, the passivation or repassivation of the UHPLC system might become necessary at some point. Therefore, instrument manufacturers are giving recommendations for passivation, which should be obeyed in all cases. Some examples for passivation are given in Table 7.10.

Passivation methods are regulated in the industry according to various standardized protocols. An important industry standard is the ASTM-967 standard, which specifies different passivation methods for stainless steel. In this standard, the interested reader can receive further information on the passivation of stainless steel [85].

Abbreviations

BSA	bovine serum albumin
cp-Ti	commercially pure titanium
DLC	diamond-like carbon
ECTFE	ethylene chlorotrifluoroethylene
EDTA	ethylenediaminetetraacetic acid

ETFE	ethylene tetrafluoroethylene
ESI	electrospray ionization
FA	formic acid
FEP	fluorinated ethylene propylene
FFKM	perfluoroelastomer
GPV	gradient proportioning valve
HSA	human serum albumin
HPLC	high-performance liquid chromatography
ICP-MS	inductively coupled plasma – mass spectrometry
ID	inner diameter
LC	liquid chromatography
MES	2-(*N*-morpholino)ethanesulfonic acid
MS	mass spectrometry
NP	normal phase
PCTFE	polychlorotrifluoroethene
PAEK	polyaryl ether ketone
PEEK	polyetheretherketon
PBC	phosphate-buffered citrate solution
PBS	phosphate-buffered saline
PTFE	polytetrafluorethylen
RP	reversed phase
PPS	poly(*p*-phenylene sulfide)
SiC	silicon carbide
TFA	trifluoroacetic acid
Tris	tris(hydroxymethyl)aminomethane
UHMW-PE	ultrahigh-molecular-weight polyethylene
UHPLC	ultrahigh-performance liquid chromatography

References

1 Collins, K.E., Collins, C.H., and Bertran, C.A. (2000) Stainless steel surfaces in LC systems, part I – Corrosion and erosion. *LC GC North Am.*, **18**, 600–608.
2 Davis, J.R. (1994) *Stainless Steels (ASM Specialty Handbook)*, ASM International.
3 Covert, R.A. and Tuthill, A.H. (2000) Stainless steels: an introduction to their metallurgy and corrosion resistance. *Dairy Food Environ. Sanit.*, **20** (7), 506–517.
4 Nickel Institute Publication No 2828 – Corrosion Resistance of the Austenitic Chromium-Nickel Stainless Steels in Chemical Environments, http://www.nickelinstitute.org/TechnicalLiterature/INCO%20Series/2828&uscore;CorrosionResistanceoftheAusteniticChromium&uscore;NickelStainlessSteelsinChemicalEnvironments.aspx (Accessed 23 September 2015).
5 Schmidt, T. (2001) Identifizierung und Untersuchung pharmazeutischer Gläser durch Laser-Ablation-ICP-MS – Dissertation, Berlin.

6 Waters Corporation (2008) Controlling Contamination in UltraPerformance LC/MS and HPLC/MS Systems, 3 März 2008, http://www.waters.com/webassets/cms/support/docs/715001307d&uscore;cntrl&uscore;cntm.pdf (Accessed 23 September 2015).
7 Dolan, J.W. (2006) Check valve problems. *LC GC North Am.*, **24** (**2**), 132–138, http://www.chromatographyonline.com/lcgc/article/articleDetail.jsp?id=302242&pageID=1&sk=&date= (Accessed 23 September 2015).
8 Dolan, J.W. (2008) Check valves and acetonitrile. *LC GC North Am.*, **26** (**6**), 532–538, http://www.chromatographyonline.com/lcgc/Articles/Check-Valves-and-Acetonitrile/ArticleStandard/Article/detail/524647 (Accessed 23 September 2015).
9 Ledtje, M.L. and Long, A.D. (1989) Check valve. US Patent US4,862,907, Sep. 5, 1989.
10 Waters Corporation (2010) Acquity UPLC H-Class System – Instrument Specifications, http://www.waters.com/webassets/cms/library/docs/720003294en.pdf (Accessed 23 September 2015).
11 Agilent Technologies (2010) Agilent 1260 Infinity Quaternary Pump VL – User Manual, 06/10, Agilent Technologies Inc.
12 Waters Corporation (2010) *Acquity UPLC H-Class System Guide*, Revision A, Waters Corporation, Milford, CT.
13 Thermo Fisher Scientific (2013) Thermo Scientific Dionex UltiMate 3000 Series SD, RS, BM, and BX Pumps Operating Instructions, Revision 1.7
14 Haeuselmann Metall Gmbh (2005) Tabellen – Chemische Bestaendigkeit, http://www.haeuselmann.ch/webautor‐data/9/Bestaendigkeit8&uscore;05Inernet.pdf. (Accessed 23 September 2015).
15 Mowery, R. (1985) The corrosion of 316 stainless steel in process liquid chromatography with acetonitrile and methanol carriers. *J. Chromatogr. Sci.*, **23**, 22–29.
16 Peters, M. (2002) *Titan und Titanlegierungen*, Wiley-VCH Verlag GmbH, Weinheim.
17 ThyssenKrupp Materials Schweiz (2013) Titan Grade 1 / Titane Grade 1, 28 Januar 2013, http://www.thyssenkrupp.ch/documents/Titan&uscore;Grade&uscore;1.pdf (Accessed 23 September 2015).
18 ThyssenKrupp Materials Schweiz (2013) Titan Grade 4 / Titane Grade 4, http://www.thyssenkrupp.ch/documents/Titan&uscore;Grade&uscore;4&uscore;A.pdf (Accessed 23 September 2015).
19 ThyssenKrupp Materials Schweiz (2013) Titan Grade 5 / Titane Grade 5, http://www.thyssenkrupp.ch/documents/Titan_Grade_5.pdf (Accessed 23 September 2015).
20 Hoar, T.P. and Maers, D. (1966) Corrosion-resistant alloys in chloride solutions: materials for surgical implants. *Proc. R. Soc. A*, **294** (**1439**), 486–510.
21 Titanium Metals Corporation (1997) Corrosion Resistance of Titanium, http://www.parrinst.com/wp‐content/uploads/downloads/2011/07/Parr&uscore;Titanium‐Corrosion‐Info.pdf (Accessed 23 September 2015).

22 Haertl, H. (2010) HPLC pumping apparatus with silicon carbide piston and/or working chamber. US Patent US 2010/0089134 A1, Apr. 5, 2010.
23 Agilent Technologies (2008) Agilent 1200 Series Quaternary Pump User Manual, 11/08, Agilent Technologies, Inc.
24 Perkin Elmer Inc. High Pressure Piston Seal Replacement Parts for Series 200 Pump, http://www.perkinelmer.com/catalog/family/id/high%20pressure%20piston%20seal%20replacement%20parts%20for%20series%20200%20pump (Accessed 23 September 2015).
25 Lemelin, M.E., Lin, T.A., Moeller, W., Koziol, S., and Usowicz, J.E. (2013) Static and dynamic seals. US Patent US 2013/0049302 A, Feb. 28, 2013.
26 Waters Corporation Waters: Vials Selector, www.waters.com/app/selector/en/vials.html (Accessed 23 September 2015).
27 Agilent Technologies Interactive Vial Selection Tool, Agilent Technologies, Inc., http://www.agilent.com/en‐us/promotions/interactive‐vial‐tool (Accessed 23 September 2015).
28 Thermo Fisher Scientific Vials and Closures Catalog 2014-2015, Thermo Scientific, http://www.thermoscientific.com/content/dam/tfs/ATG/CMD/CMD%20Documents/Catalogs%20&%20Brochures/Chromatography%20Columns%20and%20Supplies/BR20834‐sample‐handling‐lr.pdf (Accessed 23 September 2015).
29 Nichols, J.A. and Baron, B.L. (2002) Long lifetime fluid switching valve. US Patent US 6,453,946 B2, Sep. 24, 2002.
30 Idex Health&Science LLC (2013) Fluidic Products & Information for Laboratory Applications Catalog 2014.
31 Valco Instruments Co. Inc. (2014) Properties of Rotor Materials, Valco Instruments Co. Inc., http://www.vici.com/ref/mat&uscore;rotr.php (Accessed 23 September 2015).
32 Berndt, M. (2010) Shear valve with silicon carbide member. US Patent US 2010/0101989, Apr. 29, 2010.
33 Berndt, M. (2011) Shear valve with DLC comprising multi-layer coated member. US Patent US 2010/0281959 A1, Nov. 11, 2011.
34 Yasunaga, K. (2009) Flow channel switching valve. Japan Patent WO2009101695 A, Aug. 20, 2009.
35 Tower, C.R. (2011) Rotary shear valve assembly with hard-on-hard seal surfaces. US Patent US 2011/0006237 A1, Jan. 13, 2011.
36 Hochgraeber, H., Wiechers, J., and Satzinger, A. (2013) High-pressure control valve for high-performance liquid chromatography. US Patent US 2013/0284959 A1, Oct. 31, 2013.
37 Hochgraeber, H. and Ruegenberg, G. (2012) Autosampler for high-performance liquid chromatography. Deutschland Patent US 8,196,456 B2, Juni 12, 2012.
38 Collins, K.E., Collins, C.H., and Bertran, C.A. (2000) Stainless steel surfaces in LC systems, part II – Passivation and practical recommendations. *LC GC North Am.*, **18** (7), 688–692.
39 Finetubes (2012) Nickel-Legierung – Legierung MP35N, UNS R30035, Werkstoff-Nr. 2.4999, http://www.finetubes.de/uploads/attachments/g146&uscore;Legierung&uscore;MP35N.pdf (Accessed 23 September 2015).

40 Davis, J.R. (2001) *Nickel, Cobalt, and Their Alloys (ASM Specialty Handbook)*, ASM International.

41 Park, J. and Lakes, R.S. (2007) *Biomaterials: An Introduction*, Springer.

42 Valco Instruments Co. Inc. (2014) Pre-Cut Natural PEEK Tubing, Valco Instruments Co. Inc., http://www.vici.com/tube/peek&uscore;tubing.php (Accessed 23 September 2015).

43 Agilent Technologies (2013) Agilent LC Capillary Supplies Selection Guide – Confidently Build High Performance Connections From Raw Sample to Analytical Result, Agilent Technologies, Inc., https://www.chem.agilent.com/Library/selectionguide/Public/5991‐0121EN.pdf (Accessed 23 September 2015).

44 Valco Instruments Co. Inc. (2014) Nickel-Clad PEEK Tubing, http://www.vici.com/tube/ni‐clad‐peek.php (Zugriff am 23 September 2015).

45 IDEX Health & Science LLC Permanent VHP Fittings, IDEX Health & Science LLC, http://www.idex‐hs.com/products/4330/Permanent‐VHP‐Fittings.aspx?ProductTypeID=187&ProductFamilyID=3 (Accessed 13 June 2014).

46 Thermo Fisher Scientific Viper Fingertight Fitting, Thermo Fisher Scientific Inc., http://www.dionex.com/en‐us/products/accessories/reagents‐accessories/viper‐fingertight/lp‐81335.html (Accessed 23 September 2015).

47 Thermo Fisher Scientific nanoViper Fingertight Fitting, Thermo Fisher Scientific Inc., http://www.dionex.com/en‐us/products/accessories/reagents‐accessories/nano‐viper‐fingertight/lp‐81337.html (Accessed 23 September 2015).

48 Shih, C.-C., Shih, C.-M., Su, Y.-Y., Su, L.H.J., Chang, M.-S., and Lin, S.-J. (2004) Effect of surface oxide properties on corrosion resistance of 316L stainless steel for biomedical applications. *Corros. Sci.*, **46** (**2**), 427–441.

49 Andreeva, V.V. (1964) Behavior and nature of thin oxide films on some metals in gaseous media and in electrolyte solutions. *Corrosion*, **20** (**2**), 35–46.

50 Pouilleaua, J., Devilliersa, D., Durand-Vidala, S., and Mahéa, E. (1997) Structure and composition of passive titanium oxide films. *Mater. Sci. Eng., B*, **47** (**3**), 235–243.

51 Thierry, B., Tabrazian, T.C., Savadogo, O., and Yahia, L. (2000) Effect of surface treatment and sterilization processes on the corrosion behaviour of NiTi shape memory alloy. *J. Biomed. Mater. Res.*, **51** (**4**), 685–693.

52 Sundgren, J.-E., Bodö, P., and Lundström, I. (1986) Auger electron spectroscopic studies of the interface between human tissue and implants of titanium and stainless steel. *J. Colloid Interface Sci.*, **110** (**1**), 9–20.

53 Milošev, I. and Strehblow, H.-H. (2000) The behavior of stainless steels in physiological solution containing complexing agent studied by X-ray photoelectron spectroscopy. *J. Biomed. Mater. Res.*, **52** (**2**), 404–412.

54 Clark, G. and Williams, D. (1982) The effects of proteins on metallic corrosion. *J. Biomed. Mater. Res.*, **16** (**2**), 125–134.

55 Omanovic, S. and Roscoe, S.G. (1999) Electrochemical studies of the adsorption behavior of bovine serum albumin on stainless steel. *Langmuir*, **15** (**23**), 8315–8321.

56 Williams, R., Brown, S., and Merritt, K. (1988) Electrochemical studies on the influence of proteins on the corrosion of implant alloys. *Biomaterials*, **9** (**2**), 181–186.

57 Tuytten, R., Lemi'ere, F., Witters, E., Van Dongena, W., Slegers, H., Newton, R.V.O.H., and Esmansa, E. (2006) Stainless steel electrospray probe: a dead end for phosphorylated organic compounds? *J. Chromatogr. A*, **1104**, 209–221.

58 Frauchiger, L., Taborelli, M., Aronsson, B.-O., and Descouts, P. (1999) Ion adsorption on titanium surfaces exposed to a physiological solution. *Appl. Surf. Sci.*, **143** (**1-4**), 67–77.

59 Ouerd, A., Alemany-Dumont, C., Normand, B., and Szunerits, S. (2008) Reactivity of CoCrMo alloy in physiological medium: electrochemical characterization of the metal/protein interface. *Electrochim. Acta*, **53** (**13**), 4461–4469.

60 Ouerd, A., Alemany-Dumont, C., Berthomé, G., Normand, B., and Szunerits, S. (2007) Reactivity of titanium in physiological medium – I. Electrochemical characterization of the metal/protein interface. *J. Electrochem. Soc.*, **154** (**10**), C593–C601.

61 Pound, B. (2014) Passive films on metallic biomaterials under simulated physiological conditions. *J. Biomed. Mater. Res. Part A*, **102A**, 1595–1604.

62 Ouerda, A., Alemany-Dumont, C., Normand, B., and Szunerits, S. (2008) Reactivity of CoCrMo alloy in physiological medium: Electrochemical characterization of the metal/protein interface. *Electrochim. Acta*, **53** (**13**), 4461–4469.

63 Olefjord, I. (1980) The passive state of stainless steels. *Mater. Sci. Eng.*, **42**, 161–171.

64 Wallinder, D., Pan, J., Leygrf, C., and Delblanc-Bauer, A. (1998) EIS and XPS study of surface modification of 316LVM stainless steel after passivation. *Corros. Sci.*, **41** (**2**), 275–289.

65 Noh, J., Laycock, N., Gao, W., and Wells, D. (2000) Effects of nitric acid passivation on the pitting resistance of 316 stainless steel. *Corros. Sci.*, **42** (**12**), 2069–2084.

66 O'Laoire, C., Timmins, B., Kremer, L., Holmes, J., and Morris, M. (2006) Analysis of the acid passivation of stainless steel. *Anal. Lett.*, **39** (**11**), 2255–2271.

67 Pound, B.G. and Becker, C.H. (1991) Composition of surface films on nickel base superalloys. *J. Electrochem. Soc.*, **138** (**3**), 696–700.

68 Ornberg, A., Pan, J., Herstedt, M., and Leygraf, C. (2007) Corrosion resistance, chemical passivation, and metal release of 35N LT and MP35N for biomedical material application. *J. Electrochem. Soc.*, **154** (**9**), C546–C551.

69 Pan, J., Thierry, D., and Leygraf, C. (1994) Electrochemical and XPS studies of titanium for biomaterial applications with respect to the effect of hydrogen peroxide. *J. Biomed. Mater. Res.*, **28** (**1**), 113–122.

70 Nakanishi, K., Sakiyama, T., and Imamura, K. (2001) On the adsorption of proteins on solid surfaces, a common but very complicated phenomenon. *J. Biosci. Bioeng.*, **91** (**3**), 233–244.

71 Fukuzaki, S., Urano, H., and Nagata, K. (1995) Adsorption of protein onto stainless-steel surfaces. *J. Ferment. Bioeng.*, **80** (**1**), 6–11.

72 Rosengren, Å., Pavlovic, E., Oscarsson, S., Krajewski, A., Ravaglioli, A., and Piancastelli, A. (2002) Plasma protein adsorption pattern on characterized ceramic biomaterials. *Biomaterials*, **23** (**4**), 1237–1247.

73 High Performance Alloys, Inc. (2007) Mfg & Dist of Inconel, Monel, Hestalloy and Nitronic Grades, High Performance Alloys, Inc., http://www.hpalloy.com/alloys/descriptions/NITRONIC60.html (Accessed 23 September 2015).

74 Wakamatsu, A., Morimoto, K., Shimizu, M., and Kudoh, S. (2005) A severe peak tailing of phosphate compounds caused by interaction with stainless steel used for liquid chromatography and electrospray mass spectrometry. *J. Sep. Sci., Spec. Issue: Mass Spectrom. Appl. Sep. Sci.*, **28** (**14**), 1823–1830.

75 Liu, S., Zhang, C., Campbell, J., Zhang, H., Yeung, K., Han, V., and Lajoie, G. (2005) Formation of phosphopeptide-metal ion complexes in liquid chromatography/electrospray mass spectrometry and their influence on phosphopeptide detection. *Rapid Commun. Mass Spectrom.*, **19**, 2747–2756.

76 Rao, S. and Pohl, C. (2011) Reversible interference of Fe3+ with monoclonal antibody analysis in cation exchange columns. *Anal. Biochem.*, **409** (**2**), 293–295.

77 De Vijlder, T., Boschmans, J., Witters, E., and Lemière, F. (2011) Study on the loss of nucleoside mono-, di- and triphosphates and phosphorylated peptides to a metal-free LC–MS hardware. *Int. J. Mass spectrom.*, **304** (**2–3**), 83–90.

78 Kim, J., Camp, D.G., and Smith, R.D. (2004) Improved detection of multi-phosphorylated peptides in the presence of phosphoric acid in liquid chromatography/mass spectrometry. *J. Mass Spectrom.*, **39** (**2**), 208–215.

79 Castillo, A., Roig-Navarro, A., and Pozo, O. (2007) Secondary interactions, an unexpected problem emerged between hydroxyl containing analytes and fused silica capillaries in anion-exchange micro-liquid chromatography. *J. Chromatogr. A*, **1172** (**2**), 179–185.

80 Myersa, D.P., Hetricka, E.M., Lianga, Z., Haddena, C.E., Bandya, S., Kempa, C.A., Harrisb, T.M., and Baertschia, S.W. (2013) On-column nitrosation of amines observed in liquid chromatography impurity separations employing ammonium hydroxide and acetonitrile as mobile phase. *J. Chromatogr. A*, **1319**, 57–64.

81 Xu, R., Huang, X.K.K., and Hawle, M. (1995) On-column reduction of catecholamine quinones in stainless steel columns during liquid chromatography. *Anal. Biochem.*, **231** (**1**), 72–81.

82 IDEX Health & Science LLC (2014) Stainless Steel Tubing, IDEX Health & Science LLC, http://webstore.idex‐hs.com/Products/specsheet.asp?vSpecSheet=450&vPart=U‐112 (Accessed 9 June 2014).

83 Agilent Technologies (2007) Is There any HPLC Passivation Procedure Available for Sensitive Applications Like e.g. Analysis of Some Proteins,

Pyrophosphates etc. or for Applications Using Ion Exchange Columns?, Agilent Technologies, http://www.chem.agilent.com/Library/Support/Documents/Passivation%20for%20ILC.pdf (Accessed 23 September 2015).

84 Day, R. (1988) Waters Lab Highlight – Passivating Stainless Steel in HPLC Systems, http://www.waters.com/waters/library.htm?locale=en_US&lid=1524770 (Accessed 23 September 2015).

85 ASTM International ASTM A967 / A967M – 13 (2013) *Standard Specification for Chemical Passivation Treatments for Stainless Steel Parts*, ASTM International, http://www.astm.org/Standards/A967.htm (Accessed 23 September 2015).

Part II

Experience Reports, Trends

8

What a Software has to Possess in Order to Use the Hardware Optimally

Arno Simon

The answer to the question of what chromatography software has to deliver for optimal hardware usage is easily answered:

It has to be able to control all functions!

Asked in a more comprehensive way, not limited to the connection between software and hardware, the significant question has to be:

What does chromatography software need to deliver for optimal hardware use by the operator?

Parallel to the rapid development of computer technology, the applications on computers and their possibilities have changed at an extraordinary rate. Today, fitting chromatography data systems (CDSs) exist for almost every application from simple single systems to database-driven global networks.

In the past, every measuring station was connected to a particular computer; nowadays, measuring systems by diverse manufacturers are connected. Instead of decided software for each measuring system, CDSs are needed, solving diverse tasks with a various number of connected equipment. These changes have simplified many processes in the laboratory. Today, many companies are driving global-driven installations of CDSs. In this way, workflows are standardized, and the users have to manage fewer different software products.

In this chapter, I would like to discuss some thoughts on the subject divided in three aspects:

- Functionality and handling
- Data exchange
- Big installations.

8.1 Functionality and Handling

The market offers various CDSs for different types of applications: from simple file-based systems with a manageable amount of functions to global all-rounder systems based on databases.

The HPLC Expert II: Find and Optimize the Benefits of your HPLC/UHPLC, First Edition.
Edited by Stavros Kromidas.

Over 15 years, as a consultant and trainer, I have implemented a large number of different software systems in laboratories.

While none of the systems could deliver all functions users desire, all laboratories are working successfully with their CDS. To me, this indicates that there is a range of essential functions, which cover the usual requirements, both for routine laboratories or in-process control.

Nevertheless, there is (always) room for improvement, and with every new CDS version, new functions due to changing requirements are delivered: for example, a missing feature, which implements the purity of a standard component for calibration, or a really helpful new function such as the evaluation of the sample pureness according to ICH,[1] both new in Waters Empower.

Like many others, I wish that manufacturers invest in certain product areas, and I am convinced that this will happen.

8.1.1 Integration

Among the newest functions in the area of peak integration, I was most impressed by ApexTrack Integration Algorithm in Empower and by Chromeleon's SmartPeaks Integration.

To recognize the beginning and end of a peak, the ApexTrack Integration Algorithm in Empower uses the gradient of the second derivative instead of the measurement signal gradient. Because in the second derivative, a possible drift is obsolete, this is an easy way to integrate even smallest peaks reliably on drifting baselines.

Furthermore, the ApexTrack Integration Algorithm allows an easy detection of shoulder peaks and split peaks and their separation. Combined with the possibility to let ApexTrack define the parameters, an almost complete automatic integration of challenging chromatograms is feasible.

On the other hand, Chromeleon's SmartPeaks function delivers an excellent solution for assisting the integration of peaks, which are not separated on the baseline. Just choose the SmartPeaks function and select the area for integration, various suggestions for integration appear. With a simple click on the relevant graphic, the favored integration can be chosen, and the method execution is completely done by Chromeleon.

Both systems offer a relatively simple integration, even for difficult chromatograms; however, there is no guidance on how to integrate a chromatogram to detect the real peak areas. Sometimes, it is just the integration that decides about good or bad, for instance, when evaluating the area percentage to determine the purity. Here, help for a well-founded selection of the right integration would be an important feature.

Especially, help for integration of peaks, not separated on the baseline, is missing. The trend to do chromatographic separation in less time by using shorter columns and quicker methods, sometimes, leads to worse resolution. If

---

1 ICH: International Conference on Harmonisation of Technical Requirements for Registration of Pharmaceuticals for Human Use.

this happens, for example, with stability samples with many new and unidentified peaks, the amount of not baseline-separated peaks increases.

When defining the integration method, one has to determine the separating method. But is the exponential skim, which often seems to be most convenient, providing the right result? Even here, CDS manufacturers are challenged to either implement their own functions to calculate deconvolution or to open their software products to third-party providers, to retrofit the needed functions.

8.1.2 Instrument Control

To operate an analytical system these days without software connection is almost unfeasible, and specific requirements often result in the creation of "software islands." As long as not every instrument can be controlled by any software, different software products have to be implemented to control all equipment in the laboratory.

The more instruments from different manufacturers can be controlled by one software, the fewer different software systems have to be deployed in the laboratory.

Shouldn't every manufacturer of CDSs have the aim to cover as wide a spectrum as possible of supported instruments?

Yes and no! Software supporting a wide range of instruments helps a company to sell a CDS and to establish it as standard. Furthermore, software providers want to sell their own equipment, and from their point of view, compatibility with other manufacturers' instruments becomes an issue, because now customers are not bound to a certain CDS manufacturer anymore, but can choose from a wider selection without having to compromise on comfort and safety of instrument control.

A first step in the direction of open systems to control chromatography instruments has been the introduction of the Instrument Control Frameworks (ICF) by the manufacturer Agilent. The ICF was meant as a standard, to make device drivers interchangeable for different applications. In analogy to a print driver, a manufacturer delivers with the instrument, the ICF was intended to be delivered together with the HPLC or GC and to enable every CDS to control the instrument. Manufacturers of CDSs only have to implement the ICF in their software once, the connection to the equipment is done by the instrument manufacturer who also, in case of equipment changes and new versions, can quickly deliver a driver without relying on the CDS manufacturer.

Agilent is using this concept for some of their instruments and delivers an ICF driver, which not only allows the connection to Agilent's CDS but also to Waters Empower and Chromeleon by Thermo Scientific. Identical handling makes it easy for the user to switch between software.

However, this can cause problems within the workflow, because the CDSs are working with different concepts. In Empower, for instance, an injection volume can be recorded in the instrument method, though it will only be used in sample tables. Here, native device drivers have an advantage, taking the characteristics of the particular CDS into consideration.

Thermo has taken a huge step toward a unified CDS for the entire laboratory in their present Chromeleon version 7.2 (SR1) and is the first supplier that really integrates mass spectrometers (MS) in a real chromatography network solution.

In this way, the Thermo Chromeleon software managed to transfer the MS from the "expert laboratory" into the common laboratory. Until now, this integration is reduced to mass spectrometers by Thermo, but hopefully Thermo will open the integration to other manufacturers.

In the future, the changing perspective on combination of instruments and software will have an impact. If in the past, instrument specifications were the most decisive factor, today, the software is more likely to drive the process. This means that instrument manufacturers have to ensure compatibility to many CDSs to stay successful.

8.1.3 Useability

Today, almost every chromatography system is controlled by software, performing data evaluation. Fifteen years ago, the instruments' menu guidance and display size were important parameters to evaluate a new HPLC system; nowadays, HPLC systems without keyboard or display are the state of the art, and each interaction is made via computer.

Sometimes, it seems that the software as an interface between the user and system has more importance than the chromatography itself. But this is not surprising: a person working in the laboratory very often spends more time in front of computer and keypad than at the laboratory table.

In my view, this means that we need to rethink the design of user interfaces. The most common ground for complaints, in my experience, are not missing functions but the complicated handling of software packages and the time required to train new users. I do not want to suggest that functional improvement of data systems is not necessary but make clear that it is as important to provide software that can be handled by "normal" users in the laboratory.

8.1.4 Ease of Use

When designing a user front end with all its planned workflows, programmers are almost free in coding. In the 1980s and early 1990s, new software developments offered various ideas for handling and interfaces. These were largely driven by three innovations:

- The change from keypad to handling by mouse
- The development of operating systems with window technique
- Increased graphical computing power.

With the introduction of Windows NT in the mid-1990s, a standard for software handling became established.

Following Microsoft Office's development, in the past 20 years, it is obvious that there have been the same challenges we are facing with CDSs today:

Diversity of functions is growing – handling gets too complicated!

Some of you may remember the first MS-Word versions run by MS-DOS. Without mouse and window technique, all handling was done by keypad. The possibility for text formatting, for example, was shown at the bottom of the screen or opened by key combinations such as Crtl + F.

Just imagine what our screens would look like if all current formatting options in Word were shown at the bottom of the monitor!

As everybody knows, Microsoft has fundamentally changed the handling with several versions and operating systems. Now the user can choose different functions, presented in logical groups. The first toolbars appeared where functions can be selected by mouse-click. The popularity of this type of handling led to more toolbars and more keys, resulting in – for the time being – Microsoft Office's last development step: The Ribbon Design.

Instead of permanently showing all toolbars and functions, there is a selective visibility in accordance to the present work sequence. The user only sees functions that can be necessary at this stage.

These developments in interface design were more or less adopted by CDS developers. Software such as Empower by Waters – after 10 years on the market – is still using static toolbars and is therefore stuck at a development stage from the beginning of the past decade. Even if users have the possibility to choose toolbars to be shown, they are not arranged in accordance to the current work step.

In contrast, the much younger Chromeleon software version 7 from Thermo is already using the Ribbon Design. Users are normally working with one window, the relevant toolbars are shown. Besides a significant "fresher" look, a greater clarity is given.

But even the Ribbon Design cannot accommodate future requirements. Although it is orientated to the present work sequence, it does not take a user's needs or training level into account. In many cases, it will be useful, only to provide those functions the user is familiar with. While an expert would like to have access to complex or rarely used functions, a beginner will be confused by these menu items and toolbars. Even worse, there is a risk to generate an invalid result by choosing a wrong operation or to alter or destroy the data needed later by switching functions (e.g., as a calibration).

Here, it is desirable that future chromatography software allows more types of customization. User interface and available functions should be able to be increased with the user's growing training level. From a simple user interface with basic functions only, to one with all functions for the application expert, various stages should be provided!

8.1.5 User Interface

Twenty years ago, CDSs were measuring systems for chromatography specialists in the laboratory, who were familiar with the technique of chromatography as well as the measurement systems' technology.

Today, operators working in companies with CDSs have different skills and profiles:

- Specialists for HPLC and GC where the use of CDSs is daily work
- Employees in routine analysis laboratories using the CDS as measurement system
- Employees from production laboratories without specialist knowledge in chemical analysis.

Wouldn't it be helpful to have a CDS applicable in all areas, which can be adapted to the single user's needs?

Most of the products already today provide some possibilities for customization. In Chromeleon, for example, the panels for instrument control allow the configuration of the instrument functions on display; similarly, Empower includes the option to define the content of tables and keys (in each window). However, the general complexity remains.

With the very creative sector of "Open-Source Software", the so-called skins – configurable, graphical surfaces, adaptable to the needs of different users – present an interesting solution.

This graphical user interface technique (GUI) becomes particularly interesting, if it can be set up and distributed easily by users or third-party suppliers. With this kind of technology, the design of all kinds of software becomes possible – from a simple surface with two buttons for start and stop, to a high-performing and complex surface in any desired form. Ideally, such skins can be developed by third parties, such as the users themselves or service providers. In this way, it is conceivable to equip two different CDSs with the same user interface, thus making it easier for the user to switch between the systems.

8.1.6 Multilingual

In installations where CDS users from different countries are working with the same system, the language of the operating system and software often present a severe obstacle. Trying to gather locations in Japan, China, Germany, France, and the United States in one system, English is probably the only feasible system language. But can we really assume that all employees in all the connected laboratories have a sufficient working knowledge of the chosen language? Do the employees feel confident in using that language?

The decision on such an installation and the choice of language (for the common software) is often taken by employees working in multinational projects and used to communicating in foreign languages. For the employees in the laboratory, it is often a challenge to navigate in a foreign language.

Companies such as Waters supply their users with online help in several languages, but only one can be implemented into the system. In the future, it should be possible to supply users with help in their native language. Already today, CDSs provide the possibility to define user accounts and to permit certain rights. Some systems offer the possibility to define screen colors. Why not make it possible to choose a language for the CDSs at this point? The technical possibilities for software to show menu entries and help text selectively in a language chosen by the user has already been there for long time.

Software packages such as Microsoft's Office demonstrate how it works: they provide software where language packages can be integrated. In this way, it would be possible for the user to perform a basic installation of the software in English and to subsequently install the language package of his/her choice.

8.2 Data Exchange

For years, interchangeability of data and integration of different systems become more and more important. Especially, requirements for the pharmaceutical industry, regarding data transfer between different software systems, are forcing the development of interchange formats and interfaces. Furthermore, the need to read and edit data after several years makes a transfer to other and newer software packages mandatory. However, looking at all data systems in a (*n* average) laboratory, it soon becomes apparent that they, too, are grouped in single islands without sufficient data transfer between them. A holistic option to connect all information, including meta-data, does not exist. If departments inside a company or companies within research cooperation want to share data, these island landscapes become a major obstacle.

The Allotrope Foundation (http://allotrope.org) is an interesting project, trying to overcome this obstacle. Here, leading pharmaceutical companies came together to define a future standard for laboratory software:

> The Mission of Allotrope Foundation is to address the root cause of this "disease" instead of just treating the symptoms through an array of helper applications, data conversion projects, local standardization efforts of varying effectiveness, such as is done today. We believe the "cure" to the "disease" is the development of a comprehensive and innovative Framework consisting of metadata dictionaries, data standards, and class libraries for managing analytical data throughout its lifecycle. To accomplish this mission, Allotrope Foundation is working to create an open "ecosystem" through collaboration and consultation with vendors and the analytical community.

As tempting as the idea to reach complete exchange of information between diverse IT systems of the most common suppliers may be, it is unlikely that such solutions can be reached in the near future.

Development of such complex software systems as CDS, ELN, or Laboratory Information Management System (LIMS) takes several years, and the costs are enormous. Will manufacturers change their strategies and develop new products, to fulfill customer requests for a better data exchange?

At least, the biggest providers of laboratory software packages – like Agilent with the OpenLAB product, Thermo with Chromeleon/Sample Manager Integration and Waters with a combination of Empower and NuGenesis have advanced the networking capabilities between their own products.

For a company wishing to renew its general environment of lab-IT, some valuable solutions already exist today. But, the challenge is to connect existing systems, which have been implemented over the years. Here it still will be necessary for each company to develop their own solutions for data exchange.

8.2.1 Import and Export of Data

The simplest form of interaction between two software systems, such as a LIMS and a CDS, is data exchange via text files. Almost every CDS and LIMS is able to export or import these files. But this data exchange almost always requires a manual operation. Information is imported into the CDS, and, in turn, generated results are exported to a file.

The absence of a real standard of file formats is another deficit. Experience has shown that allegedly standardized formats for two-dimensional data, such as the AIA/CDF standard, are processed in different ways by different data systems. In this way, the data exchange gets complicated, incomplete, and error-prone. A standardized implementation would be easy to implement but would probably not deliver a satisfactory solution, because too little of the important metadata, such as measurement parameters and extended sample information, will be transferred with raw data.

8.3 From PCs Scalability to Global Installation

Almost every CDS on the current market can be scaled from a single PC to a global network solution. Increasingly better and efficient network connections around the globe enable the connection of sites, even far away, to a central server. In this way, many installations are driven in different time zones and languages.

These new settings represent special challenges for the system administrators but also for the user in the laboratory.

8.3.1 Software Placement

Local installations generally need less technical support than global ones, because they are more static. Once installed and configured, they run in most cases until the next update or an extension has to be installed, without interaction by the administrator. Necessary service and maintenance tasks can be arranged with the user in the laboratory and executed in certain time slots.

However, when connecting hundreds of computers in a laboratory to one network, the supply with a CDS is normally done by a group of administrators. In most cases, this results in the spatial separation between the user and administrator, meaning that the manual installation and maintenance of computers for data acquisition and evaluation in the laboratory are no longer possible. The software distribution has to be done automatically, adapted to the needs of each location. This causes a set of requirements:

- All computers that have taken part in the CDS have to be registered. Here, user access via terminal- and Citrix servers is possible and without problems; the challenge lies in servers for data collection. These servers provide the

connection between CDS and HPLC with GC systems. The Empower software from Waters, for example, requires identical software status. Where necessary, the installation of adapted drivers for (locally installed) computer hardware (e.g., extra interface cards) or connected instruments is also required.
- In case of computer breakdown, information about configuration has to be recovered quickly and completely. This requires the saving of information about the current installation status, as well as settings and configurations, which can be recalled on demand.

Each manufacturer delivers their own solutions for this:

Empower's actual version provides the possibility to directly transfer and install device drivers via a so called push installation to all integrated computers in the CDS network. In this way, administrators are free from the task to implement their own distribution mechanisms. Regrettably, this solution is at the time limited to device drivers by Waters.

The software Atlas by Thermo provides an interesting solution at least for data acquisition servers. With each booting process, the software is loaded from a central server, and the operating system is permanently stored in the ROM. To carry out an update or implement changes into the installation, an installation image has to be placed only at one central place. A computer restart is enough to distribute the software. Besides the easy distribution, this mechanism transfers each reboot to a new installation of the system. When technical problems with the server occur, a restart often can solve them. Indeed, a support-friendly solution!

The Chromeleon software package, also by Thermo, has the advantage that the computers inside the CDS network do not necessarily have to be on the same version level. In this way, a software update can be executed step by step. Installations working in several time zones and laboratories running 24-h a day, 7 days a week, thus do not have to find a timeslot for installation that is suitable for all users. Especially for computers being used for data acquisition, this is an advantage, because most software installations have to be finished with a reboot, a task that can only be executed when no data collection is in process.

Despite all advantages, these functions do not offer a solution for big installations. None of the systems has a central administration console. The focus lies on the CDS and not on complete installation including the operating system. Here, software is needed, which can – in analogy to administration tools for networks, such as a Microsoft System Center Server – provide a central administration of hardware and software, which can be integrated in customers' networks.

Perhaps said Microsoft product offers a solution, at least for the distribution of software packages.

It will be interesting to see which trends and developments will surface on the market in the coming years. Certainly, the manufacturer's development departments will try to meet customer requests. However, economical constraints, both for manufacturers and customers, will also have an impact on coming changes.

9

Aspects of the Modern HPLC Device – Experience Report of an Operator

Steffen Wiese and Terence Hetzel

9.1 Introduction

In this chapter, the authors discuss their experiences with modern Ultra High Performance Liquid Chromatography (UHPLC) systems. The focus is both on the application side and on the handling of the modules comprising a UHPLC system. The aim is to provide useful hints and tips for the most efficient use.

9.2 Determination of the Gradient Delay Volume

In Section 4.2, the importance of the gradient delay volume for the application of solvent gradients in liquid chromatography was discussed. The gradient delay volume describes the volume from the point of mixing of the mobile phases to the head of the column (see Figures 2.1 and 2.4). This volume is responsible for the isocratic step at the beginning of each solvent gradient elution. In this context, the question arises: "If the gradient delay volume is so important, why isn't it simply given by the manufacturer?" The answer is rather simple. The gradient delay volume depends on the individual configuration of the UHPLC system and changes when a different length capillary is used, the inner diameter of the capillary varies, and/or the sample loop of the autosampler is changed. Therefore, it is difficult to define a standardized gradient delay volume for a specific UHPLC system.

In this chapter, the influence of the gradient delay volume will be clarified using practical examples. To open, a method for determining the gradient delay volume of a high-pressure mixing system is introduced (see Figure 2.1 and also explanations in Chapter 4). First, the column is replaced by a zero dead volume union (ZDV). UHPLC grade water is used as solvent for channels A and B. In addition, a UV-absorbing substance is added to one of the solvents, typically channel B. Caffeine (50 mg l^{-1}), propylparaben (7.5 mg l^{-1}), and benzyl alcohol (10 mg l^{-1}) are commonly used. If solid substances are used as UV-absorbing substances, the mobile phase should be filtered after dissolving using a membrane filter ($\leq$0.2 μm) in order to prevent system clogging. Next, the chromatographic method can be

The HPLC Expert II: Find and Optimize the Benefits of your HPLC/UHPLC, First Edition.
Edited by Stavros Kromidas.

created. This could be designed as follows: an isocratic step for the duration of 5 min, followed by a linear gradient from 0% to 100% B in 10 min and an isocratic step at 100% B also for 5 min. Afterward, the system is quickly flushed back to the starting conditions, for example, within 1 min. In order to minimize errors in the subsequent data analysis, the flow rate should be chosen with respect to the expected gradient delay volume. When the gradient delay volume for a UHPLC system is anticipated within a range of 400–500 μl, a flow rate of 0.5 ml min^{-1} is recommended. In contrast, a flow rate between 0.1 and 0.3 ml min^{-1} should be applied when a gradient delay volume below 200 μl is present. In the following example, a gradient delay volume below 200 μl for the UHPLC system was expected. Therefore, a flow rate of 0.2 ml min^{-1} was chosen. However, it should be noted that a sufficiently large back pressure should be achieved in the system. This is particularly important for UHPLC systems, which have no active inlet and outlet valves. For such systems, the system back pressure should be at least higher than 15 bars in order to ensure that the inlet and outlet valves of both pumps function properly. When the desired back pressure of ≥15 bar is not achieved, the pressure can be adjusted using a back-pressure regulator or a restriction capillary with a small inner diameter. In addition, the back-pressure regulator or restriction capillary should be installed behind the detector. Here, it is important to take the pressure stability of the UV cell into consideration. A back-pressure regulator has the advantage that the pressure can be adjusted constantly and therefore independently of the flow rate. In contrast, when using a restriction capillary, the resulting pressure is flow rate dependent.

If the UHPLC system offers the possibility of "no injection," this is the preferred injection method. With a "no-injection" option, no sample is aspirated and injected. The autosampler performs the same switching procedures as a real injection of a sample. In contrast, if the "no-injection" mode is not supported by the UHPLC system, 1 μl of deionized water can be injected, for example. Furthermore, the proposed gradient should be measured several times because often the first chromatogram of the gradient profile is not suitable for the determination of the gradient delay volume. In Figure 9.1, a chromatogram for the determination of the gradient delay volume of a UHPLC system is shown using a variable flow-through autosampler. The temporal offset between the programmed (---) and experimentally (—) determined gradient profile, which is caused by the gradient delay volume, can be clearly seen.

The data analysis of the obtained gradient profiles is sometimes difficult for UHPLC systems. For conventional High Performance Liquid Chromatography (HPLC) systems, a graphic evaluation of the gradient profile is recommended. For this purpose, the chromatogram is printed on a sheet of paper. Afterward, the gradient delay volume can be determined using a ruler, pencil, and calculator. A detailed description of this procedure, which is well suited for the determination of the gradient delay volume of classic HPLC systems, can be found in the literature [1–3]. For gradient delay volumes of 1000 μl for a conventional HPLC system, an error of ±50 μl or even ±100 μl for the determination of the gradient delay volume is still acceptable under certain circumstances. This does not apply for modern UHPLC systems with a gradient delay volume of ≤200 μl. The graphic evaluation of the gradient profile for UHPLC systems usually

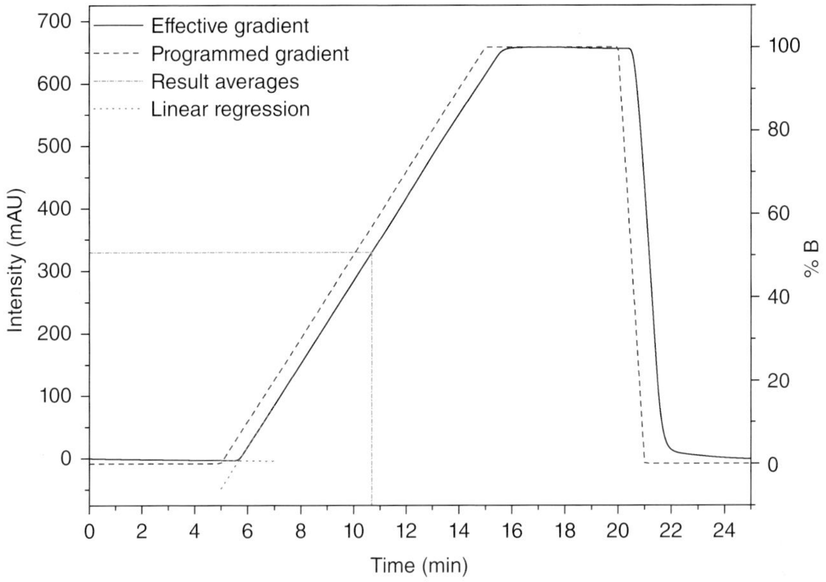

Figure 9.1 Determination of the gradient delay volume of a UHPLC system. Chromatographic parameters: stationary phase: none, replaced by a zero dead volume union (ZDV); mobile phase: (A) deionized water and (B) deionized water with 50 mg l^{-1} caffeine; flow rate: 0.2 ml min^{-1}; gradient: 0–5 min, 0% B; 5–15 min, from 0% to 100% B; injection volume: no injection, injection needle, and sample loop are switched into the flow path; temperature: 30 °C; and detection: UV at 270 nm.

leads to unacceptable errors. Therefore, in the following, an alternative method for the determination of the gradient delay volume for UHPLC systems is described.

Typically, all manufacturers of chromatography data evaluation software offer the possibility to export the chromatographic data as “Comma-separated Values” (CSV). The exported data files can be opened and edited with a spreadsheet program. The suggested methods are based on the analysis of these exported data files. Therefore, the chromatogram must first be exported as CSV or text file. On the basis of the exported values, the average signal intensity for a section of the isocratic hold at 0% B at the beginning of the chromatogram is calculated (e.g., 1–5 min, Figure 9.1). Subsequently, the calculation of the average signal intensity is determined for a section of the isocratic plateau at 100% B (e.g., 16–20 min, Figure 9.1). Then, the difference between the minimum (at 0% B) and the maximum (at 100% B) signal intensity is calculated and divided by 2 in order to obtain the half signal intensity. Afterward, the values of the minimum signal intensity and the previously calculated value of the half signal intensity are added. The result equals the half signal intensity of the linear gradient and should theoretically be achieved at the half of gradient duration.

Next, the search function of the spreadsheet program can used to find the previously calculated half of the signal intensity. An exact match of the values is not expected, but usually a high level of agreement with a difference in the first decimal place of the signal intensity can be found. The intensity value is associated with a time in chromatogram and should theoretically correlate with the half of the linear gradient. The temporal difference between these two values describes the gradient delay time and corresponds to the gradient delay volume after multiplication with the flow rate.

On the basis of the example in Figure 9.1, the half of the signal intensity should theoretically be achieved after a period of 10 min (assumption: $V_D = 0$). These 10 min consist of 5 min isocratic hold at the beginning of the chromatogram plus 5 min (10/2 min) after the half of the linear gradient section. However, based on the experimental data, the half of the signal intensity was measured at the detector at a time of 10.685 min. The difference of 0.685 min describes the gradient delay time and depends on the flow rate. This value indicates that at a flow rate of 0.2 ml min^{-1}, the effective gradient reaches the head of the column after 0.7 min and explains the differences between the programmed and experimental obtained gradient profile (see Figure 9.1).

Finally, for the calculation of the gradient delay volume, the gradient delay time is multiplied by the flow rate. With respect to the gradient profile shown in Figure 9.1a, gradient delay volume of 137 μl for the considered UHPLC system can be calculated.

The gradient profile shown in Figure 9.1 is very sharp, and the individual switching points (5–6 and 15–16 min) of the gradient are clearly visible. For a UHPLC system with a larger gradient delay volume, the observed profile at the switching points is usually not that sharp. Here, gradient curves are often observed resulting in an S-shaped profile, which may lead to difficulties with the determination of the gradient delay volume. In such cases, an alternative to the average calculation is the linear regression for the isocratic section at 0% B and for the linear increasing range at, for example, 7–14 min (Figure 9.1). Afterward, the intersection of the two lines is calculated. The straight lines (—) obtained by the linear regression for the determination of the gradient delay volume are also shown in Figure 9.1. On the basis of the presented example, the intersection of the two lines was calculated at a time of 5.669 min, resulting in a gradient delay time of 0.669 min, which leads to a gradient delay volume of 134 μl after multiplication with the flow rate. The comparison of both calculation methods shows that almost identical values are obtained, which differ only by 2%. Nevertheless, the method based on the average calculation at 0% and 100% B is more comfortable. Both previously described methods can be used to determine the gradient delay volume, although other methods are described in the literature. For example, it is recommended to replace the column by a 10-cm-long fused-silica capillary with an inner diameter of 50 μm [4]. In this case, a restriction capillary or back-pressure regulator behind the UV detector is no longer necessary. However, it should be noted that the determined gradient delay volume should be corrected for the volumes of the fused-silica capillary and the two unions.

In addition, acetone with a concentration of 0.1% in the mobile phase is often recommended as UV-absorbing substance [1–3]. At this point, it should be mentioned that the in-line degasser of modern UHPLC systems is capable of removing a certain portion of the acetone from the mobile phase. Therefore, the evaluation of the gradient profile can be difficult.

In summary, there is a variety of different methods for the determination of the gradient delay volume. With respect to the transfer of methods, the user should choose one method for the determination of the gradient delay volume and apply these to both liquid chromatography systems. If the determination of the gradient delay volume is not done personally, the user should have detailed knowledge about the experimental procedure and the UHPLC system design. This particularly applies to method transfers between different laboratories.

9.3 High-Throughput Separations

Modern UHPLC systems were optimized for the application of 2.1-mm-ID columns and sub 2 μm particles. Especially for the hyphenation of UHPLC and mass spectrometry, column dimensions of 2.1 mm ID and 50 mm length have been established, although this is not always the best column dimension for every separation problem. With electrospray ionization (ESI), flow rates between 250 and 500 μl min^{-1} are typically used. Therefore, in order to influence the elution of the early eluting analytes using the solvent gradient, a low gradient delay volume is particularly important for this hyphenation (see also Section 2.1.2.2).

The influence of the gradient delay volume for the hyphenation of liquid chromatography and mass spectrometry is shown in Figure 9.2. Figure 9.2a shows the chromatogram for a routine LC-MS/MS method for the quantification of different antineoplastic drugs. In addition, the programmed and the effective solvent gradients are illustrated. For this example, a chromatographic separation is required because ifosfamide and cyclophosphamide are isomers with at least one identical mass transition. Furthermore, ion suppression occurs when paclitaxel and docetaxel are not separated. The chromatogram in Figure 9.2a demonstrates that the elution of gemcitabine cannot be influenced by the solvent gradient and elutes from the column under isocratic conditions. The isocratic elution is caused by the large gradient delay volume of about 1000 μl of the used HPLC system. In contrast, it is possible to affect the elution of gemcitabine by the solvent gradient when the classic HPLC system is replaced by a UHPLC system with a gradient delay volume below 100 μl. The chromatogram of the optimized separation is shown in Figure 9.2b. Again, both the programmed and the effective solvent gradients are illustrated. Due to the reduced gradient delay volume, only a marginal difference between the programmed and the effective solvent gradients can be observed. In other words, the programmed gradient affects almost immediately.

The optimized separation of the antineoplastic drugs shown in Figure 9.2b was only possible due to the reduced gradient delay volume of the UHPLC system. The analysis time could be reduced by half from about 13 to 6.5 min. Further minimization of the analysis time can be achieved by reducing the column inner diameter from 3.0 to 2.0 mm. The resulting chromatogram is shown in Figure 9.2c. At a constant flow rate, by reducing the inner diameter of the column, the analysis time could be further shortened from approximately 6.5 to 4 min, and at the same time, the limit of detection and limit of quantification could be improved by a factor of 3–7. In order to further minimize the analysis time, it is possible to increase the flow rate up to 0.6 or 0.7 ml min^{-1} because the system back pressure of approximately 170 bar (Figure 9.2c) is not the limiting factor. In this case, decreasing ionization efficiency was observed, which led to a reduced sensitivity. At this point, it must be noted that this behavior strongly depends on the applied mass spectrometer ion source (see Chapter 5). In addition, it should be noted that the peak widths of the analytes are in the lower second range, and the acquisition rate of the used mass spectrometer is too low to represent the peak profile with sufficient accuracy (see Section 2.1.5 and Chapter 5).

In summary, it can be concluded that by simply replacing the classic HPLC system through a UHPLC system, the analysis time of a routine LC-MS/MS method could be reduced by a factor of 3. Furthermore, the limit of detection and quantification could also be improved by a factor of 3–7. In this context, the purchase of a new and modern mass spectrometer should be considered when decreased limits of detection and quantification are required. The presented example shows that

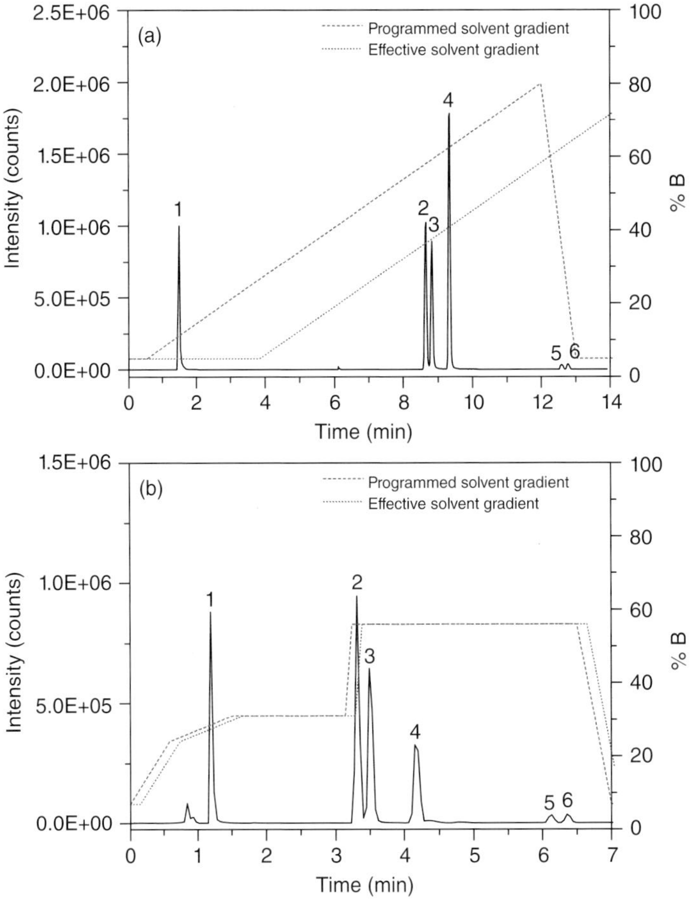

Figure 9.2 *(Continued)*

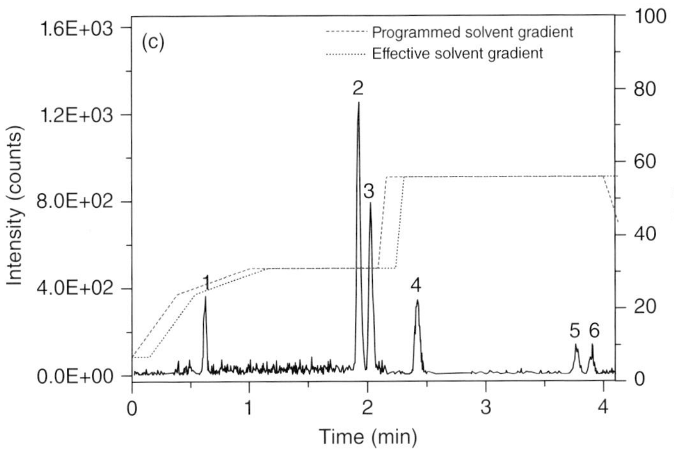

Figure 9.2 Separation of six antineoplastic drugs. Chromatographic parameters: stationary phase: Shimadzu Shim-pack XR-ODS, (a,b) 50 mm × 3.0 mm, 2.2 μm and (c) 50 mm × 2.0 mm, 2.2 μm; mobile phase: (A) deionized water + 0.1% formic acid and (B) acetonitrile with 0.1% formic acid; flow rate: 0.3 ml min^{-1}; gradient: see figure; injected absolute mass per analyte: (a) 12.5 ng, (b) 5 ng, and (c) 0.002 ng; temperature: (a) 35.0 °C and (b,c) 31.3 °C; detection: MS, MRM. Maximum pressure during the gradient: (a) 150 bar, (b) 95 bar, and (c) 172 bar; and analytes: (1) gemcitabine, (2) ifosfamide, (3) cyclophosphamide, (4) etoposide, (5) paclitaxel, and (6) docetaxel. Note to (c): differences in the peak height ratio of the analytes are caused by adjustments of the MS source parameter.

by replacing the HPLC system by a UHPLC system, significant improvements in the limit of detection and quantification can be achieved. Usually, the purchase of a UHPLC system is less expensive than purchasing a more sensitive mass spectrometer.

9.4 Method Transfer between UHPLC Systems of Different Manufacturers

As already mentioned in Section 9.2, the gradient delay volume is the decisive factor in method transfer with solvent gradient. The transfer of isocratic methods can be accomplished much more easily as the gradient delay volume is not an influencing factor. The transition of liquid chromatography methods can be done either from one HPLC system to another or from an HPLC to a UHPLC system. Furthermore, the method transfer between two different UHPLC systems is also possible. Usually, the method transfer from a system with a larger gradient delay volume to a system with a smaller gradient delay volume is less complex. Typically, this is the case for the transfer of HPLC methods to UHPLC systems. The differences in the gradient delay volume can be compensated using an isocratic step at the beginning of the gradient elution. The required calculations for the conversion of HPLC methods to a UHPLC system with respect to the column dimensions and particle diameters are well known. Therefore, the interested reader is referred to the current literature [5–10]. Furthermore, almost every manufacturer of UHPLC systems provides free calculation tools, which facilitate method transfer. The commercially available software ISET from Agilent Technologies should be noted here, which contains the system volumes of various HPLC and UHPLC manufacturers. For method transfer, the user has to choose the corresponding system, and the software performs all necessary calculations and automatically adjusts the method parameters. For method transfer to different LC systems, the maximum allowed pressure of the UHPLC systems has to be considered depending on the applied flow rate. This is particularly important when flow rates higher than 1 ml min^{-1} are used and when the method applied nears the maximum specified pressure range of the UHPLC system.

In the following, the transfer of chromatographic methods to UHPLC systems between different vendors is described. For this purpose, the gradient delay volume of both systems was determined according to the procedure described in Section 9.2. Since both systems are equipped with an integral-loop autosampler, the gradient delay volume was similar. For the UHPLC system from manufacturer A, a gradient delay volume of approximately 140 μl was determined. The gradient delay volume of the UHPLC system from manufacturer B was about 160 μl. The separation of 13 aldehyde- and ketone-2,4-dinitrophenylhydrazones on a 2.1-mm-ID and 50-mm-long column packed with sub 2 μm particles serves as an example for the method transfer. At a flow rate of 1.2 ml min^{-1}, the maximum pressure during the gradient was 1100 bar. Because the phase material and the column dimensions are kept constant, it is expected that the established method can be easily transferred to the UHPLC system of manufacturer B. In Figure 9.3a, the separation obtained on the UHPLC system from manufacturer A is shown. The separation can be achieved in 3.5 min with a critical resolution between the peak pair 6/7 of 1.6. In comparison, in Figure 9.3b, the separation of the aldehyde- and ketone-2,4-dinitrophenylhydrazones on an identical column using the UHPLC system from manufacturer B is depicted.

As expected, the separation can also be achieved on the UHPLC system from manufacturer B within 3.5 min. However, the critical peak pair and the critical resolution have changed. When using the UHPLC system from manufacturer A, the critical resolution is 1.6 between the peaks 6 and 7, whereas the peaks 7 and 8 become the critical peak pair with a resolution of 1.4 for the separation using the UHPLC system from manufacturer B. Comparing both chromatograms, the peaks elute slightly later on the system from manufacturer B (Figure 9.3b) in contrast to the separation achieved on the UHPLC system from manufacturer A. The slightly lower temperature and larger gradient delay volume of the UHPLC system from manufacturer B explain this time discrepancy. In addition, it must be considered that the dispersion volume behind the column is larger for the UHPLC system of manufacturer B because of the difference in volume for the used UV cells of both UHPLC systems. The influence of the detector cell volume is discussed in Section 9.7 in more detail. Despite the fact that the method transfer cannot be accomplished without a small loss of efficiency, it should be noted that the method achieved on the UHPLC system from manufacturer B is well suited for the quantitative analysis of the aldehyde- and ketone-2,4-dinitrophenylhydrazones.

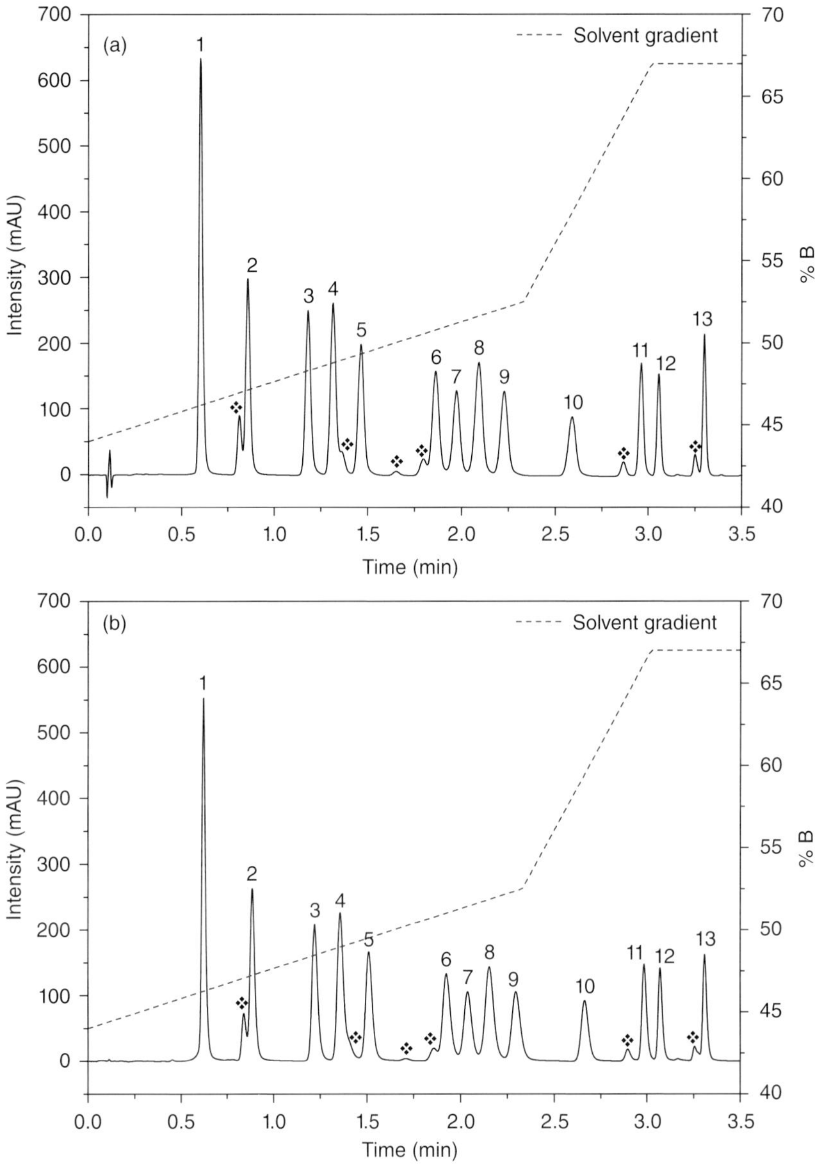

Figure 9.3 Chromatographic separation of 13 aldehyde- and ketone-2,4-dinitrophenylhydrazones on (a) UHPLC system from manufacturer A and (b) UHPLC system from manufacturer B. Chromatographic parameters: stationary phase: Agilent Zorbax SB C_{18}(50 mm × 2.1 mm, 1.8 μm); mobile phase: (A) deionized water and (B) acetone; flow rate: 1.2 ml min^{-1}; gradient: see figure; injection volume: 1 μl; temperature: (a) 33.4 °C and (b) 33.0 °C; detection: UV at 360 nm. Maximum pressure of the method: (a) 1100 bar and (b) 1080 bar. Analytes: (1) formaldehyde-2,4-DNPH, (2) acetaldehyde-2,4-DNPH, (3) acrolein-2,4-DNPH, (4) acetone-2,4-DNPH, (5) propionaldehyde-2,4-DNPH, (6) crotonaldehyde-2,4-DNPH, (7) methacrolein-2,4-DNPH, (8) 2-butanone-2,4-DNPH, (9) butyraldehyde-2,4-DNPH, (10) benzaldehyde-2,4-DNPH, (11) valeraldehyde-2,4-DNPH, (12) *m*-tolualdehyde-2,4-DNPH, and (13) hexaldehyde-2,4-DNPH. Ψ: impurities of the standard which are not considered during method development.

Furthermore, this example clearly illustrates that a method transfer between different UHPLC systems from different manufacturers renders almost one-to-one results despite the slightly different gradient delay volumes.

However, this is not always the case. Method transfer is particularly problematic when a method from a UHPLC system with a smaller gradient delay volume has to be transferred onto a system with a larger gradient delay volume or when the gradient delay volumes of both UHPLC systems significantly differ.

For this kind of method transfer, it is possible that a method cannot be transferred, and this should be considered when developing new methods. This particularly applies when the method development is carried out on a UHPLC system with a gradient delay volume of, for example, 200 μl, and afterward, the new method should be transferred to a classic HPLC system with a gradient delay volume of, for example, 1000 μl. Therefore, it is useful to consider larger gradient delay volumes of HPLC systems during method development using a UHPLC system. This can be easily done by a programmed isocratic step at the beginning of the separation. In this case, the isocratic step can be used to compensate different gradient delay volumes.

9.5 Application of Elevated Temperatures

With the introduction of UHPLC, high-temperature liquid chromatography was established in order to exploit the advantages of the reduced system back pressure. This is particularly advantageous when highly viscous solvents such as methanol or 2-propanol are used in combination with long columns (≥10 cm), small inner diameters (≤2.1 mm), and sub 2 μm particles. By increasing the temperature, the system pressure is significantly reduced. As a consequence, higher flow rates can be applied in order to accelerate the separations. For instance, the separation of 13 aldehyde- and ketone-2,4-dinitrophenylhydrazones at 20 and 50 °C is shown in Figure 9.4. By increasing the temperature by 30 °C, the system pressure can be reduced roughly 240 bar from 567 bar at 20 °C to 320 bar at 50 °C. Furthermore, the analytes elute earlier from the column and display narrower peak shapes.

The application of elevated temperatures in liquid chromatography requires sufficient temperature stability of the analytes and the columns. A list of different stationary phases from different manufacturers, which can be used at higher temperatures, can be found in the literature [11]. Usually, modern UHPLC systems offer the possibility to work at temperatures up to 100 °C. However, it should be noted that the back pressure behind the column must be sufficiently high to prevent evaporation (phase transition) of the mobile phase in the column.

In addition, it has to be ensured that the mobile phase is preheated before entering the column. Furthermore, as spectroscopic detectors are very sensitive to slight changes in temperature, the mobile phase should be tempered to a defined temperature after leaving the column. Otherwise, fluctuations of the peak area and height can be observed.

In comparison, an insufficient preheating of the mobile phase usually leads to less efficient separation because the peak shape is negatively affected [12].

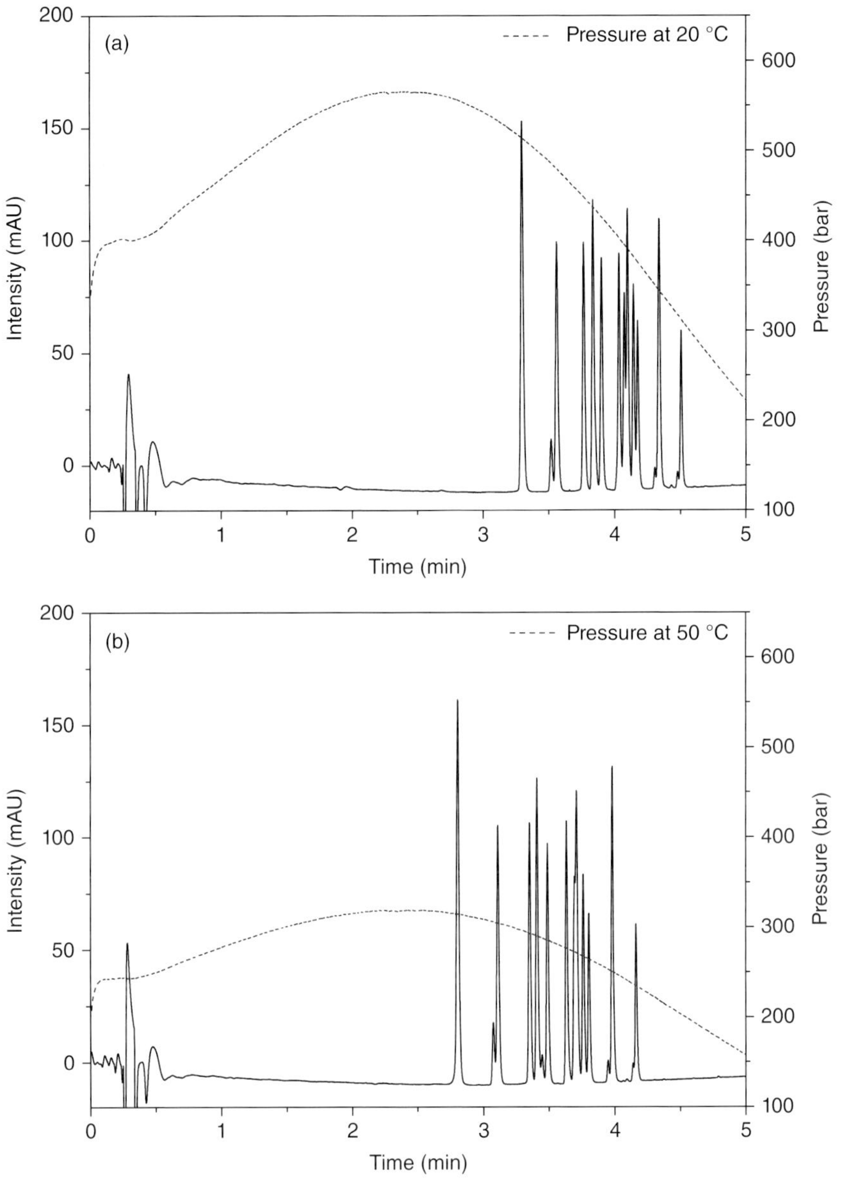

Figure 9.4 Chromatographic separation of 13 aldehyde- and ketone-2,4-dinitrophenylhydrazones. Chromatographic parameters: stationary phase: Agilent Zorbax SB C_{18} (50 mm × 2.1 mm, 1.8 μm); mobile phase: (A) water and (B) acetone; flow rate: 0.5 ml min^{-1}; gradient: 5–95% B in 5 min; injection volume: 1 μl; temperature: see figure; and detection: UV at 360 nm. Maximum pressure during the gradient: (a) 567 bar and (b) 320 bar.

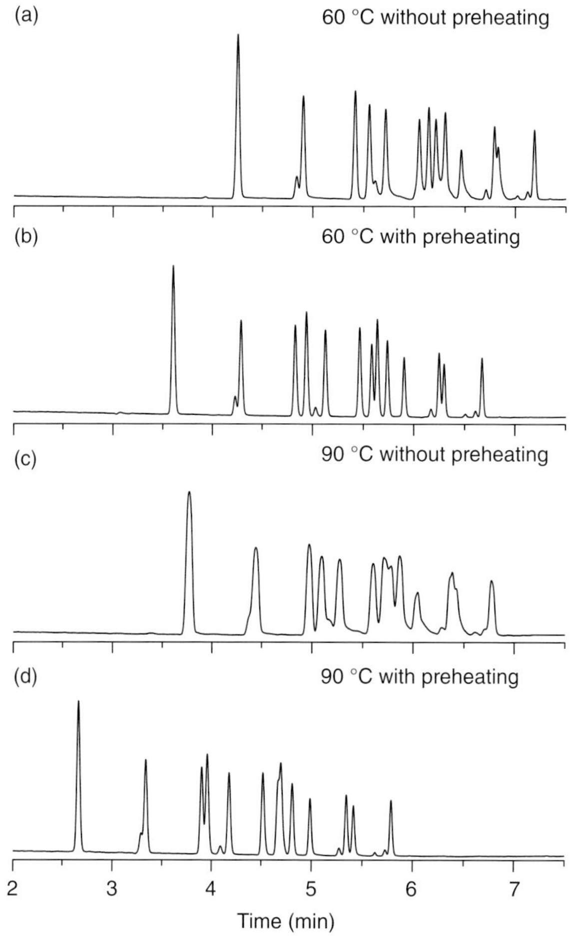

Figure 9.5 Chromatographic separation of 13 aldehyde- and ketone-2,4-dinitrophenylhydrazones. Chromatographic parameters: stationary phase: Agilent Zorbax SB C_{18} (50 mm × 2.1 mm, 1.8 μm); mobile phase: (A) water and (B) acetone; flow rate: 0.5 ml min^{-1}; gradient: 5–95% B in 10 min; injection volume: 1 μl; temperature: (a,b) 60 °C and (c,d) 90 °C; and detection: UV at 360 nm. (a,c) No preheating and (b,d) passive preheating in a volume of 1.6 μl.

In Figure 9.5, two chromatograms are shown at temperatures of 60 and 90 °C with and without preheating of the mobile phase.

It can be clearly seen that even at a relatively low temperature of 60 °C, the peaks are broader and elute later from the column (Figure 9.5a,b). This is due to the formation of axial and radial temperature gradients in the column. If the temperature is increased to 90 °C (Figure 9.5c,d), the loss of chromatographic efficiency is even more pronounced when the mobile is not preheated before entering the column (also see remarks in Section 2.2).

In summary, it should be noted that increasing the temperature in UHPLC offers a variety of advantages such as the reduction of the system back pressure, which allows higher flow rates in order to accelerate the separation. However, this is only possible if the used columns and analytes have sufficient temperature stability and the mobile phase is preheated before entering the column. Otherwise, a significant loss of chromatographic efficiency can be expected.

9.6 Large-Volume Injection (LVI)

In the field of water and wastewater analysis, polar substances and metabolites are of increasing interest over the past years, and clear steps to regulate the allowed concentration of certain environment-related substances have been taken. However, different laboratories are confronted with the problem that many of these polar compounds are present in a very small concentration. Therefore, sample preparation and sample enrichment procedures are necessary prior to the actual analysis in order to obtain valid results in the range of the regulated values. One of the most applied techniques is the solid-phase extraction (SPE), which can be done either online or offline to enrich the analytes. However, SPE can be problematic due to the selectivity for specific groups depending on the used phase material. Therefore, a large number of polar compounds can be discriminated during enrichment. Furthermore, it has to be considered that the final step of SPE includes the elution or desorption of enriched analytes from the SPE material with a small volume of an organic solvent, often methanol. The obtained organic extract also represents the injection solution. Thereby, the user is faced with the problem that only a small volume of the extract can be injected. Otherwise, nonacceptable peak shapes can be observed [13]. This particularly applies to polar substances, which show weak or no retention on classic reversed-phase materials. As an alternative to reversed-phase chromatography, hydrophilic interaction chromatography (HILIC) can be used for injection of highly organic extracts. Nevertheless, the injection of extracts with a high amount of methanol can also lead to nonacceptable peak shape using HILIC [14].

Another alternative to online or offline SPE is the direct injection of high volumes onto the column. This technique is known as "large-volume injection" (LVI). As a rule of thumb, in reversed-phase chromatography, the injection volume should not be higher than 10% of the column void volume [14]. With respect to a 50-mm-long column with an internal diameter of 2.1 mm and a porosity of 0.7, the injection volume should not be higher than 1.2 μl. In the following, the nomenclature "LVI" is used when the injection volume is higher than 10% of the column void volume. The decisive advantage of the injection of large volumes is that nonpolar analytes can be enriched at the column head using an aqueous standard and elute only from the column at a certain amount of organic solvent in the mobile phase. For this enrichment technique, usually classic C_{18} reversed-phase columns are used, which are suitable for the enrichment of nonpolar substances [14, 15]. In contrast, the enrichment of polar compounds cannot be achieved with reversed-phase materials. For this purpose, different stationary phases such as graphitic carbon (porous graphitic carbon, PGC), better known as *HyperCarb*, should be used. By means of this phase material, it is possible to retain strongly polar substances [16]. It has to be considered that nonpolar compounds are also heavily retarded. As a consequence, sometimes these components are irreversibly bonded or elute very late from the column. Due to this, in the following, a coupled phase system is described for the enrichment of polar as well as nonpolar substances. For this purpose, a short HyperCarb precolumn (10 mm × 2.1 mm, 5 μm) and a classic C18 reversed-phase column (50 mm × 2.1 mm, 3.5 μm) are coupled. The advantage of this column combination is that by using the HyperCarb precolumn, sufficient retention for polar compounds can be achieved. At the same time, the retention of nonpolar substances is not that high. Thereby, the risk of no or late elution can be significantly reduced.

First, the influence of the HyperCarb precolumn is illustrated for the retention of polar, semipolar, and nonpolar analytes. In Figure 9.6, two chromatograms for the separation of four pharmaceuticals are shown. Figure 9.6a shows the chromatogram for the separation without the HyperCarb precolumn, thus without online enrichment, whereas Figure 9.6b demonstrates the separation using the HyperCarb precolumn. Comparing both chromatograms, it can clearly be seen, that there are no differences for the nonpolar fenofibrate nor for the semipolar ifosfamide and cyclophosphamide. For these analytes, the injection volume of an aqueous standard can be increased up to 1000 μl without any negative influence to peak shape. In comparison, the strongly polar gemcitabine elutes are broad, nonfocused bands from the C_{18} column, if no precolumn is used. The retention of gemcitabine on a C_{18} material is so weak that under these conditions an enrichment cannot be achieved. However, when a combination of HyperCarb precolumn and C_{18} reversed-phase column is used (Figure 9.6b), it is clear that the retention on the HyperCarb precolumn is responsible for the focusing of the very polar gemcitabine on the column head. Afterward, the focused gemcitabine can be eluted as a narrow band using the solvent gradient. With respect to the optimum injection volume of $\leq$10% of the column void volume ($\leq$12 μl), this corresponds to an enrichment by the factor of approximately 83.

Finally, it is shown that the injection of large volumes is not necessarily associated with a significant loss of chromatographic efficiency. Because in practice, it is more important to securely measure a regulated or predefined LOQ.

Therefore, it is less important in UHPLC to use the recommended column dimension (50 mm × 2.1 mm) and injection volume of $\leq$5 μl in order to reduce the influence of the injection volume on the separation efficiency [18]. In Figure 9.7, an overlay chromatogram for the separations of pharmaceuticals using an injection volume of 5 and 1000 μl is shown. The differences in retentions times can be explained by the different volumes of the applied sample loops. For the injection of 5 μl, a sample loop of 5 μl was used, whereas for the injection of 1000 μl, a sample loop with a volume of 1000 μl was installed. The difference in time of about 2 min corresponds to the time that is needed to flush the 1000 μl sample loop at a flow rate of 0.5 ml min^{-1}.

It can be concluded that the injection of large volumes in UHPLC is usually possible when solvent gradients are used, and sufficiently high retention of the analytes can be achieved on the applied column.

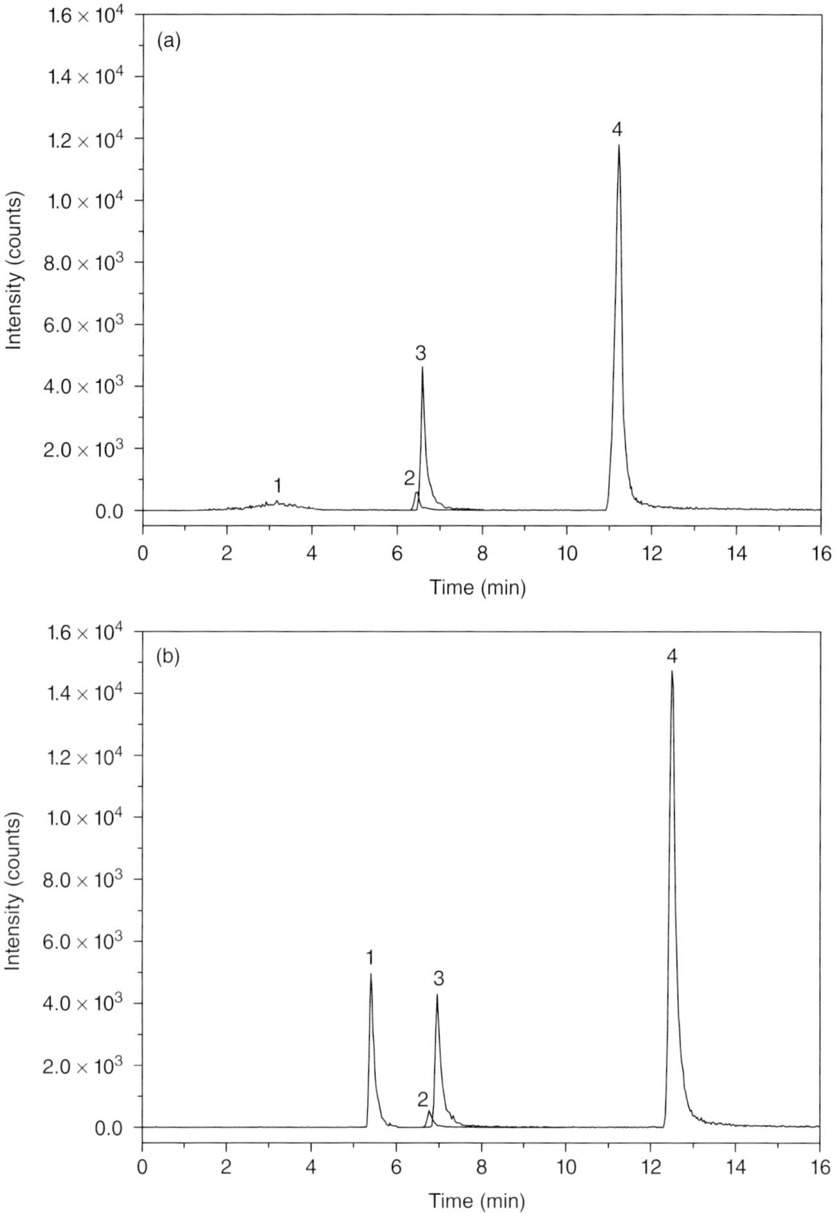

Figure 9.6 Separation of four antineoplastic drugs (a) without and (b) with online enrichment [17]. Chromatographic parameters: stationary phase: (a) Waters XBridge C_{18} (50 mm × 2.1 mm, 5 μm) and (b) ThermoHyperCarb (10.2.1 mm, 5 μm) coupled to a Waters XBrdige C_{18} (50 mm × 2.1 mm, 3.5 μm); mobile phase: (A) deionized water + 0.1% trifluoroacetic acid and (B) acetonitrile + 0.1% trifluoroacetic acid; flow rate: 0.5 ml min^{-1}; gradient: 0–99% B in 10 min, 10–20 min, 100% B; injection volume: 1000 μl; temperature: 35 °C; detection: MS. Analytes: (1) gemcitabine, (2) ifosfamide, (3) cyclophosphamide, and (4) fenofibrate.

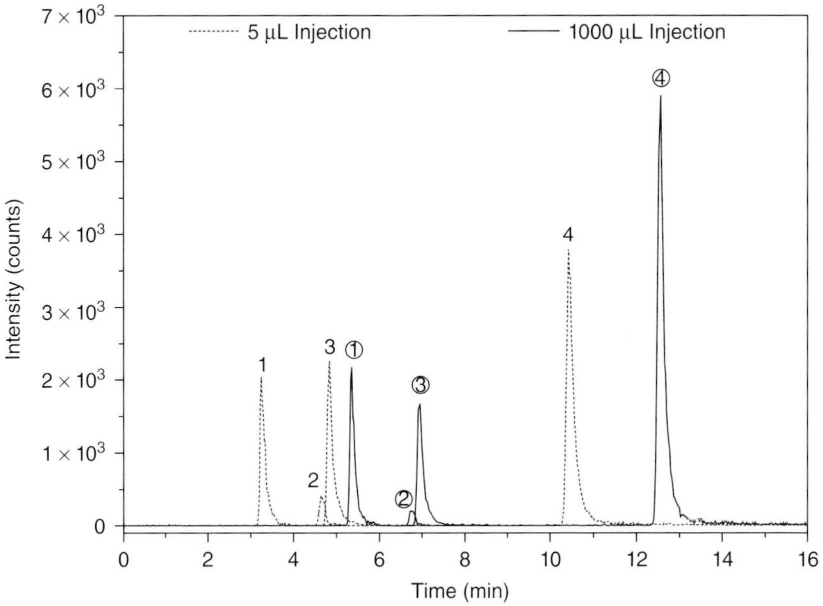

Figure 9.7 Separation of four antineoplastic drugs [17]. Chromatographic parameters: stationary phase: ThermoHyperCarb (10 2.1 mm, 5 μm) coupled to a Waters XBrdige C_{18} (50 mm × 2.1 mm, 3.5 μm); mobile phase: (A) deionized water + 0.1% trifluoroacetic acid and (B) acetonitrile + 0.1% trifluoroacetic acid; flow rate: 0.5 ml min^{-1}; gradient: 0–90% B in 10 min, 10–20 min, 100% B; injection volume: 1000 μl; temperature: see figure; and detection: MS. Injected absolute amount of substance: 25 ng. Analytes: (1) gemcitabine, (2) ifosfamide, (3) cyclophosphamide, and (4) fenofibrate.

9.7 UHPLC Separation with 1 mm ID Columns

Since the introduction of UHPLC in 2004, there have been many further developments of this technology such as the commercialization of column with an inner diameter of 1 mm. By changing the inner diameter from 2.1 to 1.0 mm, the solvent consumption can be reduced at constant column length by a factor of 4.4. In addition, the compatibility to mass spectrometry can be increased when 1-mm-ID columns are used. However, in the scientific literature, the applicability of 1-mm-ID columns is nowadays not recommended and critically discussed [18–21]. The reason for this is that the efficiency of 1-mm-ID columns is not comparable to the efficiency achieved on 2.1 – mm-ID columns. Generally, the chromatographic efficiency of 1-mm-ID columns is significantly lower than for 2.1-mm-ID columns due to the resulting peak dispersion caused by the extracolumn volume of the applied UHPLC system. This issue is discussed in detail in the academic research. From a practical point of view, the question that should be addressed is whether the efficiency of 1-mm-ID column is sufficient to solve the individual chromatographic separation problem or not. In this context, one example is described assuming that a chromatographic separation requires a theoretical plate number of 10,000. In order to solve this problem, a 2.1 – mm-ID column is available, which provides a theoretical plate number of 25,000. A corresponding 1-mm-ID column only generates 15,000 plates. Why should not the 1-mm-ID column be used, although the efficiency is lower than that for the 2.1-mm-ID column?

In the following, it is shown that modern UHPLC systems can easily be used with 1-mm-ID columns for simple separation problems using solvent gradients. For this purpose, the method described in Section 9.4 for the separation of 13 aldehyde- and ketone-2,4-dinitrophenylhydrazones is transferred to a 1-mm-ID column.

In order to transfer the separation illustrated in Figure 9.8a achieved on a 50-mm-long column with an inner diameter of 2.1 mm without any loss of efficiency to the corresponding 1-mm-ID column with the same length, it is necessary to further reduce the system volume of the UHPLC system. Therefore, the flow-through autosampler configuration was adapted to a fixed-loop injection (see Section 2.1.3). This way, the flow path of the sample band as well as the gradient delay volume could be significantly reduced. As the used columns differ only in their inner diameter, it was only necessary to reduce the flow rate from 1.2 to 0.272 ml min^{-1} for the method transfer. The profile of the solvent gradient needed no further adjustments. The results for the method transfer to the 1-mm-ID column are shown in Figure 9.8b.

The separation can be achieved as expected within 3.5 min with a maximum pressure during the solvent gradient of 825 bar. Nevertheless, it can be observed that a significant loss of efficiency occurs. This is illustrated by the marked impurity of the standard highlighted by a star. For the application of the 2.1-mm-ID column (Figure 9.8a), the impurities are partly separated from the main components. In comparison, only peak shoulders (Figure 9.8b) can be observed using the 1-mm-ID column. The decreased efficiency can also be appreciated by comparing the critical resolution. The resolution between the peak pair 7/8 is 1.41 and decreases to 1.01 using the 1-mm-ID column. It can be assumed that the observed band broadening is caused by the large dispersion volume behind the column. In order to reduce this volume, the high-volume and highly sensitive detector cell with a cell volume of 9 μl was replaced by a cell with a volume of 1 μl. The resulting chromatogram is shown in Figure 9.8c. By reducing the dispersion volume behind the column, a significant improvement of the efficiency was achieved. The critical resolution for the peak pair 7/8 could be increased from 1.01 to 1.43. Moreover, the marked impurities highlighted by a star are again partly separated from the main components.

However, the reduction of detector cell volume also decreases the limit of detection. The measured peak areas of the analytes are about a factor of 9–10 smaller using the 1 μl cell compared to the cell with a volume of 9 μl.

On the other hand, the application of 1-mm-ID columns in UHPLC is advantageous in terms of solvent consumption. With respect to the shown example in Figure 9.8, the consumption of acetone can be reduced by a factor of 4.4 using the 1-mm-ID column.

Another advantage of 1-mm-ID columns concerns the hyphenation to mass spectrometry using electrospray ionization (ESI-MS). Usually, for an optimal ionization in LC/ESI-MS, a flow rate between 250 and 500 μl min^{-1} is used. If 2.1-mm-ID columns are used, the flow rate is 1.2 ml min^{-1}, which corresponds to an increase in the flow rate by a factor of 3–5. Nevertheless, the resulting limit of detection is decreased. In contrast, the flow rates using 1-mm-ID columns are in the optimal range for the LC/ESI-MS hyphenation.

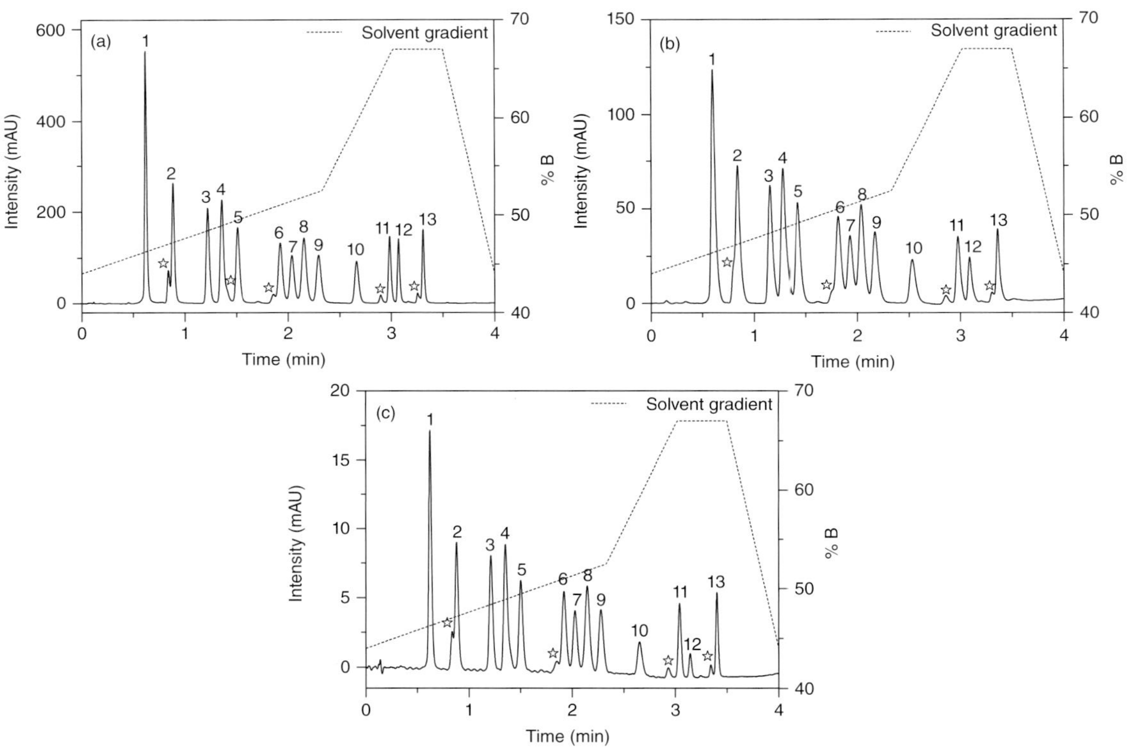

Figure 9.8 Chromatographic separation of 13 aldehyde- and ketone-2,4-dinitrophenylhydrazones. Chromatographic parameters: stationary phase: (a) Agilent Zorbax SB C_{18} (50 mm × 2.1 mm, 1.8 μm) and (b,c) Agilent Zorbax SB C_{18} (50 mm × 1.0 mm, 1.8 μm); mobile phase: (A) deionized water and (B) acetone; flow rate: (a) 1.2 ml min^{-1} and (b,c) 0.272 ml min^{-1}; gradient: see figure; injection volume: 1 μl; temperature: 33.0 °C; and detection: UV at 360 nm. Maximum pressure of the method: (a) 1080 bar and (b,c) 825 bar. Volume of the detector cell (a,b) 9 μl and (c) 1 μl. Analytes: (1) formaldehyde-2,4-DNPH, (2) acetaldehyde-2,4-DNPH, (3) acrolein-2,4-DNPH, (4) acetone-2,4-DNPH, (5) propionaldehyde-2,4-DNPH, (6) crotonaldehyde-2,4-DNPH, (7) methacrolein-2,4-DNPH, (8) 2-butanone-2,4-DNPH, (9) butyraldehyde-2,4-DNPH, (10) benzaldehyde-2,4-DNPH, (11) valeraldehyde-2,4-DNPH, (12) *m*-tolualdehyde-2,4-DNPH, and (13) hexaldehyde-2,4-DNPH. Ψ: impurities of the standard.

In summary, it should be noted that modern UHPLC system can easily be used with 1-mm-ID columns for simple separation problems using solvent gradients. Thereby, the costs for solvents and their disposal can be significantly reduced. Moreover, the resulting flow rates when using 1-mm-ID columns are perfectly suited for the hyphenation to ESI-MS. However, UHPLC systems need to be optimized in terms of the minimization of the dispersion volumes prior and behind the column.

Acknowledgment

The authors would like to thank Dr Kathryn Monks for the fruitful discussions and her help during the preparation of this chapter.

References

1 Gilroy, J.J. and Dolan, J.W. (2004) *LC GC Eur.*, **17**, 566.
2 Snyder, L.R. and Dolan, J.W. (2007) *High-Performance Gradient Elution: the Practical Application of the Linear-Solvent-Strength Model*, John Wiley & Sons, Inc., Hoboken, NJ.
3 Snyder, L.R., Kirkland, J.J., and Dolan, J.W. (2010) *Introduction to Modern Liquid Chromatography*, John Wiley & Sons, Inc., Hoboken, NJ.
4 Fountain, K.J. and Iraneta, P.A. (2012) in *UHPLC in Life Sciences* (eds D. Guillarme and J.L. Veuthey), Royal Society of Chemistry, Cambridge, pp. 29–66.
5 Ryan, T.W. (1995) *J. Liq. Chromatogr.*, **18**, 51.
6 Swartz, M. and Krull, I. (2006) *LC GC North Am.*, **24**, 1204.
7 Guillarme, D., Nguyen, D.T.T., Rudaz, S., and Veuthey, J.L. (2007) *Eur. J. Pharm. Biopharm.*, **66**, 475.
8 Guillarme, D., Nguyen, D.T.T., Rudaz, S., and Veuthey, J.L. (2008) *Eur. J. Pharm. Biopharm.*, **68**, 430.
9 Webster, G.K., Cullen, T.F., and Kott, L. (2013) in *Ultra-High Performance Liquid Chromatography and Its Applications* (ed. Q.A. Xu), John Wiley & Sons, Inc., Hoboken, NJ, pp. 31–54.
10 Debrus, B., Rozet, E., Hubert, P., Veuthey, J.L., Rudaz, S., and Guillarme, D. (2012) in *UHPLC in Life Sciences* (eds D. Guillarme and J.-L. Veuthey), Royal Society of Chemistry, Cambridge, pp. 67–101.
11 Teutenberg, T. (2010) *High-Temperature Liquid Chromatography: A User's Guide for Method Development*, RSC Publishing, Cambridge.
12 Wolcott, R.G., Dolan, J.W., Snyder, L.R., Bakalyar, S.R., Arnold, M.A., and Nichols, J.A. (2000) *J. Chromatogr. A*, **869**, 211.
13 Kozlowski, E.S. and Dalterlo, R.A. (2007) *J. Sep. Sci.*, **30**, 2286.
14 Busetti, F., Backe, W.J., Bendixen, N., Maier, U., Place, B., Giger, W., and Field, J.A. (2012) *Anal. Bioanal. Chem.*, **402**, 175.
15 Hogendoorn, E.A., Verschraagen, C., Brinkman, U.A.T., and Vanzoonen, P. (1992) *Anal. Chim. Acta*, **268**, 205.

16 West, C., Elfakir, C., and Lafosse, M. (2010) *J. Chromatogr. A*, **1217**, 3201.
17 Reeh, A. (2011) Master thesis. Entwicklung einer Onlineanreicherung für die Multikomponentenanalyse von Arzneimitteln mittels HPLC-MS unter Verwendung einer Chromatographiesimulationssoftware, Fachbereich Chemie, Hochschule Nierderrhein.
18 Wu, N.J., Bradley, A.C., Welch, C.J., and Zhang, L. (2012) *J. Sep. Sci.*, **35**, 2018.
19 Lestremau, F., Wu, D., and Szucs, R. (2010) *J. Chromatogr. A*, **1217**, 4925.
20 Heinisch, S., Desmet, G., Clicq, D., and Rocca, J.L. (2008) *J. Chromatogr. A*, **1203**, 124.
21 Fekete, S., Kohler, I., Rudaz, S., and Guillarme, D. (2014) *J. Pharm. Biomed. Anal.*, **87**, 105.

10

Experiences of an Independent Service Engineer – Hints and Recommendations for an Optimal Operation of Agilent and Waters-Devices

Siegfried Chroustovsky

10.1 Introduction

This section aims to provide a brief description of the various methods that can be used to localize and eliminate errors. The absence of drawings is intentional, and these can be found in the user manuals issued by the manufacturers and on the Internet. It is assumed that the reader has an in-depth knowledge of the functional principles of all components in an high-performance liquid chromatography (HPLC) system, as well as the software used by the various manufacturers.

10.2 The Degasser, Principles

All of the mobile solvents and solvent mixtures used in HPLC (including those with added inorganic or organic compounds) contain various quantities of dissolved gases. If these gases, which are essentially made up of oxygen, nitrogen, and carbon dioxide, are not removed to an adequate extent, they can exert an adverse effect on the quality of separation or make it difficult to evaluate the chromatograms. Under certain circumstances, they can even cause malfunctioning of the pump if a vacuum is produced as a result of long suction distances or clogged intake frits upstream of the pump, causing the aforementioned gases to be released. The resulting gas bubbles cannot be dissolved again during the pressure phase of the pump, however, and can give rise to brief interruptions in the flow of solvent and cause retention time fluctuations, separation problems, or error messages (with the Waters 2690, 2695, etc.).

Nearly all of the older types of degassers that use coiled bundles of Teflon tubes have long suction distances. There are two methods of removing the gases from the eluent as completely as possible, which have proven to be effective – helium sparging and gas osmosis – and these can also be combined. The practice of heating the eluent with or without a vacuum and the ultrasonic degassing technique should only be used as a last resort if there is no efficient degassing equipment

The HPLC Expert II: Find and Optimize the Benefits of your HPLC/UHPLC, First Edition.
Edited by Stavros Kromidas.

available as both of these methods are relatively complicated, and they are not particularly effective.

10.2.1 Different Manufacturers, Different Concepts

The most well-known HPLC equipment manufacturers have decided in favor of gas osmosis and have been adopting this technique for many years now. Agilent initially launched a four-channel degasser – the G1322 – with four tubular plastic membranes in a large vacuum chamber, whereby the vacuum is controlled by means of a pressure transducer, an electronic evaluation unit, and a diaphragm pump. However, a defect in the vacuum chamber causes a failure of the complete degasser. The advantage of this device is its large channel volume, which continues to work very effectively even if there are large quantities of gas dissolved in the eluent. The successor model – the G1379 – works with four individual chambers, which are accommodated in two enclosures of two chambers each. Each of the two enclosures can be replaced individually. Each chamber contains a membrane made of gas-permeable material. This means that the volume of a channel is considerably smaller than that of a bundle of tubular membranes. Both degasser models can be used in all Agilent HPLC systems, from Agilent 1100 to Agilent 1290. These degassers can also be used with any other HPLC systems as they do not need to be coupled to the system electrically.

Although the manufacturer recommends the use of a sub-D connector to link either of these two degasser models to the rest of the system, this can cause the entire system to come to a standstill if a fault occurs in the degasser during a run (e.g., overnight). This connection should not be used at all and, if connected, should be simply disconnected. A red LED at the top, on the right-hand side, already lights up via a fiber-optic conductor in the event of a degasser failure.

Waters, the other major HPLC equipment manufacturer, uses four individual degasser chambers, which can be replaced individually, in both the older modular systems and in the more modern, compact models including the 2690 and 2695 and so on. The older modular systems operated with a so-called inline degasser, which was a separate unit with four chambers.

The degassing function can be checked by generating an air bubble in the intake hose to each channel, for example, by lifting the intake frit briefly. This air bubble must then be divided into several smaller bubbles by pressing the hose between your fingers several times as a large air bubble would require a much longer dwell time in the degasser (less surface area!) than several smaller bubbles. If the flow rate amounts to approximately 1 ml/min, no further air bubbles may be visible downstream of the degasser, that is, upstream of the pump, under these conditions.

All of the hose connections between the eluent and the degasser, and between the degasser and the pump inlet, must be checked for signs of porosity from time to time. Eluents containing buffer can make hoses brittle as a consequence of the pH value. If so-called modifiers are used, they may be caught up here and affect subsequent chromatograms.

10.3 The Pump, Principles

The high-pressure pump of an HPLC system is of paramount importance as far as the chromatographic result is concerned, as the accuracy and long-term stability of eluent delivery affect the retention time and, with it, the reproducibility of the chromatograms.

Modern HPLC pumps are piston pumps, which comprise one or several pump heads. Each pump head contains one piston, whereas a double-head pump contains two pistons, which are made of different extremely hard materials. Pistons made of sapphire or ruby are commonly used in the majority of pumps. Both materials are varieties of corundum (aluminum oxide). Some pump manufacturers also use pistons made of zirconia oxide. Sintered silicon carbide is a new material used in high-pressure pumps.

Each pump head is equipped with an inlet valve and an outlet valve. The inlet values may be electromagnetically controlled, whereas the outlet valves take the form of ball valves or double-ball valves.

Depending on the manufacturer, the valve may comprise a cartridge, which can simply be replaced in the event of a malfunction, using the outer valve body again, or a complete valve, which can certainly be given a new lease of life by an experienced HPLC operator in the event of a malfunction.

Contamination can adversely affect the functional capability of cartridges and complete valve units alike. Outlet valve malfunctions may be attributable to piston abrasion, whereas contamination from the eluent can cause inlet valve malfunctions. In most cases, the functional capability of such a valve or cartridge can be restored by putting it in a glass beaker filled with acetone and treating it in a warm ultrasonic bath for around 15 min. Before commencing this procedure, a syringe should be used to inject acetone into the valve or cartridge in the forward direction to eliminate the possibility of air bubbles, which could interfere with the cleaning process. A spray bottle is subsequently used to flush the valve or cartridge thoroughly with methanol or isopropanol in the direction of flow, and it can then be tested by blowing into it in the forward and reverse directions.

If eluent has escaped from an Agilent outlet valve (part number G1312-60012), but the threaded capillary connections are dry, the eluent is escaping through the relief well under the threaded ring, indicating that one of the ceramic guide tubes inside the valve has broken.

There is no need to despair, however. If an old valve removed from service following a similar incident has been kept for spare parts, it is only necessary to open both valves carefully and transfer the ceramic tube that is still intact (there are two in each valve) from the old valve to replace the defective one. These are double valves, which means that only one of these components is broken. The procedure for dismantling the valve is as follows:

1) Remove the valve from the system.
2) Pull the plastic ring off upward.
3) Remove the threaded ring.
4) Remove the gold seal.
5) Remove the strainer, if fitted.

6) Clamp the valve body in a vice by the flattened surfaces with the inlet pointing downward. You will need the usual 5/16 in. double open-end wrench. Applying a little force and a great deal of care, loosen the two parts of the valve, but do not unscrew them completely (or the entire content will fall out).
7) Transfer the valve to a suitable supporting surface (soup plate!) and separate the two parts, being careful to note the order of the components. Replace the defective component, and put everything back together without exercising too much tightening force with the valve clamped in the vice, and then replace the strainer and the gold seal.

Fit the valve into the system, but do not switch the pump on yet as something must have caused the valve to become defective. It is now time to undo the capillaries downstream of the valve in order to find the blockage. Switch the pump on, then, beginning at the valve, start to tighten the capillaries, one after another, while monitoring the pressure. If the pressure does not rise, you can tighten the capillary completely, then continue with the next one until you have found the blockage. It goes without saying that the Rheodyne valve must also be included in the check. If a capillary is blocked, apply pressure carefully in the opposite direction, and the blockage will usually be cleared accompanied by a noise. This procedure should also be adopted if the valve is replaced by a brand new one, as the new one could otherwise become defective immediately afterwards.

A pressure limit (approximately 300 bar) can be entered as another means of restricting excessively high pressure. This course of action requires the user to accept the fact that the pump switches off every time.

A new improved outlet valve is now available for the Agilent 1100, 1200, and 1260 systems (part number G1312-60067). This outlet valve has no strainer or the usual gold seal. The internal structure is also more robust, eliminating the need for the relief well described earlier.

A malfunction in an Agilent system may also be attributable to a defective active inlet valve (AIV) (part number G1312-60025). If the valve still looks good on the outside, the AIV cartridge is probably defective or blocked. Pressure and retention time fluctuations are also manifestations of this fault. The AIV cartridge is very expensive, and the cleaning treatment mentioned earlier should be carried out here as well before reaching for a new cartridge (part number 5062-8562). An open-end 14 mm wrench is used to remove the valve unit after switching the pump off and disconnecting the gray plug connector. You can simply pull the cartridge out with your fingernails.

If a pump of any make fails to deliver the necessary volume or the delivery rate gradually deteriorates, this may be due to acetonitrile polymerization. This has already been the subject of many reports, which also discuss a means of avoiding such polymerization by adding 5% water. However, nobody can describe what these polymerization products really resemble or how they can be prevented.

If a situation such as the one described arises, the system should be flushed with a circulating mixture of 50% methanol and 50% acetone for 24 h. At the end of this procedure, a fresh mixture should be used to rinse the system clean. This cleaning process has proved to be successful in the majority of cases. If little time is available for the cleaning process, the flushing period may be shortened. The user can also resort to a slightly more radical alternative in the form of a mixture of isopropanol or methanol and tetrahydrofuran (THF). Caution must be exercised here, however, as THF is an excellent solvent for plastics (and PVC in particular). Appropriate exposure times and THF concentrations must therefore be chosen to ensure that degasser membranes and high-pressure seals are not damaged.

It may be advisable to perform such a cleaning process from time to time where acetonitrile and buffers of any description are required for gradient elution, rather than waiting until it is too late and the system ceases to function.

10.3.1 Different Manufacturers, Different Concepts

Pump manufacturers adopt two different concepts when it comes to gradient elution. One of these involves mixing two to four eluents together in the low-pressure section, that is, upstream of the high-pressure pump. This is the method adopted by Waters, not only in the company's modular systems (with Waters 600 pump), but also in such compact models as the Waters 2690, and 2695. In the other concept, the two eluents are not mixed together until they are downstream of at least two high-pressure pumps. The ability to process four eluents is ensured by connecting a combination of four valves upstream of one or two high-pressure pumps. One manufacturer (Agilent) even offers the alternative use of both principles in one of its systems.

This means, for example, that the Agilent 1100 system can be operated with either a quaternary pump (G1311) or a binary pump (G1312).

The G1311 pump has just a single pump head with multichannel gradient valve (MCGV) (part number G1311-657701) fitted upstream. Although the G1312 pump has two pump heads, it can be upgraded by fitting a supplementary four-way MCGV, whereby attention must be given to ensuring that two channels are used for just one pump head.

Another possible eluent delivery and mixing variant was adopted in the Hewlett Packard 1090 system. In this case, up to three separate low-pressure pumps were used to produce the gradient mixture. The finished mixture was routed to a high-pressure diaphragm pump. The technology was highly complex, and, this being the case, maintenance also proved to be very expensive.

A fundamentally different principle is adopted in the modern Waters systems (2690, 2695, and subsequent models). A four-channel gradient proportioning valve (GPV) assembly is fitted upstream of the high-pressure pump, and its four inputs are connected to the four outputs of the four degasser chambers. The shared GPV output is connected to the inlet valve (cartridge) of the high-pressure pump – the primary piston drive – which has no outlet valve. From here, the eluent flows via a pressure transducer (primary transducer) to the inlet valve (cartridge) of the accumulator piston drive. The eluent leaves the accumulator via another pressure transducer (system transducer) and passes through a purge valve and sintered frit on its way to the injector. The electronic system detects the minutest changes in pressure and controls the two piston drives independently of one another with such precision as to balance out any pressure fluctuations.

The test software provided by the various manufacturers offers numerous means of checking the pressure stability and leak-tightness of the different systems. These leak and pressure tests are generally performed automatically by the software with guidance. A restriction capillary should be used to generate pressure without having to use an old column. This capillary is available from Agilent under part number 5022-2159. The capillary can be used with Agilent systems directly as Agilent uses standard Swagelok connections in all of its units. An adapter comprising a Waters fitting with ferrule on one side and a Swagelok fitting with ferrule on the other side must be used for a Waters system. An attempt to screw a Waters fitting to a Swagelok seat causes so much damage to both components that it would probably be impossible to make the connection leak-tight, or extreme force would be required. The head of the Waters ferrule is likely to be torn off in an attempt to undo such a connection again. The 1/4- and 5/16-in. open-end wrenches are the only wrenches that may be used for such threaded connections. If metric wrenches are used in spite of this, the heads of the fittings will be rounded off, making it impossible to undo the connections, even if the proper imperial (inch) tools are used.

10.4 The Autosampler, Principles

Until a few years ago, valves that were looped in between the pump and the column were used to enter a sample into an HPLC system. Most of these valves were manufactured by Rheodyne, and they essentially comprised a rotatable unit – the rotor – and a rigid component – the stator – which usually had six ports. The rotor included a so-called rotor seal, which connected two of the ports in the stator together. It was therefore possible to steer the flow of liquid in two directions as there were three small grooves milled into the rotor seal. The stator, which is also referred to as a *lapped stator*, had a thread cut into it above the holes that made contact with the rotor seal, and Rheodyne produced special ferrules and fittings of various lengths, which fitted onto this thread. The ferrules were designed for steel capillaries or PEEK capillaries. Although this type of simple valve (model 7010) was unsuitable for injecting in itself, it could be upgraded by a port and a sample loop (for 5 μl or more). Model 7125 could be used to inject samples directly. A sample loop (for 5 μl or more) was permanently installed at two of the valve's inputs, and this was filled by means of a normal syringe, but with a special needle, in the "load" position. At least five times the loop volume should be allowed to flow through the loop before switching the valve to "inject." This was how the content of the sample loop was introduced into the circuit, that is, applied to the column. The HPLC system received its start signal via an additional contact (optional!). Assuming that it had been well maintained, a Rheodyne 7125 valve was even capable of an injection accuracy of less than 1.0%.

The development progressed rapidly, supplementing the valve with mechanical and electronic components, which offered a means of processing a larger number of samples for each program. A simple but highly reliable autosampler of this nature is the G1313, manufactured by Agilent.

This autosampler is also available with a thermostat (G1330) under the number G1329. Both autosamplers are equipped with a special Rheodyne valve with a 100 μl sample loop. The sample loop is connected to an injection needle, which has a mechanical raising and lowering facility. Its travel path terminates in a needle seat which, as the needle, needs to be replaced from time to time. A handling unit (robot) transports the vial from the programmed position under the raised needle, which travels into the vial, whereby a so-called metering pump draws the required quantity of liquid into the sample loop. The needle releases the vial again, and the handling unit ensures that it is precisely returned to its original position. The needle is then pressed into the needle seat, and the Rheodyne valve switches over. Driven by a stepper motor, the metering pump presses the required sample quantity through the needle seat and its capillaries, via the Rheodyne valve, into the flow of eluent, which carries the sample to the column where the actual chromatography takes place.

The sampling and application processes are fairly simple and straightforward thanks to the use of a Rheodyne valve. The system works very accurately and reliably if maintained properly.

10.4.1 Different Manufacturers, Different Concepts

Other HPLC equipment manufacturers do without this Rheodyne valve and use valves of a proprietary design to control the eluent. The autosamplers produced by Waters many years ago contained motorized valves with switchover heads, for example. Two of these valves were required in order to control the streams of liquid. They were referred to as *V1 (sample inlet valve)* and *V2 (syringe valve)*. They were installed in modular HPLC systems with autosampler models 717 and 717+. The same valves are also found in Waters compact models 2690 and 2695, and so on. In recent years, the die-cast supporting component actually responsible for the switchover has been replaced by a supporting component made of plastic. Originally designed with thread screws, the four screwed connections are now designed with thin, self-tapping screws, which can strip the thread and lose their grip if tightened a little too much in the course of maintenance work. In this case, the valve is no longer leak-tight. The supporting component is not available as a spare part, which means that the entire valve has to be replaced with a very expensive new (or possibly refurbished) valve.

One economic way of solving this problem involves squeezing a small amount of a two-component epoxy adhesive (UHU Plus) into the damaged thread, inserting the slightly lubricated screw and tightening it as far as the stop, and leaving it to cure. The screw can then be undone again, and the valve-reconditioning process can be continued. Caution must be exercised in spite of this, however, as the new plastic material is not as robust as the original plastic. Nevertheless, this represents a significant cost saving.

The sample is metered into the aforementioned autosamplers by means of a glass syringe (which are available in various sizes for different applications) controlled by a stepper motor. There is another small solenoid valve (V3 – waste valve) fitted downstream, which is used to replace the content of the syringe during cleanup flushing and for sample metering. The V3 valve must close properly as part of the measured quantity of liquid would otherwise escape into the waste container, which is on the other side of V3. The positive aspect is that there are testing options available for every valve, and these can be performed directly via the touch screen. The entire injection system can be checked by means of the compression check, for example.

Every time a needle or seal is replaced, an applet (seal pack adjustment) must be used to exactly adjust the motor-controlled injector needle with the side port to ensure that it is positioned in the correct chamber for the respective stage of the injection process, that is, in the wash position, the injection position, and in the stream position. If a problem occurs during the adjustment process, the range must be corrected by means of an adjusting screw on the injector.

The injection accuracy of an autosampler is checked by filling a vial with water and sealing it with a septum and an aluminum cap. The complete vial must then be weighed on the analytical balance. The vial is placed in the autosampler, and 50 injections of 10 μl each are made over a run time of 1 min. The direct injection function of a Waters autosampler offers a means of entering this type of activity via the display directly, whereas the program of an Agilent unit should be used to access this function.

At the end of the process, the vial must be weighed on the balance again, and the difference between the weighed values gives a good indication of whether the injector is still working accurately.

The vial containing the sample to be analyzed is not moved into position under the injection needle in the systems of some manufacturers, which move the needle into position over the vial in order to withdraw the sample. This technique requires that electric cables and lines filled with liquid are constantly moved along with the needle, which can make the system more susceptible to faults.

If the accuracy of the entire HPLC system is to be checked, this can be done by running through the process with a caffeine solution or a mixture of ethyl benzoate and isopropyl benzoate diluted to an appropriate extent in methanol or water with a 5 μ C18 column, eluent:water/methanol 1 : 1 at 254 nm. The Relative Standard Deviation can be determined after five to eight injections.

10.5 The UV Detector, Principles

The most frequently used detectors are the ultraviolet absorption detectors. They generally cover a range of 190–700 nm. Almost all manufacturers of HPLC systems have produced their own proprietary detectors. Some detectors are virtually or absolutely identical but are marketed under different brand names. Detectors are always stand-alone devices, which are not integrated into compact units as other functional groups.

HPLC systems can also be combined with devices produced by other companies. One example of a popular combination consists of an Agilent 1100 LC system and a Waters DAD 996 or 2996 detector. In this case, a SAT/IN A/D converter and a LAC/E acquisition server with Empower software are used to preprocess the output signal from the detector.

A various wavelength detector (VWD) essentially comprises a deuterium lamp, a focusing device, a reflective mirror, and an electrically rotatable grating downstream. A beam splitter is installed behind the grating, which diverts half of the radiant energy to a suitable photodiode, which passes it to the evaluation electronics as the reference value. The other half of the energy is routed through a flow cell to another photodiode – the sample diode – and the measured value is compared with the reference value to obtain the value for the signal in the cell.

The rotatable grating (which replaces the prisms that were originally used) enables selection of the required wavelength by means of the electronic system.

In a diode array detector (DAD), all of the light from the deuterium lamp is allowed to pass through the flow cell, and it is absorbed at many points of the spectrum. This light is broken down into its individual wavelengths in the monochromator and then irradiates an array made up of several hundred diodes. The electronic system scans the array of diodes in quick succession to produce an image of each wavelength's intensity. Scanning must take place very quickly as the flow of liquid in the cell includes brief changes. The user can also select a specific wavelength and call the pertinent values to the display.

Apart from the deuterium lamp, Agilent's G1315 diode array detector (DAD) and G1365 multiple-wave detector (MWD) are also equipped with a small flashlight bulb, which shines through a hole in the deuterium lamp and increases the intensity above 370 nm.

10.5.1 Different Manufacturers, Different Concepts

In practice, detectors that use deuterium lamps may also prove to be potential error sources, which are not immediately identifiable.

When changing a lamp, the user should actually enter all of the necessary information requested by the system. This is very easy when it comes to Agilent detectors. Regardless of whether the controller is an older one (model G1323B) or a more recent instant pilot (model G4208), the lamp change can be entered in the controller along with subsequent tests. The user can also call up a display of the current status of the lamp across the entire wavelength range, as well as the running time of the lamp, whether the lamp has been changed or not.

The service life of a lamp and the decision to replace it depend on the wavelength used to a great extent. Lamps operated at between 190 nm and around 230 nm will have the shortest service life. This is the range at which the lamp energy diminishes most quickly, requiring more frequent lamp replacement. The service life of a lamp operated in the upper range may amount to 3000 h.

Attention must be given to the fact that lamp noise may increase considerably as the energy diminishes.

If a lamp is too old, the plasma discharge may exhibit sudden changes, which are evident as steps in the baseline. This is often interpreted as being an indication of a defect in the pump, but this possibility can be easily ruled out by recording the pressure.

If the chromatographic peaks have been diminishing continuously when working in the range between 190 and 220 nm, and the lamp test in this range does not produce the easily recognizable energy peak, the cell has probably become impermeable to a large extent. If conventional washing methods (isopropanol) have proved ineffective, there is another tried and trusted, if somewhat unusual, method of cleaning the cell window, rather than having to replace it unnecessarily.

Fill a small glass beaker or a small Erlenmeyer flask with glacial acetic acid, insert the intake frit for channel A into the acid, and secure the arrangement with Parafilm to keep the odor pollution at a low level. Remove the column, and connect the column input and column output together.

Undo the screw fitting at the detector output, allowing the liquid to escape. Start the pump at 1.0 ml/min, and use absorbent paper to check the detector output until such time as the glacial acetic acid is discharged (odor). Then tighten the screw fitting again. Leave the pump to run with glacial acetic acid for 3 min, and then transfer the inlet frit for channel A to a liter bottle of water. Check the cell from time to time to establish whether the energy is becoming visible again in the range from 190 to 220 nm. It could take some time to flush the system with water, and you can observe the way in which the energy gradually increases again within this range.

This washing process has been frequently carried out on detectors manufactured by Agilent and Waters and has always produced good results. Before washing the Waters DAD, only one line running parallel with the *X*-axis was visible in the 190 to 220 nm range. The process was rendered visible by means of the diagnostic function with Empower software.

11

The Analyte, the Question, and the UHPLC – The Use of UHPLC in Practice

Stefan Lamotte

11.1 Introduction

The requirements on liquid chromatographic separations depend on the individual separation task. Indeed, there are applications where UHPLC methods are the method of choice. On the other hand, there are quite a number of applications where conventional HPLC methods with maximum pressure up to 600 bar are clearly more beneficial. For achieving the ideal condition for a liquid chromatographic separation next to the questionable HPLC/UHPLC, the choice of the ideal column dimension (length and inner diameter) is crucial, depending on the individual separation task. This chapter gives a guidance when and which UHPLC separation conditions should be used and when they are superior over conventional HPLC conditions.

11.2 When Does It Make Sense to Use UHPLC and When Should I Better Use Conventional HPLC?

UHPLC columns are packed with particles with particle sizes ≤2 μm. In contrast to conventional HPLC columns, UHPLC columns are shorter and have generally narrower inner diameter. The narrower inner diameter is necessary to get rid of generated heat stemming from higher linear velocity and higher column back pressure in UHPLC. UHPLC columns can be shorter because the required plate number for a separation can already be achieved in a shorter column. The practical consequence of the reduction of column inner diameter and length is a reduced volume of the column. Therefore, it is necessary that the UHPLC instruments used are also volume optimized. In some chapters of this book (Chapters 2.1 and 3), this is discussed in detail. For the application of UHPLC in a liquid chromatographic laboratory, this means that it is necessary to dissolve the sample in a solvent with a weaker or at least the same solvent strength as the chromatography started. On a conventional HPLC column, elution bands (peaks) are wider, and because of the larger volume, a larger dilution of the elution bands occurs. Therefore, in conventional HPLC, stronger solvent

The HPLC Expert II: Find and Optimize the Benefits of your HPLC/UHPLC, First Edition.
Edited by Stavros Kromidas.

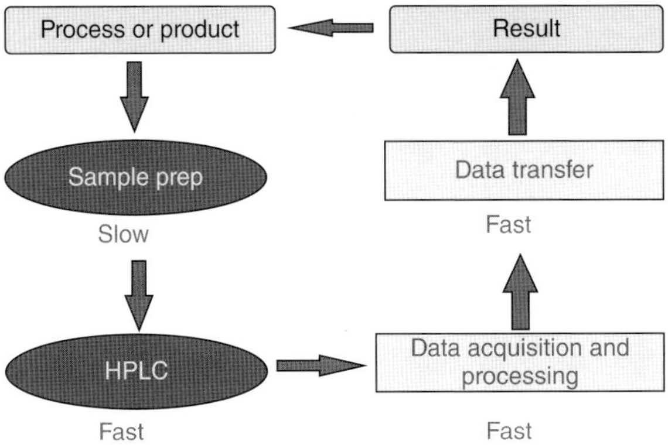

Figure 11.1 Flowchart of a typical HPLC analysis.

strengths can be easily compensated by the larger column volume. For the daily work in a liquid chromatographic laboratory, this means that UHPLC is harder or even impossible to apply if more complex sample preparation, which results in higher solvent strength solvents, such as it is generally in solid-phase extraction, should be necessary. But in situations such as this, UHPLC is senseless anyway because the sample preparation is slow and is therefore the speed-limiting step in the entire separation process. Consequently, fast chromatography and therefore UHPLC are not required. Before the decision for or against UHPLC is done, the entire separation process should be analyzed. In Figure 11.1, the flowchart of a typical HPLC analysis is illustrated. If the HPLC separation is not the speed-limiting step, UHPLC makes no sense, except saving solvents or higher mass sensitivity is an issue. When do the two factors solvent saving and mass sensitivity become critical? This is if the solvent price is going up. This was the situation we faced during the acetonitrile shortage in the years 2008/2009. Sometimes, it is required to use expensive solvents for the chromatography due to the solubility of analytes, which are not soluble in classical HPLC solvents such as methanol or acetonitrile/water mixtures. Sometimes, the analytes are only available in small amounts, or they are very expensive. In these cases, a high mass sensitivity is required. Here, UHPLC next to Nano-HPLC is a solution because the injection volume can be dramatically reduced compared to conventional HPLC.

It can be concluded that the advantages of UHPLC over conventional HPLC are higher speed of chromatography, lower solvent consumption, and higher mass sensitivity.

Misleadingly, it is often claimed that the use of UHPLC would lead to higher plate numbers and lower limits of quantification (LOQ). This is wrong, and exactly the opposite is true. Lower LOQs are only achievable with UHPLC if the sample amount is limited. because loadability of the column is directly proportional to the mass of the stationary phase in the column in conventional HPLC, higher amounts of sample (either mass or volume or both) have to be injected on HPLC columns. This fact is sometimes forgotten if one transfers a method from HPLC to UHPLC. The injection volume needs to be adapted by the factor of the column volumes between the HPLC and UHPLC columns.

Higher plate numbers are always achieved by packing large particles in long columns. The limit here is of course the back pressure, and the back pressure rises linearly with the column length but quadratic with the particle diameter. The rule is as follows: the longer the columns and the larger the particles, the higher the plate number. The higher plate number is paid by longer analysis time. For further discussions, it is referred to the literature [1, 2] and to the discussions in Chapter 1.

UHPLC offers great benefits for all separation where approximately 10 000 plates in a minimum time of analysis are required. Restrictions may be a very complex matrix where a sample preparation may be required and very high solvent strengths are needed to dissolve the sample or dirty samples where small particle diameter may be blocked. Besides these restrictions, UHPLC is beneficial over HPLC.

UHPLC is not useful if very high peak capacity (>1000) or plate numbers >100 000 plates should be required. Such high plate numbers are not achievable in UHPLC. Neither the less it is still useful to have a UHPLC instrument available in these cases. Longer columns packed with 5 µm particles coupled in series can be used in these cases, and these columns can be used even at pressures higher than 600 bar if required. Plate numbers as in gas chromatography can be achieved this way, but also the time of analysis is low.

In the following section, the application of UHPLC on different fields of liquid chromatography is discussed with the special focus on the column dimensions that are recommended for use.

11.3 Dissolution Tests in Pharmaceutical Industry

In these tests, the release of active pharmaceutical ingredients (APIs) in a formulation or support under defined conditions is monitored. These conditions should simulate the release of the API in or on different parts of the human body. In general, these samples show a low complexity. Often, only one or two components need to be separated from a matrix and quantified. Because the kinetics of the release is the aim of those studies, fast separations are necessary. Therefore, often short columns (20–30 mm length) are used because a high number of samples need to be measured in a short period of time. In these cases, UHPLC columns packed with sub 2 µm particles are not necessarily required. Also, short columns packed with conventional 3 or 5 µm particles or monolithic columns can be used. Especially, monolithic columns show an excellent tolerance against polymeric matrices, which are often used in combination with APIs in pharmaceutical industry. These matrices might be critical if columns with sub 2 µm particles are used.

11.4 Method Development and Optimization

Especially, in method development in HPLC, a large number of different parameters such as different selective columns, different pHs of the mobile phase, different solvents and gradient steepness, different column temperatures should be investigated. To get a full picture of the critical parameters of the separation, it is important to know what happens to the separation if one of these parameters is changed. Further, the investigation of these parameters ensures a robust and optimum method, which is able to be validated. For checking all these parameters, a lot of chromatographic runs are necessary. To achieve an equilibrium state in chromatography takes a relative long time, and if parameters such as different solvents or pH of the mobile phase are varied, it takes a long time, which cannot be used for measurements. The equilibration time strongly depends on column volume. If the column volume is lower as in UHPLC, equilibration is faster. Additionally, in gradient elution, peak capacity strongly depends on the volume of mobile phase, which passed during the gradient through the column. If the column volume is lower, peak capacity is increasing if the gradient steepness is kept constant [3, 4]. Therefore, UHPLC is a big advantage in this application. Further, the sub 2 μm UHPLC columns can be run at higher flow rates without losing efficiency. Therefore, even more column volumes per time unit can be run compared to conventional HPLC, leading to higher peak capacity or faster separations, faster equilibration, and finally, an enormous boost in the time required for the whole method development process.

The typical column dimensions for the sub 2 μm particles are UHPLC columns of 50 mm length and 2 mm inner diameter. These UHPLC columns offer the typical plate number of an HPLC column (depending on the analytes: 10 000 – 15 000 theoretical plates) in the shortest possible time of analysis. Typical run times are 5 – 10 min.

The biggest advantage of UHPLC in method development and optimization remains the short equilibration time in between the different parameters, which boosts up the entire method development times dramatically.

11.5 Typical "Classical" Liquid Chromatographic Analysis

Also, in typical liquid chromatography in a research laboratory, UHPLC can be beneficial. Some claim there would be a big advantage due to higher sample throughput. This is not true in a typical research laboratory. Only rarely you have sample series of larger than 50. If you have less number of samples, it is no difference if you work with HPLC or with UHPLC because you will have an autosampler, and it makes no difference if the last sample of series will be injected at 8 p.m. in the evening or next morning at 6 a.m. In both cases, the results will be reviewed in the morning next day and going out to the researchers.

The real advantage of UHPLC over HPLC in these laboratories is that you will see much earlier if something goes wrong with your sample sequence, that is, a sample lies outside your calibration curve and needs further dilution or the largest calibration point is outside the linear range of the calibration curve. If you have typical cycling times of UHPLC of approximately 10 min, you can still interact within the same working day, and the sample series can still be started again overnight. In conventional HPLC, the failure cannot be recognized on the same day, and the sequence needs to be repeated next morning.

Therefore, if UHPLC can be applied in research laboratories, it saves time and even more important resources, which can be used for other samples the next day.

11.6 Fast (Most Second) Dimension of Multidimensional Chromatography

In 2D-HPLC, it is necessary to differentiate between the purposes for the corresponding chromatography (see Chapter 6). In heart-cut 2D-HPLC, it is not necessary to use fast HPLC. Here one individual analyte should be separated from a very complex mixture of components. To do this, the part of the chromatogram in which the analyte of interest is eluting from the column is switched to a second-dimension column. This column provides a retention mechanism as orthogonal as possible to the first dimension. In addition, it is beneficial if the second-dimension column has a high peak capacity. Therefore, the use of a conventional HPLC column in the second dimension for heart-cut 2D-LC is often superior to an UHPLC column.

In comprehensive 2D-LC, the situation is completely different. Here a fast second dimension is compulsory. Each fraction, which is eluting from the first separation column, is immediately transferred to the second dimension and can only be parked for a short period of time before it is separated. During the separation in the first dimension, the second-dimension column is equilibrated, and a third column collects the eluting band from the first dimension. Therefore, short columns packed with sub 2 μm particles are the ideal choice for the second dimension of comprehensive 2D-LC. The fractions eluting from the first dimension are preseparated, cleaned, and diluted and represent therefore the ideal sample for UHPLC columns. The big advantages of UHPLC columns – high mass sensitivity and high plates per time unit (fast time of analysis) – can be fully exploited in this application. In most of the cases, the eluting band from the first-dimension separation is eluting in a mobile-phase solvent with a compatible weaker solvent strength for the second dimension. If this is difficult, the design of the separation strategy has to be adapted that way that the starting mobile phase for the second dimension has a weaker solvent strength. In this case, the elution bands of the first-column dimension are focused on the column head of the second-dimension column.

Due to increasing sample complexity, 2D-LC applications will gain more importance in the future, and consequently, the role of UHPLC as second-dimension column in comprehensive 2D-LC will increase.

11.7 Separation of (Bio)polymers

The separation of polymers is often complex. Due to multiple interaction sites of the analytes with the stationary phase, the interactions are much stronger compared to small molecules. As a matter of consequence, the elution profiles from those columns are broad and asymmetrical. Therefore, polymers need to be eluted from the column in gradient elution mode. In contrast to small molecules, the elution behavior of polymers is "digital." Between stop and go are often only small differences in solvent strength of the mobile phase [4]. Therefore, the columns used for polymer HPLC should be as short as possible. Every millimeter longer column leads to band broadening. Due to smaller diffusion coefficient of larger analytes and therefore slower mass transfer, the advantage of small particles as stationary phases in polymer HPLC is even higher than for small molecules. Mass transfer is getting more critical with slower diffusion [4]. But due to possible size exclusion issues, large pore size or nonporous supports need to be used as stationary phases. The disadvantage of wide-pore supports is the lower mechanical strength, which makes it difficult to use them under higher back pressure. Therefore, especially in biopolymer applications, often monolithic columns are used. Another excellent alternative are nonporous supports with sub 2 μm particle size.

11.8 Process Analysis (PAT)

Fast chromatography offers new opportunities to process analysis. It is possible to monitor fast chemical reactions and their kinetics. This allows a fast interaction in a chemical process if necessary. The big advantage of a chromatographic method over other spectrometric and optical methods is a higher selectivity and therefore less interference. In the past, HPLC was often not the method of choice due to the speed limitation. UHPLC offers now cycling times between two analyses of less than 5 min, and the robustness of new UHPLC instruments allows the use of the technique in process analysis. However, to establish a robust UHPLC method in a real process environment is a quite complicated task. In this area, a strong demand on flawlessness and redundancy must be ensured.

11.9 Conclusion

UHPLC offers a broad variety of solutions for chromatographic separation tasks. Whether it can or should be applied for an individual separation depends on the individual analytical problem.

References

1 Kromidas, S. (ed.) (2016) When should I use my UHPLC as a UHPLC? in *The HPLC Expert II – Find and Optimize the Benefits of Your HPLC/UHPLC*, Wiley-VCH Verlag GmbH & Co. KGaA.

2 Halàsz, I., Endele, R., and Asshauer, J. (1975) Ultimate limits in high-pressure liquid chromatography. *J. Chromatogr.*, **112**, 37–60.
3 Poppe, H. (1997) Some reflections on speed and efficiency of modern chromatographic methods. *J. Chromatogr. A*, **778**, 3–21.
4 Snyder, L.R. and Kirkland, J.J. (2010) *Introduction to Modern Liquid Chromatography*, John Wiley & Sons, Inc.

12
Report of Device Manufacturers – Article by Agilent, Shimadzu, and ThermoScientific

12.1 Agilent Technologies

Jens Trafkowski

During the past years, many developments in High-Performance Liquid Chromatography (HPLC) have taken place, leading to the extension of the term HPLC to UHPLC. Nowadays, "Ultrahigh-Performance Liquid Chromatography" (UHPLC) is seen as a new class in analytical separation sciences. Many users think mainly about the fact that these HPLC systems are characterized by a high operating pressure, enabling the use of columns with small particle sizes (typically sub 2 μm). As a result, the much higher efficiency and plate number lead to a better chromatographic resolution on the same separation length or/and a faster separation. In fact, not only the offers but also the needs of modern UHPLC systems are much more:

- Faster separations require higher data rates in the detectors. These needs are fulfilled from the later generation of diode array detectors (DADs), easily offering data rates of 100 Hz and more. Here, Agilent Technologies offers market leading data rates of up to 240 Hz.
- Further developments in the latest DAD techniques enable to reach new levels of sensitivity, which were not possible before. With the implementation of the Max-Light Cartridge, a flow cell with up to 60 mm length at the lowest dispersion volume and highest light transmission, Agilent Technologies was the pioneer in the elimination of refractoric and thermal effects.
- New autosampler techniques increase the flexibility and allow faster injection cycles, leading to higher sample throughput. This follows the need that new autosamplers have the necessary design and intelligence to handle much larger sample numbers. In future, this need may even grow, and so will the requirements to the upcoming autosamplers. Here, Agilent Technologies has set a benchmark with the 1290 Infinity II Multisampler:
  - Maintaining the standard bench size of an Agilent UHPLC-System, the sample capacity and the flexibility have been dramatically increased, without the need of additional bench space or an additional and costly module. In many laboratories, the bench space is already a limiting factor, and this issue shall not be increased.

The HPLC Expert II: Find and Optimize the Benefits of your HPLC/UHPLC, First Edition.
Edited by Stavros Kromidas.

- Depending on the need, different sample vessels can be used – starting with 384 microtiter plates going up to 6 ml vials. The flexibility of setting up the autosampler with up to 8 drawers, hosting 16 vial, or microtiter plates allows working off more than 6000 samples – in a single system of only 40 cm width!
- The Multiwash technology enables both cleaning of the injection needle from the outside in a separate wash port and the additional backflushing of the needle seat and the seat capillary with up to three different solvents. In addition, the needle and the injection loop are cleaned by the LC gradient. As a result, the growing demands to the highest detection sensitivity, especially when using high-end triple-quadrupole mass spectrometers, were taken into account by reducing carryover to a single digit parts per million value, which has never been reached before (see Figure 12.1.1).
- The unique use of two injection needles (Dual Needle Technology) is high flexibility and allows two different benefits, depending on requirements: on the one hand, the use of two injection needles with similar loop sizes enables the minimization of an injection cycle down to 5 s, and on the other hand, two different sample loops for large and small volumes can be used without any compromises in terms of delay volume, analysis time, and sample carryover.

- Besides providing highly stable thermal conditions for the separation, column ovens are expanding possible temperature ranges. Furthermore, they are becoming more and more established as functional units, making it possible to use increasingly more different methods in a single system, without any manual intervention. The 1290 Infinity II Multicolumn Thermostat (Figure 12.1.2) allows the simultaneous use of eight columns up to a length of 150 mm, including dispersion-optimized heat exchangers by means of a (removable) column switching valve for all eight columns. As a result, extensive method development and multimethod projects can run smoothly on one fully automated system without any manual interaction, optimizing the measurement time of the system to a maximum usage.
- Agilent Technologies not only provides valves for automatic column selection but also offers also a large portfolio of different valves, which can be used for other workflow applications such as solid-phase extraction or automatic column regeneration. With the introduction of the Quick Change Valves, the possibility to variably populate the column oven or an external valve drive with the different valves was created. These valves can be changed easily, within 10 seconds, enabling the user to switch between different workflows very fast.

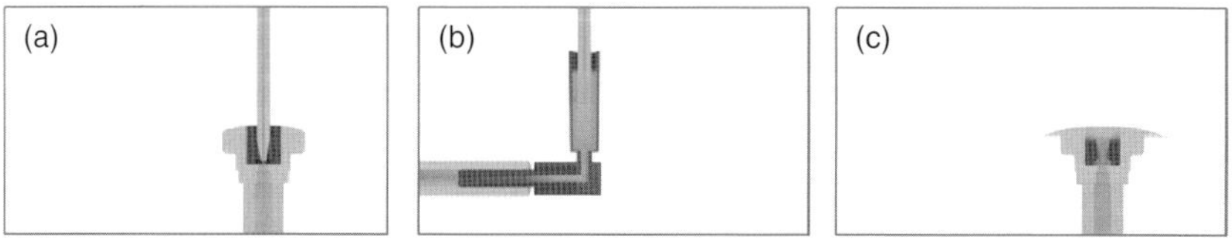

Figure 12.1.1 Schematic representation of the extended rinsing procedure: (a) rinse the needle by means of gradients, (b) rinse the outer needle surface in the wash port, and (c) backflushing of the needle seat and the needle seat.

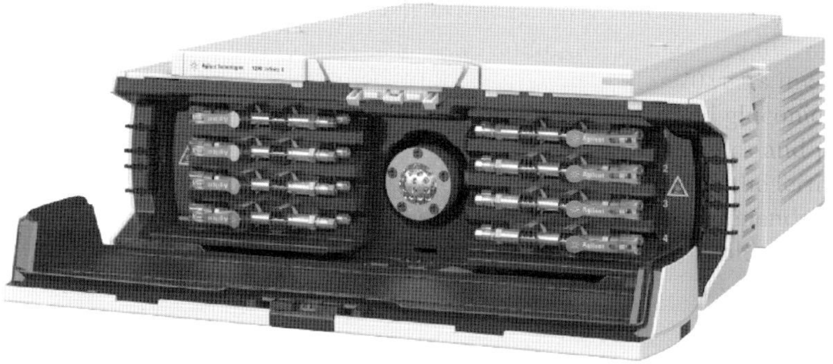

Figure 12.1.2 Agilent Infinity II Multicolumn Thermostat with eight built-in columns, low dispersion heat exchanger, A-Line Quick-Connect Fittings und column-switching valve.

Now the upcoming question is whether future developments will have the goal to outperform existing specifications ("higher, faster, wider") or if other criteria will be more in focus. Similarly to the developments in consumer electronics, where customers no longer decide only on the basis of large data rates or high optical resolutions, the experienced analytical chemists do not only ask for good paper specifications or extravagant product names but rather want a product that fulfills their analytical tasks the best. The analytical chemists' needs aim in different directions, all of which have the goal of increasing the laboratory's efficiency:

- Enlarging the analytical range with alternative and additional chromatographic techniques
- Increasing the chromatographic performance
- Maximizing the ease of use and thereby minimizing possible sources of error
- Determining ways to perform dedicated applications in special niche areas.

These different goals can be approached in different ways, which is explained in the following:

The enlargement of the analytical range can be achieved to a large extent with actual techniques, which only need to find their way into the existing laboratories. It runs in parallel with the increase in chromatographic performance, which will be discussed subsequently.

The prevailing standard of HPLC is still Reversed-Phase Chromatography (RP). By using orthogonal separation techniques, an increase in selectivity can be achieved, which can resolve previously poorly separable substances. In addition to exploiting the entire width of RP column materials, normal-phase chromatography (NP), ion-exchange chromatography (IEX), size-exclusion chromatography (SEC), and gel permeation chromatography (GPC) can be used as highly orthogonal techniques. Still relatively new is the range of Hydrophilic Interaction Liquid Chromatography (HILIC), which represents an intermediate region between NP and RP. Due to the strong interaction with polar analytes, the stationary phase corresponds therefore rather to NP, while known eluents from the RP act as mobile phase.

SFC (Supercritical Fluid Chromatography) has experienced a boom in recent years, and with supercritical CO_2 as mobile phase, it offers another orthogonal separation technology as alternative. With its SFC-UHPLC hybrid system, Agilent Technologies offers the ability to apply SFC and to use (U)HPLC in one system. The switch between both two techniques can be made without manual interaction. This can support the start of the new technique from two sites: the application aspect itself and the financial aspects, as a single system can be used for maintaining the standard LC workflow, but can also be used to start with the new technique. As no manual interaction is required to switch between both chromatography modes, the hurdle to try the additional technique may be more easily overcome.

As listed in the previous two paragraphs, there are already now a large number of alternatives to conventional HPLC, all distinguished by their own advantages and disadvantages. It raises the question whether the development of additional systems with higher pressure specifications is the best way to develop the "HPLC of the future." Increasing the pressure specifications can theoretically increase the resolution of the system to a certain extent but also brings up further system challenges such as dealing with material fatigue and maintenance requirements of the critical UHPLC system parts. These problems can be solved, but it is questionable whether the benefits of a slightly improved theoretical resolution are in meaningful proportion to the costs. Agilent Technologies has therefore decided to concentrate on the much more promising technique of 2D-LC.

A drawback of many analytical workflows is the need to perform some analysis with two different separation technologies, requiring two separate analysis runs. Two-dimensional HPLC (2D-LC, discussed in Chapter 6) allows the coupling of two (often highly orthogonal) techniques in a single run and therefore offers a unique opportunity to increase the peak capacity in orders of magnitude, which is not possible with one-dimensional HPLC. In 2D-LC, many of the above described techniques can be coupled together.

Agilent Technologies does not only offer the possibility to increase the pressure of a one-dimensional system but also enables the user to integrate the second chromatographic dimension and therefore to significantly raise the separation power using different 2D-LC techniques.

While Heart-Cutting 2D LC has already been established for some time and has found applications in different analytical problems, where the substances of interest are usually known (but incompletely resolved), this is not the case for more complex and still growing technique of Comprehensive 2D-LC. To enable the entry for the broadest possible range of users, Agilent Technologies provides a complete high-end system, which minimizes any difficulty for the user in setting up a system and designing advanced methods. The most important advantages of the Agilent 1290 Infinity II 2D-LC Solution are as follows:

- A pump with highest precision and accuracy, which provides highest performance also at the typically low flow in the first dimension, but also at the ultrafast run times in the second dimension (usually 20–30 s [1]). In all 2D-LC modes, retention times and quantitative results must be highly accurate and reproducible.

- A specially designed 2D-LC valve, which is characterized by inner symmetry leading to two identical flow paths in both valve positions. This eliminates the risk of switching interferences, which often occur, when standard valves are used (typically 2/10 or two 2/6 valves are used). Furthermore the unique Muiltiple Heart-Cutting Valves allow the easiest and fasted hardware setup for this advanced technique (see below).
- A smart software interface, which enables to easily program and vary complex 2D-LC gradients combinations. The use of "shifted gradients" offers a big gain in resolution [2]; therefore, it is of highest interest that the software simplifies the method setup also for this analytical technique. This full software support minimizes the risk of manual mistakes when programming even demanding gradients and has therefore a direct positive impact on the full 2D-LC system performance.

As the descriptive name already describes, comprehensive 2D-LC gives a complete overview about the whole sample, but the ultrashort gradients on the second dimension are also a demanding task. Longer gradients on the second dimension enable a better separation, the big advantage of Heart-Cutting 2D-LC. As a 2D gradient blocks the second dimension for the next 2D analysis, as long as it is running, the next injection is only possible, when this run is finished. To address the manifold cases, in which users would like to analyze multiple spots also in a second chromatographic dimension, as it is often the case in quality control, Agilent Technologies launched in 2014 a fully automatized and software supported Multiple Heart-Cutting 2D-LC system. In this system, the two sample loops are exchanged by two column selection valves, which are connected to the unique 2D-LC valve. The six-column selection valves are themselves equipped with six loops each. In total, there are now 12 loops in the system for taking cuts from the first dimension and storing them before the injection onto the 2D column. This technique is also known as *peak parking* and enables to analyze cuts, which are taken from the first dimension, while the second dimension is still occupied and running (Figure 12.1.3).

Multiple Heart-Cutting 2D-LC offers the possibility to cut various peaks out of the first dimension within a short amount of time and to analyze them subsequently in the second dimension. The decoupling of the two chromatographic dimensions enables Heart-Cutting 2D-LC "upon need" (Figures 12.1.4 and 12.1.5).

In addition to the purely technical aspects, which aim to enhance the chromatographic resolution and data quality, thus opening up more applications, it has already been mentioned that a higher user friendliness can minimize the possibility of errors, and thereby, the analytical results can be achieved faster and better. Many users aim for this efficiency enhancement, especially as more and more users are facing the task of increasing sample throughput.

The development of new analytical methods with the goal of improving the chromatographic quality further is only one part of the analytical work. As HPLC has already been established for many years as one of the most important gold standards in analytical laboratories, it is also of highest importance to safely reproduce and apply established methods in different laboratories. This

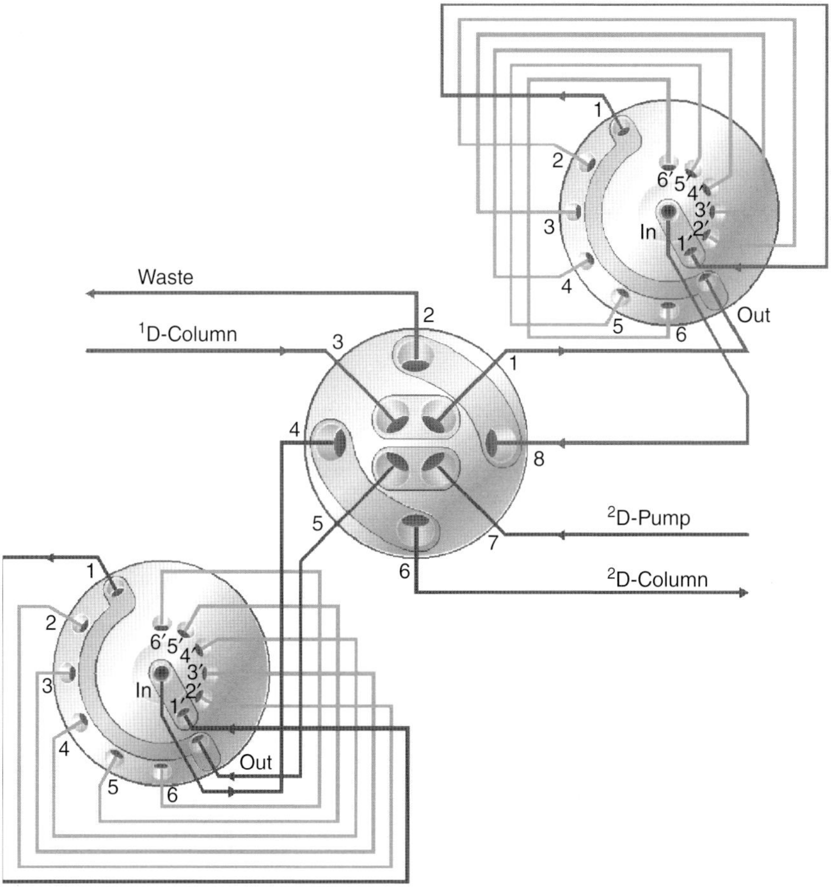

Figure 12.1.3 Agilent Multiple Heart-Cutting Solution Valve Setup. The two sample loops are replaced by two complete six-column selection valves with six sample loops each.

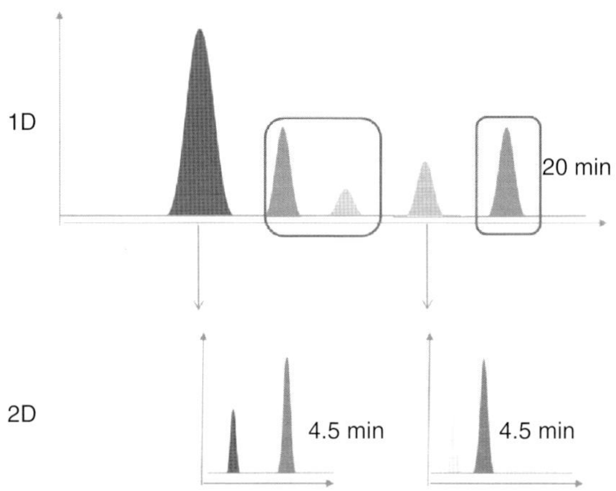

Figure 12.1.4 Heart-Cutting 2D-LC without "peak parking." The red marked peaks cannot be analyzed in the second dimension, as the actual 2D run is not finished.

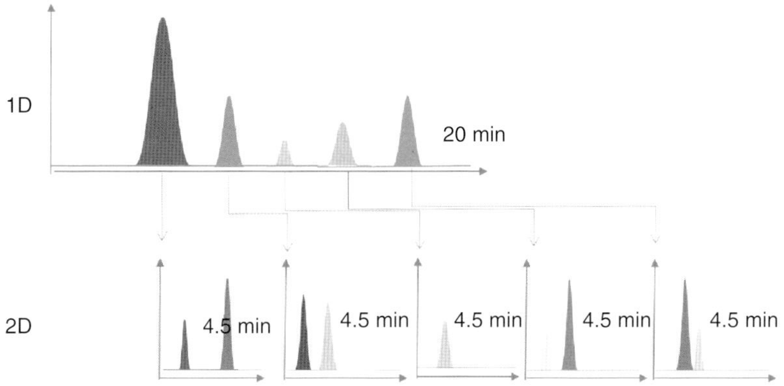

Figure 12.1.5 Heart-Cutting 2D-LC with "peak parking." All peaks can be analyzed in the second dimension, peaks are worked off subsequently.

repeatability shall also be given, when existing systems, which have been doing a good service in routine analysis over many years, need to be replaced. Here, the user expects a seamless transition from the current to the new systems. Agilent's unique software solution ISET (Intelligent System Emulation Technology) enables the emulation of other systems by the Agilent 1290 Infinity and Infinity II LC systems and thus the introduction of the next-generation UHPLC while maintaining established standard HPLC methods – without the need of time-consuming and complex reprogramming and revalidation of the existing methods.

When emulating a system, not only the gradient start is corrected by using the delay volumes, but also the typical characteristics in the gradient (especially, the mixing behavior) are taken into account to run the gradient. In addition to the myriad of measurements, which are required to characterize the different HPLC or UHPLC systems, for a good working emulation, an intelligent software algorithm and especially a system that clearly exceeds the emulated systems performance are needed – only a (much) better system can emulate another system, and vice versa, this is not possible. The Agilent 1290 Infinity II LC with ISET offers the possibility of replacing HPLC systems by UHPLC systems and benefitting from the advantages of future-proof UHPLC technology in addition to the use of already validated methods. Besides the Agilent (U)HPLC systems (1100, 1200, 1220, and 1260 series), the most established systems from other manufacturers can be emulated: Waters Alliance, Waters ACQUITY, Waters ACQUITY H-Class, and Shimadzu Prominence. More systems are to come to this list to further simplify the method transfer for any laboratory.

In general, all solutions that reduce manual work and increase flexibility will gain more and more importance in the future:

- Quaternary gradient systems already offer the possibility to replicate binary gradients by blending the mobile phases with stock solutions of the additives (acids, bases, buffers, etc.), so that no manual preparation of the final eluents is necessary anymore. The Agilent 1290 Infinity Quaternary Pump and 1290 Infinity II Flexible Pump offer the hardware prerequisites in terms of accuracy, repeatability of gradient composition, and gradient flow and have the software solution Blend Assist already implemented. This solution makes it easy to minimize your manual work of mixing the mobile phases manually and limits the possibility of mistakes.

- Future-proof plays a more and more important role in many purchasing decisions. UHPLC systems, which are not fixed to a defined system setup, allow higher flexibility and also allow the later upgrade to systems with higher performance. Agilent Technologies has not only followed the philosophy of a modular system design already for years but also guarantees the possibility to combine modules from different systems and different system generations together. Existing systems can be easily upgraded step by step, for example, to take advantage of new technologies such as automated method development, online SPE or 2D-LC.
- On the other hand, also dedicated and "specialized" UHPLC systems could be developed for dedicated analysis or analysis areas, covering only their limited field of analysis with highest efficiency. Such solutions can help users getting started in dedicated analysis techniques, especially when only a limited spectrum of analytes must be covered. In the area of LC-MS, already complete sets with standards, columns, and customized methods, which can be implemented on a new system very easily and quickly, can be purchased, for example, in food, environmental, or toxicological analysis. This development may go even further to facilitate the start for new users.
- Even small improvements in the handling of mechanical parts and compounds, which significantly reduce or even eliminate the need for tools, can help users who work directly on the systems themselves. Agilent offers the possibility to completely connect capillaries without any tools. With the A-Line Quick-Connect Fittings, highest tightness at pressures up to 1300 bar at zero dead volume are achieved with highest ease of use (Figure 12.1.6).
- Software-controlled solutions with intelligent valve circuits can perform complete workflows, such as method development or (online) solid-phase extraction. This supports laboratories to further increase efficiency, precision, and accuracy of their daily work and minimize unproductive phases and manual errors. Here Agilent Technologies offers not only suitable software solutions but also a complete portfolio of special valves for dedicated applications (see

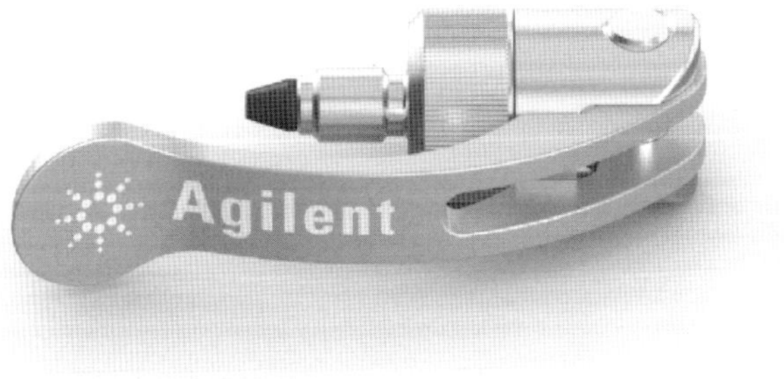

Figure 12.1.6 A-Line Quick-Connect Fitting.

above) as required hardware. The ability to manually change valves quickly and to use them in various different modules increases flexibility and possible application width.

- The overall efficiency of a laboratory can be also raised by increasing the detector linearity: If, despite big ranges of the analyte concentrations, samples need to be analyzed only once instead of twice, measuring time and solvent consumption can be halved. This increase in efficiency will also gain more and more importance in future in order to reduce the total number of measurements while still maintaining the analytical quality and meaningfulness. Agilent's High Dynamic Range (HDR) solutions enable the coupling of two differently sensitive DADs or ELSDs to one detector with the maximum linearity and sensitivity and support the reduction in analysis time and solvent consumption.

In order to further increase lab efficiency in the future, intelligent software systems such as OpenLab Laboratory Software Suite will play an important role, taking over the entire data management. Instead of using different software packages for system control, pure data acquisition, data analysis, and archiving of these data, in future decentralized overall data systems, will take over all tasks in order to control and manage the entire lifecycle of important (scientific) company data. To achieve this goal, several obstacles must be overcome, and an architecture must be created, bringing together all tools and data under one roof: all the different applications themselves, but also a variety of data records and formats (different manufacturer, different software, different systems, etc.) and any other information about samples (internal ID number, date, etc.). Of course, it is of highest importance that the safety and security of all these decentralized data are guaranteed.

In addition to the many already established applications (such as QA/QC in pharmaceutical and chemical industry, residue analysis, toxicological and clinical analysis, and many more), in future, especially the number of biopharmaceutical applications will grow. The past few years have already experienced a rapid growth in the biopharmaceutical market, and the majority of important newly approved drugs came from this area. Complete analytical solutions for this major pharmaceutical future area are needed, meeting the dedicated biopharmaceutical requirements in terms of sample stability and complexity of sample matrix and analyte spectrum:

- Truly bio-inert HPLC systems have been developed to avoid that samples get in contact with metals on the entire chromatographic way, and thus, possible interaction of sensitive (intact) proteins with metal surfaces is prevented. The 1260 Infinity II Bio-Inert LC System is characterized by a completely metal-free flow path. In this system, there is no risk of corrosion, which might come up due to large pH ranges and the high buffer concentrations in the typical bio applications.
- Special columns for biopharmaceutical applications are designed to optimize the systems' overall performance.
- The sensitive samples require special sample preparation. Through automation of elaborate method development and sample extraction, the highest level of

reproducibility and the lowest possible error rate can be guaranteed (see the previous discussion).
- Due to the high sample complexity and their broad analyte spectra, 2D-LC will directly address many questions with its enormous increase in chromatographic performance [3–5].
- Also, capillary electrophoresis (CE) as orthogonal separation technology, often coupled with (Q-)ToF mass spectrometers, will offer a better solution than UHPLC for many applications.

In summary, it can be said that the future of HPLC is going to be more and more UHPLC. In future, techniques will have to balance between diversity and specializations in individual analytical areas, whereby both the existing standard analysis can be maintained and analytical niche techniques will keep a safe place, so that new chromatographic challenges can also be solved. From a user perspective, innovations in the field of HPLC hardware and software operation will make the work even safer, more reproducible, and easier and will for sure underpin the status of a gold standard in analytics.

References

1 Vanhoenacker, G., David, F., Sandra, K., and Sandra, P. (0000) Determination of Taxanes in *Taxus sp.* with the Agilent 1290 Infinity 2D-LC Solution, Application Note, https://www.chem.agilent.com/Library/applications/5991-3576EN.pdf.
2 Li, D. and Schmitz, O.J. (2013) Use of shift gradient in the second dimension to improve the separation space in comprehensive two-dimensional liquid chromatography. *Anal. Bioanal. Chem.*, **405**, 6511–6517.
3 Sandra, K., Moshir, M., D'hondt, F., Tuytten, R., Verleysen, K., Kas, K., François, I., and Sandra, P. (2009) Highly efficient peptide separations in proteomics: part 2: Bi- and multidimensional liquid-based separation techniques. *J. Chromatogr. B*, **877** (**11–12**), 1019–1039.
4 Vanhoenacker, G. *et al.* (0000) Analysis of Monoclonal Antibody Digests with the Agilent 1290 Infinity 2D-LC Solution, Application Note, https://www.chem.agilent.com/Library/applications/5991-2880EN.pdf.
5 Vanhoenacker, G., Sandra, K., Vandenheede, I., David, F., and Sandra, P. (0000) Analysis of Monoclonal Antibody Digests with the Agilent 1290 Infinity 2D-LC Solution Part 2: HILIC × RPLC-MS, Application Note, https://www.chem.agilent.com/Library/applications/5991-3576EN.pdf.

12.2 HPLC Current Status and Future Development

Björn-Thoralf Erxleben

Actual HPLC market reflects the trend started with the first release of UHPLC systems, increasing demand for high (>100 MPa) or medium high-pressure system, even the actual applications and method do not require that. Customers would like to be prepared for future tasks and be ready to change toward UHPLC type of methods. Multipurpose use and flexible configurations of an "all-round" system are not just wanted; the target is to use systems for more than one analytical task. In future, this will reduce the number and types of instrument and their variety at user side, not talking here about the ongoing trend to replace existing HPLC systems by UHPLC systems.

This sounds somehow a contradiction as existing, mainly classic (conventional) HPLC methods are still in use and will be replaced only step by step the UHPLC methods in accordance to specific demands and rules of the company or laboratory.

Reasons for this are different: on one side, routine analytical tasks are first of all based on current available instruments in the laboratory, and on the other side, still users and management in many laboratories are concerned about robustness of UHPLC systems in daily use. This is finally more important than economic advantages such as time-saving or less operation costs.

From manufacturer's point of view, the trend toward UHPLC systems is clear, and on medium term, such systems will replace the conventional HPLC. Right now, already medium high-pressure (~60 MPa) systems are seen a "standard." Development of systems nowadays focuses on medium as well as high-pressure UHPLC. Market introduction and acceptance of those also depend on development and availability of stationary phases, which has to prove the claims about pressure stability under routine conditions and small particle size distribution.

Talking about detectors, especially the mass spectroscopic detectors play the most important role whenever we consider detection of small, fast eluting peaks. Acquisition of peaks with enough data points to allow accurate quantitative analysis is a challenge and a key specification of modern MS and MS/MS systems. Shimadzu's LCMS-8050 offering a scan speed of 30 000 amu s^{-1} in 0.1 u steps and fast switching between positive and negative mode with maximum of 555 MRM per seconds is one of those instruments suited well to a modern UHPLC system (Figure 12.2.1).

Important for the further use of a UHPLC for all LC-related applications will be that the robustness and reliability improve compared with conventional LC

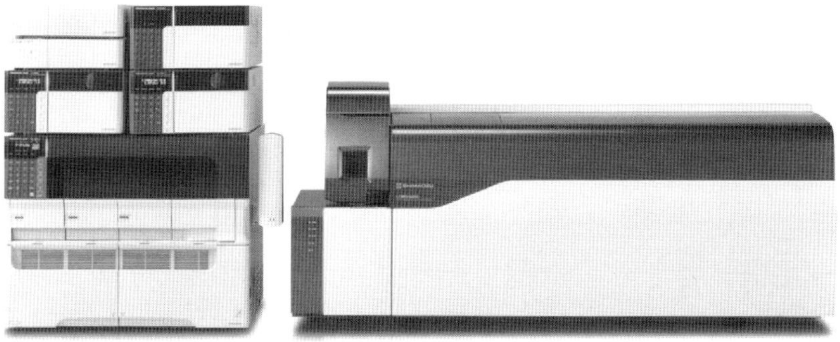

Figure 12.2.1 Nexera MP system with LCMS-8050 triple-quadrupole mass spectrometer.

systems. During the past decade, HPLC became more and more reliable and robust, which despite improved performance and sensitivity was the key for the growing wide use of LC for many separation tasks.

Use of new materials for plunger and rotor seals and special coatings could contribute for longer lifetime of parts; however, still there are differences to be observed when compared to conventional LC systems. Generally, looking at analysis run time and number of samples in a cost of ownership calculation, the picture is a bit different, but the final goal should be that UHPLC replaces conventional systems also in terms of robustness and reliability.

Too often, UHPLC is just taken by talking about high system pressure, but this is just a part of improvements and innovations on Liquid Chromatographs. Low dwell volume, fast injection, and ultralow carryover are meanwhile key specifications for a state-of-the-art autosampler, which no one would like to miss.

Higher temperature separation, preheater for mobile phase, and postcolumn cooler to avoid air bubbling in the detector cell are features allowing changes in separation mode, which also influence the existing and established LC methods. Last but not least, small cell volume and higher data acquisition speed are not only the result of faster separation by UHPLC, they also stand for a real step forward in terms of HPLC development and set milestones.

These standards, of course, could improve conventional LC systems, but just adding a UHPLC detector on a "normal" system will not automatically improve separation performance with an old method.

The combination of the mentioned new features with high efficient mixing, low system volume, and state-of-the-art column technology is, from our point of view, essential to get the best possible results with the system.

The classic detectors, especially photodiode array detectors, profit mostly out of the new trend: new cell design, small cell volume, novel optical coating, and high-tech parts for the detection, and electronics has dramatically reduced the sensitivity gap between standard UV detector and PDA, for some examples even turn around. Use of a total reflection cell with a long path length – for Shimadzu SPD-M30A 85 mm – and low cell volume and lower peak dispersion could improve the sensitivity significantly. PDA-based multiwavelength detectors are more often used, as they are nowadays equivalent to standard multichannel UV in terms of sensitivity.

New detectors and technologies are required, which enable shorter analysis time and detect a broad range of chemical structures and classes of substances in different concentration ranges qualitatively and quantitatively. Thanks to the price drop over the past years, MS detectors are getting more attractive and affordable for smaller laboratories too. Looking into the details of MS, as every other detection principle, there are advantages and limitations. New concepts, such as Ion Mobility, offer new opportunities but also set new requirements onto a UHPLC system. More and more operation cost, analytical environment, and infrastructure are considered, as well as environmental aspects and use of resources.

Differentiation between research use and use in routine instruments is one aspect; on the other side, method development intends to cover both application areas with one instrument, as this enables an easier transfer of methods and avoids adaptation. In practical terms, a system for "high-end" analytic needs to be robust enough to run under the standard condition of a QA laboratory.

From technical point of view, a dual approach looks attractive: a flexible modular concept for research and a compact system with fixed specifications concerning system volume and detector type and flow cell design for routine use. Such systems would be easy to operate and support a fast and easy method transfer between systems.

Current operators of an HPLC/UHPLC system are less experts in chromatography and do not care much about HPLC in particular. Such systems shall operate by single key operation and shall offer troubleshooting and technical advice in case of error. Especially, this is difficult to realize on a modular system, that is why a compact system for routine gets more importance.

Various log files about operation, system conditions, and errors are required and will support in future dynamic adaptation of maintenance intervals, as we know this already from automobiles.

From developer's perspective, a compact system is an ideal platform for special application systems. Fixed system volume and compact design allow with only a few technical modifications to create a "ready-to-use" solution.

Regarding optimization of system volume, the compact design has the advantage that flow line can be lowered and the connection between the different parts shortened. Considering lower flow rates and injection volumes, band broadening and dispersion and long delay times can be avoided.

Special configuration such as comprehensive LC × LC and other multidimensional system have their place for special applications and will be further developed. A closer cooperation between the user and the manufacture/developer is needed to incorporate demands and practical experiences into new developments. Advantages of multidimensional separation after a single injection and the resulting high chromatographic resolution need to be transferred into standard applications, and such methods are continuously improved. The beauty of this approach is that various substances/classes of substances are analyzed in only one run. Especially, for complex matrices such as natural products, synthesis products, and food, time saving could be achieved, under the condition that a multidimensional LC method can run under routine condition in a testing laboratory, and operation is easy. Data processing may still require specialists concerning interpretation of results and judgment of relevance.

On-line solutions will be more important in future, in two directions: first, direct sample intake from production or synthesis flow stream with an LC system at the side; second, automated sample pretreatment reducing manual work and failure risks. Sample dilution or reagent adding is possible already with standard autosamplers, while filtering and centrifugation are sometimes requested, as well as on-line solid-phase extraction and liquid–liquid extraction.

Besides complex control and operation logic of such a system, the goal is to create a platform for different application fields. Requirements are the same as for UHPLC systems: robustness, reliability in daily operation, and less downtime.

New developments in terms of HPLC and UHPLC are competed by other chromatographic techniques such as supercritical fluid chromatography. Combining different techniques, their strength into one-solution system provides new possibilities for the user.

Recently, Shimadzu released the Supercritical Fluid Extractor (SFE) and an SFC completing the Nexera product line, which can be a novel approach for on-line sample pretreatment and online analytic in food control, as it is supposed to reduce manual work and time for sample preparation.

Besides real hardware novelties, software tools are of interest, which support method development and optimization. Based on a predefined number of scouting runs with different columns and system parameters such as run time, gradient composition, and temperature, several quality-by-design (QbD) models can be created, modeling the separation and the robustness of the method. This helps to eliminate a big number of experiments to run with the system to get these data. In addition to the software, hardware solution supports column selection and switching and enables an "overnight" processing. The Nexera Method Scouting System offers the combination of a binary high-pressure gradient with two quaternary gradient pumps, column switching, and either photodiode array or MS detection. Embedded software tools support an easy creation of the sample table (Figure 12.2.2).

Peak Tracking by 2D and 3D detector signals, here the MS detector will become more and more the unit of choice, completing the modeling of the separation. Software supports each step in this chain of scouting, column and solvent selection, method optimization, and validation. From our point of view, for such complex tasks, cooperation with strong partners such as the Molnar Institute of Applied Chromatography and company ICD is very useful.

Automated qualification of the system and reporting will help the user to reduce the time for this procedure and the downtime of the system. Special

Figure 12.2.2 Nexera method scouting system.

in-built software for system check, performance tests prior to and after a service intervention, and intelligent feedback are just parts of such a concept. Web-based application can help the service for getting the system status and information about spare parts, give an overview about system use and planning of resource, and enable a remote overview about work in progress.

Use of mobile equipment and support from various operating systems are more and more in focus. Control of a complete system from a tablet PC, such as an iPAD, including the transfer of methods and sample tables in the WLAN area, is realized with the i-Series. All interactions and changes of the method are tracked in the software and transferred between the iPAD, instrument display, and the control software.

To meet the requirements of regulated environment, user access and log files are applied (Figure 12.2.3).

Often, users ask for compatibility of new hardware to the existing Chromatography Data Systems (CDS). Since several years, therefore, we have developed and maintained, besides our standard software, drivers for Empower®, Chromeleon®, and OpenLAB®.

Finally, besides the current development in Europe and America, especially in the emerging markets, conventional systems are still widely used and requested. Furthermore, several separation tasks cannot be easily transferred onto UHPLC or will not show significant improvements on analysis time and performance.

In the coming years, HPLC and UHPLC will coexist and complete each other and not just replace.

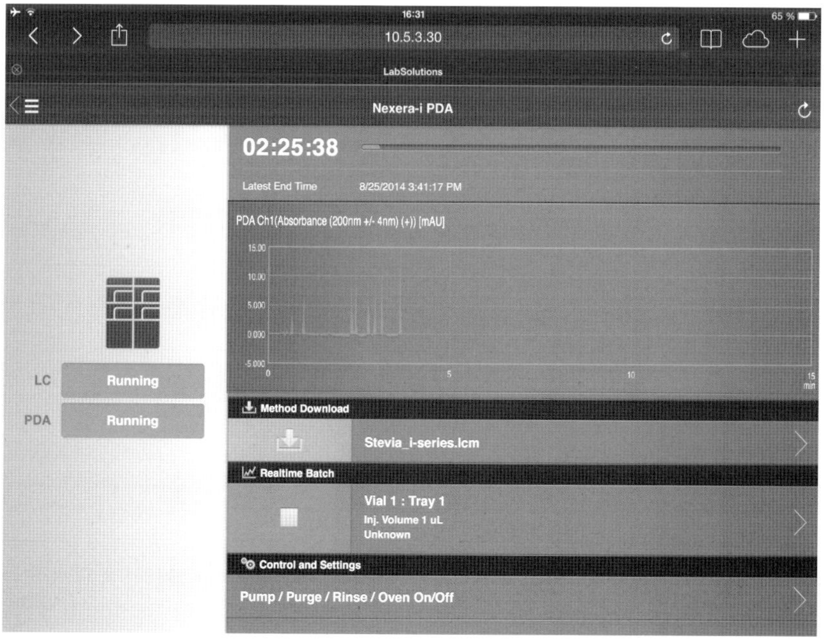

Figure 12.2.3 Shimadzu i-Series/Controller screen and i-PAD screen.

12.3 Thermo Fisher Scientific, Germering

Frank Steiner

It dates back to the mid-1980s that HPLC had already reached the status of a robust and automated routine analysis method with even the first automated systems on the market. Since that time, it set standards that are hard to match for any other emerging analytical technique. Further development over the last 30 years was mainly addressing improvements in productiveness and usability. The goal is to even facilitate less educated users to successfully work with this technique. Nevertheless, the evolutionary improvement over the past decades has been quite substantial and is still continuing at a similar pace.

The increased productiveness resulted from a higher degree of automation of the entire workflow from sample to results, but also from increased separation speed and improved separation efficiency. For all improvement steps of the kinetic potential of this technique, the combination of column and instrument technology was a crucial factor, while instrument control and data processing software were other important contributors. The strive to achieve the optimized separation efficiency in the shortest time possible in a robust way with the lowest operation cost led to smaller particles packed in smaller volume columns that are run at higher linear velocities. This required instruments that enabled higher operating pressures, reduced system volumes, and faster detectors. Superficially porous particles achieve impressive separation efficiency even with particle sizes between 2 and 3 μm, which reduced the ratio of operating pressure per plate count substantially and often gave rise to the false assumption that respective columns could also be successfully operated in conventional HPLC systems. Meanwhile, the smallest particle size of superficially porous materials is even smaller than that of fully porous particles on the market. The need for improvement of instrumentation in all respects continues, and there is no slowdown in sight. Besides the separation performance focus, there is a continuous improvement of detection devices. Improvements are addressing trace-level detection, quantification accuracy across the widest possible concentration range, universal detection capabilities as well as specific information on peak identity. These developments involve many different techniques beyond the widespread UV absorbance detection and include the impressive progress in MS detection. The following sections report on the recent HPLC key developments in Thermo Scientific and give very brief outlook.

12.3.1 Total System Requirements and Related Key Experiences

Some key factors for instrument design decisions are clearly related to user-group-specific preferences. While HPLC addresses a very diverse range of user groups, there are several substantially different requirements between stand-alone HPLC and LC-MS users, as well as between users in research and development versus users in routine analysis and quality control. Finally, the preferred column diameter is of importance for the design of the instrument. In this respect, the systems to operate nanoLC columns have an exceptional position. One reason is that they are almost exclusively used as MS front ends, and the other reason is that column diameters of less than 100 µm require extraordinary technologies for the lowest flow generation and highly miniaturized fluidics. For these reasons, nanoLC instrumentation will be briefly discussed separately, before the main emphasis will be put on the developments in instrumentation for the normal flow range.

12.3.1.1 NanoLC

Thermo Scientific offers two markedly different nanoLC systems, the UltiMate 3000 RSLCnano and the EASY-nLC, and both have an independent history. Both these systems are always connected to mass spectrometers and applied for protein identification and quantification. The main markets are proteomics and biopharmaceutical characterization. The roots of the UltiMate 3000 product line lie in the mid-1990s when a company named LC Packings introduced the first commercial nanoLC system to the market. LC Packings was acquired in 2000 by Dionex, and Dionex came into Thermo Fisher Scientific in 2011. While the UltiMate system of the 1990s was based on a rather rudimentary but yet routine proof passive flow splitting, the modern RSLCnano variant uses an advanced direct nanoflow generation technique. There has been an intermediate UltiMate 3000 system based on a patented active flow-splitting technique as well. The EASY-nLC system has its roots in one of the first generations of direct nanoflow systems that had been introduced in 2006 by a company named Proxeon. Proxeon was acquired by Thermo Fisher Scientific in 2010.

The UltiMate 3000 RSLCnano and the latest version of the EASY-nLC are fully UHPLC-compatible nanoLC systems that support pressures of up to 1200 bar in routine applications. Equally important as the operation pressure is how both systems empower the user to make nanofluidic connections that are not affected by any dead volumes. This is a true challenge in nanoLC and the Viper technology (see Section 12.3.1.3) that is also available in the nanoViper version is a true game changer in this respect. The EASY-nLC1200 system is characterized by a compact fully integrated design and can be operated by a touch screen at the system front. This screen enables to read operation parameters and to call typical nanoLC workflows to run routine applications that also include some multidimensional capabilities. The compact design also facilitates to easily move the system close to the ion source of any mass spectrometers in the lab. The UltiMate 3000 RSLCnano system is a partially integrated system, where the flow generation is combined in one module with a very effective column thermostatting and column switching device, while the autosampler is a separate module. The UltiMate 3000 system is characterized by the wide flow range and the outstanding workflow flexibility up to fully automated off-line 2D-LC. The wide flow range results from the combination of a direct-flow binary nanogradient pump and a ternary gradient pump to cover micro-to-normal flow range. The flow rate range of the nanopump can easily be extended by changing flow meters, so that it can also cover capillary and microflow rates.

12.3.1.2 HPLC and UHPLC on Two Instrumental Platforms (UltiMate 3000, Vanquish)

While nanoLC is a fascinating technology and ideal for LC-MS applications with a very limited sample size, the majority of modern liquid chromatography users still operate in a flow rate range of 0.2–2 ml min^{-1}. More recent UHPLC methods focus in spite of the applied high linear eluent velocities on the flow rate range below 1 ml min^{-1} generated on lower internal column diameters of mainly 2.1 mm, but also 1 mm to a certain extent. Thermo Fisher Scientific offers the UltiMate 3000 Standard (SD) line with a maximum pressure of 620 bar and on the same platform the Rapid Separation (RS) for up to 1000 bar operating pressure.

With the introduction of the Vanquish system in 2014, a new platform and generation of Thermo Scientific UHPLC instruments were introduced with a strong focus on the high-end market. Eight years of experience with the UltiMate 3000 line and almost 30 years of experience in developing and manufacturing HPLC systems at the site in Germering, Germany, enabled the development of the technology built in Vanquish. Its future proof system concept deviates quite substantially from the UltiMate 3000 system, while some of the successful core technologies of this very mature predecessor system were still sustained. The key design goal of Vanquish was to build the system around the column as the central separation device and to design it for optimal use of column performance and best possible operator friendliness with emphasis on performance, throughput, and ease of use. This resulted in a system concept of the so-called integrated modularity (Figure 12.3.1). Integrated modularity is firstly the mechanical design of a system that is composed of individual modules for full flexibility, but these modules attach to each other to an integrated system with automatically connecting system drainage and optimized fluidic connections. Integrated modularity

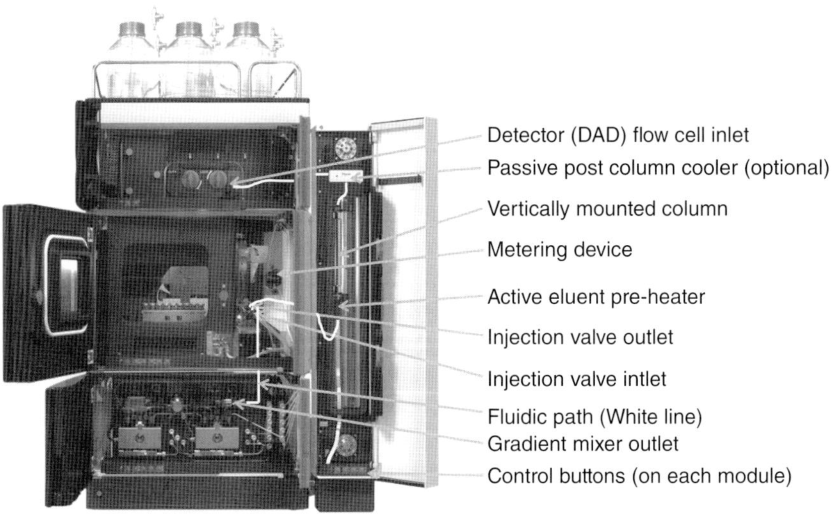

Figure 12.3.1 Setup of the Thermo Scientific Vanquish Horizon UHPLC system following the concept of integrated modularity. The fluidic path is marked with a white line for better traceability.

does, however, also refer to the fact that the individual modules communicate with each other via the intrasystem link connection, and the operational concepts are matched so that certain features of one module positively influence the performance of another module. The mechanical integration of the system allows for easy and quick access to internal parts for exchange of wear parts and special maintenance without dismantling the stack. The individual modules can be drawn outside the front only after unbolting some screws. All electric connections of the main modules are now located at the rear part, but they point to the side of the system for easier access. It is an important design element of the UltiMate 3000 systems to show the main operational parameters (e.g., pressure, flow rate, detector signal, elapsed retention time) easily readable on large displays integrated in the front door of each module. These displays also allowed to access menus to set values for all important functions with the help of a magnetic pen. While the large displays were appreciated by many users to quickly receive information on the key parameters, it turned out to be sometimes a robustness challenge to have electricity in moving parts as the front doors and exchange of damaged doors required a service expert. The user friendliness of the magnetic pen was limited, although it avoided erroneous change of settings that may occur with a touch screen. The new Vanquish system abandoned electricity in the front doors that can be unhinged if required for better access during setup changes. The front uses a more modern concept of a light conductor band across the entire door width. This light conductor guides the light from an LED array in the module front behind the door to the system front once the door is closed. This light band can encode several operational states by the use of four different colors and additional swell, blink, and scan functions. A blue scan across the detector band indicates, for instance, the recording of data, the blue scan across the autosampler the sample injection process. Behind the front doors of each module are buttons with several basic functions located. They allow actions such as switching on the UV lamp in the detector, purging the pump fully automated, and turning the sample carrousel. Another element for the instrument control is the optional touch screen tablet that is detachably mounted to the railing of the solvent rack. This device communicates wirelessly with the instrument controller and provides a mobile Chromeleon software app that allows the control of all instrument functions and the launch of prepared sequences packaged in the so-called eWorkflows. It also displays the key parameters in the sense of the UltiMate 3000 front door displays. The wireless controller can also be detached from the system, making the aforementioned functionalities fully portable to every place with mobile Internet access. Software control of the individual modules in the Vanquish system occurs via USB, as this is the only way to enable real-time control of all functions. This is a prerequisite for certain triggers, for example, a detector-based collection of fractions. There is the option to have an embedded industrial PC in the system base, which performs the USB control but is connected via Ethernet to the network installation of the respective laboratory. This combines the advantages of the USB real-time control with the easy connection to any remote PC in the lab, and the user is not restricted by the lengths of the cable connections as it is the case for USB connection between the control PC and the instrument.

The concept of building the system around the column is mainly reflected in the design of the fluidic path as well as the mounting of the column and its thermostatting. Unlike the UltiMate 3000 stack, the Vanquish stack hosts the pump as the bottom module sitting directly on the system base. The eluent degasser is integrated in the pump module not only for the quaternary but also for the binary pump. The outlet of the gradient mixing device at the upper right corner of the pump is shortly connected through an internal pipe fairlead with the injection valve of the autosampler. This injection valve is placed directly at the bottom of the module. The column thermostat is located on the side of the main stack and holds the column vertically. This allows a short horizontal connection from the injection valve into the column inlet thanks to the free positioning of the active eluent preheater that is directly integrated into the connection capillary. The position of the preheater can hence be easily adapted to the column length without significant additional capillary length. The detector module is located on top of the main stack above the autosampler. In this respect, the column bridges the distance from the autosampler to the detector flow cell, and it is operated with a bottom-up flow direction. Columns are mounted at the upper end of the thermostat regardless of the column length. This enables a relatively short predominantly horizontal connection to the detector flow cell inlet. An optional postcolumn eluent cooler can be used, which is mounted outside the heated compartment and is located at the height level of the detector flow cell. This instrumental setup allows for a total system height of only 75 cm and enables even the operation of long columns with relatively short capillary connections.

12.3.1.3 Viper-Based System Tubing

All fluidic connections in the Vanquish system are based on Thermo Scientific Viper™ capillaries. Viper is a unique tip-sealing fitting concept that works with a seal ring exactly at the fluidic contact between the capillary and its counterpart (see Figure 12.3.2). Relative to ferrule-based systems, this provides the massive advantage of achieving virtually zero dead volumes as soon as the fluidic connection is tight. It is also tool-free and can be easily and safely tightened for pressures up to 1500 bar, which is an essential prerequisite for successful operation in the Vanquish high-end system. Another advantage of Viper is the small design of the end fittings with the removable knurl that makes it pass even narrow holes and allows it to fit to any 10-port switching valve. The Viper technology was already applied in the UltiMate 3000 platform and has been appreciated by users for many years. Over time, it has been further improved to make its use even safer and better compatible with a wide variety of HPLC and UHPLC columns in the market place.

12.3.2 The Contribution of the Individual Components to the Success of a System

This section presents all relevant technical information on the individual instrument components (or modules) along the fluidic path of the system from the pump to the detector.

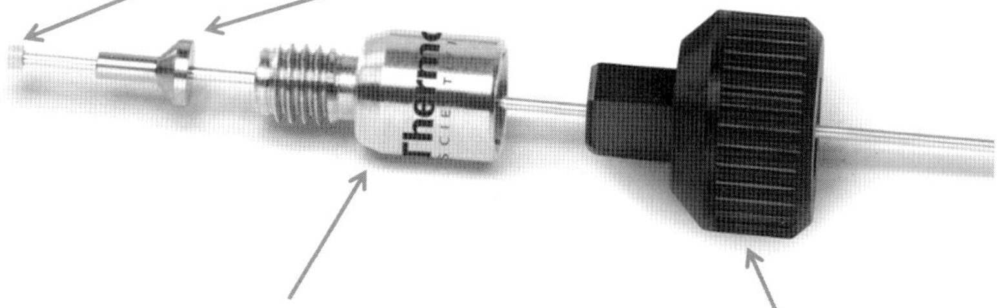

Figure 12.3.2 Design of the universal and tool-free Thermo Scientific Viper UHPLC fittings. The photo shows the first version of Viper for a clearer view on the different parts. In the current version, the PEEK ring, the force adapter, and the cylinder with the screw are connected for functional reasons.

12.3.2.1 The Flow Delivery Device – Much More Than a High Pressure Pump

It has been described in detail in Chapter 4 that requirements on a modern UHPLC pump go far beyond the generation of a constant flow and exact gradient formation at even elevated pressure levels. The striving to fully meet these requirements is reflected in the developments from the UltiMate 3000 pumps to the Vanquish high-end device. The UltiMate 3000 pump platform offers a range of 620 and 1000 bar pumps that exist in low pressure gradient (LPG), high pressure gradient (HPG), and DGP variants, next to isocratic variants, dedicated nanoLC pumps and a biocompatible semipreparative gradient pump. DGP stands for dual-gradient pump. This pump comprises two ternary low-pressure gradient pumps in a standard size enclosure and facilitates 2D-LC next to other automation and productivity increasing workflows that would otherwise require an additional pump. UltiMate 3000 pumps follow the solvent delivery concept of serial 1½-cylinder blocks with shaft-coupled cam drives. This concept is characterized by relatively low purchase and operation costs combined with operational quietness and robustness. The inherent patented isokinetic operation technology named SmartFlow features an effective principle to minimize residual pulsation through a continuous measurement and consideration of the actual solvent compressibility. This also deliberates the user from entering compressibility values into the control software and is useful for LPG pumps where the compressibility in the pump cylinder changes during the gradient run. Under challenging conditions of delivering a very compressible solvent such as methanol at pressures of 1000 bar or higher, the coupled cam-driven delivery principle reaches its limitations when minimal residual pulsation is required. It is particularly critical with HPG pumps where the pulsation of the individual blocks translates into mixing ripples that can disturb the detector baseline.

With the development of the Thermo Scientific Vanquish H pump, an ultrahigh-performance HPG, the delivery principle of the normal-flow UltiMate 3000 pumps has not been further pursued. The binary Vanquish H pump employs linear drives with independent motors for both cylinders in each block and a parallel instead of a serial principle. It is able to deliver a maximum flow rate of 5 ml min^{-1} at the maximum pressure of 1500 bar and is optimized for the lowest residual pulsation even under extreme conditions. This is due to an advanced flow delivery control and enables to apply the pump with a low mixing volume of 35 μl for demanding gradient methods even under ultrahigh pressure. The parallel delivery concept decouples the delivery procedure of the two pistons in one block even beyond the completely independent drives. With that, it is facilitated to compensate for thermal effects that stem from solvent compression under ultrahigh pressures more effectively by intelligent control algorithms. At the same time, does the parallel principle distribute the workload effectively between the alternately delivering cylinders and thus halves the mechanical stress on the pistons and seals over time. This helps to make the pump even under extreme conditions very robust, and it extends the endurance of wear parts by approximately factor 2. To boost the application flexibility, the binary Vanquish H pump is equipped with ternary eluent selection valves that enable to switch between three different solvents on each block. A switch is also possible during the run, which allows for accomplishing very effective column cleaning steps by solvent changes. The ease of operation was an important design goal in the development of the pump. It is characterized by a clearly structured fluidic path, easy access to all fluidic elements that can be dismantled without tools thanks to the Viper finger-tight fitting technology. Even the check valves can be screwed in and out tool-free within seconds. This accounts for best possible ease of use of this parallel pump that features twice the number of check valves than a serial pump.

12.3.2.2 The Injector and Liquid Handling Devices for Robust and Ultra-Precise Sample Dosage Even in High-Throughput Workflows

The UltiMate 3000 product line that preceded the Vanquish instrument platform was determined by a wide range of autosamplers to cover the different specific flow and pressure ranges. The majority of the UltiMate 3000 autosamplers apply the split(-loop) or flow-through needle (FTN) principle, while the samplers for the nano range but also the integrated autosampler column compartment module (ACC-3000) use the pulled-loop principle. Another representative of the pulled-loop principle is an integrated autosampler and fraction collector (WPS-3000 TFC). This standard-sized autosampler module enables fractionation and reinjection to support special workflows such as automated off-line 2D-LC. The UltiMate 3000 autosamplers feature a relatively open design for easy access to the fluidic parts, which significantly facilities, for instance, the swapping of a sample loop. The sample compartment is protected by a cover that is automatically opened and closed by the injection needle arm. With this concept, it is inevitable to expose the sample vessels during the sample draw process for a couple of seconds to the laboratory atmosphere. In some parts of the world where typical labs are by far not at the standard of those in long-established industrialized countries, this could lead to serious problems with condensed water and sample contamination from the lab atmosphere. To operate safely and robustly even under such conditions, the UltiMate 3000 autosamplers can be equipped with a transparent front cover, which must be dismounted to access the sample carousel. The change to the Vanquish autosampler was characterized by significant modifications, among which the closed design is possibly the most obvious. It features a hinged front door that effectively closes the sample compartment and provides an efficient thermal insulation. In order to completely shield the interior from the environment, an optional transparent cover can also close the needle arm and injection valve section where the user does not need to manipulate unless parts such as the sample loop have to be exchanged. While the UltiMate 3000 autosampler already featured space for three trays or well plates, the on-board capacity of the Vanquish instrument was increased to four tray positions. The concept of the sample carousel was inherited from the UltiMate 3000 product, and with that the very fast access to distinct positions, as rotation movement can be accomplished faster than translation movement. Next to the advantage of fast injection cycles down to 12 s, the carousel enables a fast movement back and forth, which can be employed to shake the sample vessels for mixing of their content.

The need for an injection device that can operate up to pressures of 1500 bar gave rise to the development of a proprietary Thermo Scientific valve and valve drive. The material that is in contact with the mobile phase and the sample is a metal-free ceramic that is coated on one side with diamond-like carbon (DLC) to account both for sealing and to protect against abrasive wear. The development of a proprietary valve with a radically new design allowed the switch to the concept of a service-free valve. All established injection valves in the market are based on using the rotor seal as a wear part, and the manufacturers have adopted the need for regular maintenance into their business model, which is automatically inherited by the manufacturer of the autosampler that uses these valves. With the development of the Vanquish autosampler, this paradigm was changed, and the user can take advantage of no longer needing service on the autosampler valve due to wear under standard use. Another major difference between the UltiMate 3000 and the Vanquish autosampler relates to the sample dosage principle. In contrast to the conventional dead-end syringe in the UltiMate 3000 instrument, the Vanquish autosampler features a flow-through syringe in the high-pressure part, also known as a *metering device*. This metering device is not only used for the sample dosage but is also able to perform a precompression of the sample loop prior to switching the injection valve. The metering device comprises both a fine drive for exactly positioning the piston to achieve the best possible dosage down to a 10 nl resolution of the injection volume and a course drive that can apply the required force to precompress the liquid in the loop up to 1500 bar. This precompression avoids the pressure dip in the fluidics at the time of the sample valve switch. A longer column lifetime is only one positive consequence, but it also has a strongly positive effect on the retention time precision. With avoiding the flow interruption and suppressing random solvent compositions that result from different ratios of compressed aqueous and organic solvents in the pump blocks of an HPG, the retention time precision can be improved by a factor of

10, and the Vanquish instrument is able to achieve a precision level of 0.005% in numerous applications. The concept of the sample-loop precompression also enabled to achieve such precision with relatively large stroke volumes on the pump of 120 μl in standard operation, which increases the lifetime of the pump thanks to a lower stroke frequency. Finally, it also enabled a relatively large maximum injection volume of 100 μl without adverse effects at the point of the valve switching. In order to deal with the relatively large dead volume of the metering device that supports such high maximum injection volumes, an intelligent control of the metering device piston enables to set a so-called idle volume parameter. This can be used to place the metering device piston in a favorable position if only small injection volumes are applied. With that, it can reduce the gradient delay volume (GDV) contribution significantly. In addition, this parameter can be used to tune the GDV contribution of the autosampler in order to mimic other instruments for successfully transferring methods from other instruments. It enables a seamless adaption of the GDV beyond the change of mixers in the pump. Taking advantage of a minimized idle volume, the GDV contribution can be as low as 80 μl when the Vanquish autosampler is equipped with the smallest sample loop that serves for a maximum injection volume of 10 μl.

The sample capacity of the Vanquish autosampler was increased relative to the UltiMate 3000 predecessor by implementing a maximum of four trays or well plates. The tray design was also changed as a consequence of the altered air-to-air sample cooling concept that avoids the use of cooling liquids and eliminates any risk of coolant leakage. The Vanquish vial trays are no longer metal blocks with significant heat capacity but very light and open devices that give full access to the circulating cooled air for fast thermostatting. This light design reduces space for the individual vial bores and thus increases the capacity of the trays for standard 2 ml-vials from 40 (UltiMate 3000) to 54 vials per tray. Thanks to the capacity of four trays with up to 54 vials, the total on-board sample capacity could be increased by 80% from 120 to 216 vials. If higher sample capacity is needed, the Vanquish autosampler can be equipped with an additional rack loader module, the so-called Vanquish Charger. This module can host up to 9 vial trays or alternatively up to 20 standard well plates. With this extension, the maximum capacity of the system comprises 23 well plates, because one space in the autosampler must be left empty for shuffling between the two modules. Using 384-well plates on 23 positions, the total sample capacity is 8832. For high-throughput workflows that handle such extreme numbers of samples as well as for open-access solutions, where different users can add samples to the system via the Charger module, it is essential that the different racks are properly tracked by the system. This is achieved by integrated bar-code scanners both in the Vanquish Charger and in the Vanquish autosampler module.

12.3.2.3 New Ways of Column Thermostatting to Combine Highest UHPLC Column Efficiency and Best Method Transfer Capabilities

All UltiMate 3000 column thermostatting compartments applied the so-called forced-air principle, which accounts for fast thermal equilibration and very effective column temperature control by a fan that internally circulates the thermostatted air. In this respect, the UltiMate 3000 column compartments deviate from the still-air concept that most other vendors employ for their high-end UHPLC instrumentation. With still-air thermostats, only minor heat dissipation occurs between the interior of the separation column and the column compartment space. UHPLC methods operating at elevated pressure generate substantial frictional (or viscous) heat that increases the internal column temperature relative that of the incoming eluent. Forced-air thermostats dissipate at least a significant portion of that frictional heat, which inevitably results in a radial temperature gradient across the column. This phenomenon was extensively discussed in Chapter 2.2, and it leads to distorted band profiles inside the column as the migration velocity is faster in the center of the column relative to the layers closer to the column wall. This effect substantially reduces the separation efficiency below the theoretical capability of the respective UHPLC columns. In contrast to that with the still-air principle, the frictional heat mostly remains inside the column, which implies that the column heats up from the inlet to the outlet and induces a longitudinal temperature gradient instead of a radial one. The advantage is that due to the absence of radial temperature gradient and resulting distorted band profiles, the efficiency under frictional heat is significantly better than in forced-air thermostats and much closer to the theoretical column efficiency. The disadvantage is that the temperature in the column evolves relatively uncontrolled by frictional heating. This has an impact on analyte retention and possibly even on selectivity relative to the result at the given column temperature in the absence of frictional heating. In any case, the chromatogram recorded under still-air thermostatting will substantially deviate from that recorded with the same method, but in a forced-air thermostat, this can be a significant challenge for method transfer between instruments with different column thermostats. The Vanquish column compartment was developed to support both still-air and forced-air thermostatting. This design allows in the forced-air mode, effective heat dissipation with the best possible control of the internal column temperature for fast thermal equilibration and smooth method transfer from UltiMate 3000 or other forced-air instruments. Alternatively, the still-air mode allows for the best possible efficiency under frictional heat conditions. In fact, the fan speed can be controlled in seven steps in order to adjust the amount of heat dissipation and adopt the thermostatting conditions as good as possible to those occurring in other instruments.

Another important feature is the flexibility with respect to mobile-phase prethermostatting in front of the column. For a very effective preheating up to the maximum Vanquish column compartment temperature of 120 °C, a feedback-controlled active preheater can be used. It even allows setting the inlet temperature of the mobile phase to a different value than the compartment temperature. This feature enables (under frictional heat conditions) to generate similar temperatures inside the column in still-air mode by entering the column at lower temperature as it would result in forced-air mode with setting the preheater to the compartment temperature. In this way, the retention factors from forced-air mode can be combined with the peak efficiency of the still-air mode by a compensation of the temperature increase due to frictional heating. The optional passive eluent preconditioner would be the way of choice for subambient column temperatures as it can more effectively cool down the mobile phase than the active preheater device that can only heat up the mobile phase.

This broad range of column and eluent thermostatting possibilities offer a plethora of method optimization options and close a common gap with respect to method transfer that other vendors tend to neglect.

12.3.2.4 How to Detect Fast and Ultraefficient UHPLC Separations

Thanks to its wide range of different detection techniques such as variable-wavelength detection, multiwavelength detection, diode array detection, fluorescence detection, electrochemical detection, and the proprietary charged aerosol detection (CAD), the UltiMate 3000 product line provided an unmatched range of possibilities. With the Vanquish products becoming another UHPLC platform, the clear goal is to provide an equivalent detector range. The Vanquish Horizon system started with diode array detection as the most widespread optical detection for top-level UHPLC instruments and the unique CAD that is increasingly gaining share among nebulizer-based detectors. This detection principle is based on exhaustive adsorption of ionized nitrogen to the surface of the dried aerosol formed from the peak zone that elutes from the column. Eventually, a current resulting from the charge is measured, which is proportional to the compound mass. This leads to the most universal response behavior of all LC detectors independent of the chemical structure of the analyte. It also generates response curves that can be easily transformed to linearity by applying a simple power function to compensate for the systematic reduction in surface-to-volume ratio of the aerosol particles with increasing concentration. The fact that small aerosol particles provide a relatively higher surface that absorbs more charged nitrogen helps to boost the sensitivity of this detector for trace-level concentration. This sensitivity combined with the ease of linearity transformation and the most universal response factors are clear advantages of the CAD over the competing evaporative light scattering detectors (ELSDs). With the UltiMate 3000 Corona Veo model introduced in 2012, the CAD technology already reached a very mature state. This enabled to integrate it with only minor modifications into the Vanquish system platform. The Vanquish CAD exists in the high-end (H-) version with a maximum data rate of 200 Hz and mid-tier (F-) version with a maximum data rate of 100 Hz. In contrast to the CAD, the Vanquish DAD (H-version) was a complete technical redesign that has virtually nothing in common with the UltiMate 3000 DAD. While the DAD-3000RS with its two light sources (deuterium and tungsten) represents a very flexible and economic diode array detector based on conventional flow cell design, the Vanquish H-DAD is a highly optimized detector with focus on detection in the UV wavelength range. The flow cell is based on fused-silica light conductor technology, and the optics have been optimized for a single-deuterium high-intensity light source. This provides clear performance advantages over the alternative design with two lamps. Together with the focus on temperature-stabilized optics and fluidics as well as the minimization of stray light, the result is a combination of excellent signal-to-noise ratio and an extended linearity range up to 3000 mAU for distinct applications. The fused-silica light-pipe technology is based on the total light reflection at the outer surface of the silica tube against the surrounding air. This differentiates substantially from the Teflon AF technology where the total reflection occurs at the transition from the mobile phase to the Teflon AF layer, thanks to its low optical density. Total reflection is required to guide the light as completely as possible through a relatively long and narrow detection channel, and it always occurs at the transition from the optically denser to the less dense medium. The advantage of the fused-silica concept is the favorable total reflection angle and the independence of this from the mobile-phase composition whose optical properties are not involved in the total reflection process. These properties help for optimal detection performance and justify a single disadvantage of the fused-silica concept, namely the risk of breaking under pressure or mechanical shocks. The Teflon AF technology is less prone to mechanical damage but more susceptible to contamination of the inner surface of the plastic tube and changes of the optical properties of the mobile phase impacting on the total reflection behavior.

For very fast transient signals that have to be recorded in speed-optimized UHPLC, the detector electronics are equally important as the optics. Most important is an intelligent filtering algorithm to account for the lowest possible baseline noise without compromising (or damping) the recording of very fast signal changes resulting from extremely sharp peaks. This is possible with filter time constants below 0.1 s, which is not a specification that is commonly discussed with detectors for fast chromatography. The specification in the focus of interest and thus the mainly advertized one is normally the maximum data collection rate with values of 200 Hz and more. To understand the relevance of such data collection rates, it is helpful to calculate time-based peak widths resulting from ultrafast methods. The following example will illustrate this:

- On a short column of 20 mm length with an efficient stationary phase that achieves a theoretical plate height in practice of 4 μm, there will result a geometrical 8σ peak width (which are typically the integration limits) of 2.5 mm.
- Pumping the mobile phase at a velocity of 10 mm s^{-1}, which enables a gradient time of only 20 s (at $10t_G/t_M$) such a 2.5 mm peak width translates into 0.25 s time-based peak width.
- With a date rate of 100 Hz, one can still record 40 data points across this peak, which is generally considered sufficient.

With a maximum data collection rate of 200 Hz, the Vanquish H DAD provides even capabilities to detect peaks from 10 s gradient runs with extreme linear velocities of 20 mm s^{-1} on such a column while it still guarantees for sufficient data points.

More important than the data rates are the detector baseline noise of less than $\pm 3\ \mu$AU (standard flow cell under specified conditions) and the optimized dispersion volume of flow cell (specified 1σ dispersion volume of 0.8 μl). An optional high-sensitivity flow cell with optical path length of 60 mm is also available and allows enhancing the signal-to-noise ratio of the Vanquish H DAD further in applications where the main focus is not on recording very small peak volumes.

In addition to the DAD and CAD detectors, the Vanquish platform also offers a variable-wavelength detector with a maximum data rate of 250 Hz that presents an economical alternative to the DAD and offers a wide range of conventional detector flow cells down to a cell volume of 3 nl. The Vanquish detector range is completed by a fluorescence detector that is build on the UltiMate 3000 FLD technology using a xenon flash lamp and the option for a second photo multiplier that records long emission wavelength (700–900 nm) with optimal sensitivity.

12.3.3 2D-LC and Alternative Ways to Increase Productivity for Analyzing Complex Samples – and Outlook to Changing Paradigms

The UltiMate 3000 product line was already in the past decade capable of running multidimensional workflows based on the dual-gradient pump principle, the integration of two switching valves in the column compartment, and the capabilities of the Chromeleon software. These workflows range from on-line solid-phase extraction (SPE) over 2D-LC up to possibilities for throughput increase based on overlapping gradient runs with simultaneous column equilibration on the second channel (tandem mode), automated switching between two methods that are both kept equilibrated (application switching), and the simultaneous operation of two methods with use of two detectors (parallel mode). The so-called automated off-line two-dimensional LC as a particular variant of 2D-LC is well supported by the dual UltiMate 3000 system and the Chromeleon software. This technique is accomplished by a unique autosampler module (WPS-3000TFC) that can collect fractions from the first dimension and reinject them into the second dimension. The software automates the generation of the sequence for the second dimension and the presentation of the 2D-retention map as a so-called gel plot. In comparison to the very common fraction transfer by one or several collection loops (on-line 2D-LC), the fractionation from the first dimension into well plates provides the following advantages over the use of loops:

- The fractions from the first dimension do not have to be completely reinjected, and thus, they can be used for several different methods in the second dimension and provide a spare volume in case something fails with the second-dimension runs in the first place.
- The injection volume for the second dimension can be adapted to the column dimension used there, and the problem of volume overloading that often enforces the use of larger column diameters and elevated flow rates are avoided.
- The fractions can be manipulated with respect to the solvent composition to make it better compatible with the phase system of the second dimension, and even a chemical derivatization prior to the run in the second dimension is possible

The controlled manipulation of the solvent composition can be crucial for the success of the separation in the second dimension, as a too strongly eluting solvent may completely distort the peak shape there. If the solvent can be modified to provide strong retention, it can help for zone focusing on the column inlet of the second dimension, similarly to the cryo-focusing technique in 2D-GC. Beyond the automated off-line method, the dual UltiMate 3000 and Chromeleon also support on-line 2D-LC and heart-cut techniques.

Although the peak capacities of successfully optimized 2D-LC methods with good orthogonality of both dimensions can never be matched by 1D-LC, it cannot be ignored that 2D-LC implies several disadvantages and challenges. 2D method development can be very time-consuming, and the analysis time of a comprehensive LC × LC is generally quite extended. The instrumental implementation is always difficult, and not only the peak capacities multiply between the two dimensions, but also the sources of error and robustness gaps.

The primary approach for analyzing complex samples with the Vanquish Horizon UHPLC system is to achieve the highest possible peak capacities in one separation dimension and to achieve this in the shortest time possible. In addition to the optimized fluidics for the best possible peak efficiency and the advanced column thermostatting options, the extended pressure capabilities up to 1500 bar support this approach. The application of long and even coupled columns packed with 2.2 μm Acclaim stationary phases or 2.6 μm Accucore superficially porous materials helps to generate total column lengths of 50 cm and more. Such setups achieve plate numbers of 150 000–180 000, and this generates peak capacities of 600–800 in full-range gradients with analysis times of less than 1 h. The use of 1.5 μm Accucore Vanquish solid core materials packed in 10–15 cm columns enables peak capacities of 300–400 in only 10–20 min. Such approaches may not always be appropriate to separate challenging complex samples, but they should be considered as a primary attempt for the sake of simplicity. The inferiority relative to 2D-LC is mainly because it cannot take advantage of the combination of two orthogonal selectivities. Efficiency only can account for high theoretical peak capacities, but it can never replace selectivity, which is the intrinsic origin of the separation in chromatography.

It is hence a natural next step to further extend the capabilities of the Vanquish platform for respective dual capabilities known from the UltiMate 3000 platform that enable 2D heart-cut, multi heart-cut, and even comprehensive LC × LC solutions. These further evolutions are not only about transferring existing techniques to the other platform, but to rethink and fundamentally improve them. Alternative fluidic setups or automation capabilities could, for instance, replace column switching techniques that always require a certain effort for method development. The Vanquish autosampler design is ready for a second completely independent needle arm with separate metering device for independent fluid connections of two LC systems in one. One example to use this technology with a dual-gradient pump is a new way of the so-called Parallel LC, a throughput-increasing solution known from UltiMate 3000 Dual instruments. Instead of switching the autosampler between the fluidic lines of the two methods running in parallel (as on UltiMate 3000), this new design enables to run both methods with fully decoupled fluidics, so that the mechanical access to the sample container (vial/well plate) is the only limitation in the method combination. Other than this, the two methods need no further match for solvent compatibility, and no further method development is required to set up parallel sequences on one instrument.

Another significant potential to increase throughput in laboratories is the integration of automated sample preparation into the HPLC. The Vanquish Charger module that currently only serves as tray feeder was developed together with the Thermo Fisher Scientific automation experts, and it is prepared to interface with additional devices for sample preparation techniques. In this respect, it can act as a sample handler that moves trays and well plates from other devices into the Vanquish UHPLC system. Next to such developments outside the main Vanquish instrument, there are also ongoing developments at the Vanquish autosampler itself beyond the dual needle arm technology. In this context, innovative methods for automated detection of liquid levels and even liquid–liquid phase boundaries are successfully under investigation. One way to achieve this liquid boarder detection is by oscillating the injection needle and measuring the influence of the surrounding medium on the oscillation frequency. This together with advanced pipetting abilities can be used for automated liquid–liquid extraction, as a very useful alternative technique to SPE for sample preparation and cleanup purposes. The one-way use and disposal of the cleanup media can be more easily automated with liquid–liquid extraction than with SPE, and the phase polarities can be easily adjusted.

In open autosamplers, the concept of mechanically independent injection and fractionation devices with separated fluidics is used to provide extended capabilities for workflow integration and throughput enhancement. A future Vanquish dual-arm autosampler could implement such technical possibilities in a stackable standard size autosampler and would significantly reduce bench space requirements relative to advanced open autosamplers that span over a substantial bench width. In addition to such an autosampler, the reduced height of the Vanquish stack with the column compartment at the side enables powerful pump combinations beyond the use of the dual-gradient pump that can still be mounted in a single stack. Automated off-line techniques for comprehensive LC × LC and multistep chromatography workflows are under investigation based on these possibilities.

All the described ongoing developments are not meant to be an outlook for the Vanquish product pipeline but should inform the interested reader on the Thermo Fisher Scientific philosophy of how to meet the current and future requirements of demanding users in modern analysis laboratories.

About the Authors

Stavros Kromidas Stavros Kromidas studied chemistry in Saarbrücken, completing his PhD thesis on optically active phases for HPLC. He subsequently worked as a Sales Manager for Waters, when he founded NOVIA GmbH, an independent consultant company for analytical chemistry. Since 2001 he has been working as a consultant and has given lectures and training courses on HPLC and Validation. Stavros Kromidas has authored, coauthored, and edited numerous of books and articles on HPLC, validation, and quality in analytical chemistry.

Frank Steiner Frank Steiner is currently Scientific Advisor in the HPLC organization of Thermo Fisher Scientific. In this function, he coordinates scientific collaborations with external partners to advance UHPLC technologies and applications. Frank Steiner received his PhD degree in Chemistry in 1995 from Prof. Dr Heinz Engelhardt at the Saarland University in Saarbrücken, Germany, working on development of stationary phases for IC. After a postdoctoral research stay at the CEA, Saclay in France, he returned in 1997 to Saarland University to conduct research on electro-driven separation (nonaqueous CE and CEC), LC purification, and MS coupling technologies and became an assistant professor in 2003. In 2005, Frank Steiner joined the Dionex Softron GmbH in Germering, Germany, now a part of Thermo Fisher Scientific. He played a significant role in developing and launching the UltiMate 3000 UHPLC systems and solutions, as well as the recently introduced Vanquish UHPLC system.

The HPLC Expert II: Find and Optimize the Benefits of your HPLC/UHPLC, First Edition.
Edited by Stavros Kromidas.

Stefan Lamotte Stefan Lamotte studied Chemistry at Saarland University and received his PhD in 1998. The thesis issued by Prof. Heinz Engelhardt was entitled: Synthesis, characterization, and application of nonporous 1.5 μm silica particles for fast HPLC. After working for Bischoff Analysentechnik und geräte GmbH for more than a decade being responsible for HPLC columns and stationary phases, he entered BASF SE in 2011. He is a team leader in the HPLC laboratory in the Competence Center Analytics in Ludwigshafen, Germany. As an expert in chromatography especially for HPLC, he is also part of the chromatographic separation techniques group of the EDQM.

Monika Dittmann Monika Dittmann received her PhD at the Mining Research in Essen with a work on phase equilibrium thermodynamics. She has been working for 2 years as a Humboldt scholar (postdoctoral research) in the Department of Chemical Engineering, University of California, Berkeley. Her work was focused on statistical thermodynamics. Since 1988 Monika Dittmann has been working for Agilent Technologies in Waldbronn, Germany, in the development of analytical instruments. Her area of work encompasses research and optimization of liquid-phase-based separation techniques such as HPLC, UHPLC, LC/MS, capillary electrophoresis, and separations in microstructures.

Michael Heidorn Michael Heidorn has been working as laboratory assistant in the R&D department of Honeywell Specialty Chemicals in Seelze, Germany. After that, he studied chemical engineering in Lübeck, including one practical semester in the R&D department for HPLC columns at the company Merck in Darmstadt. In 2010, he graduated with a work on the influence of frictional heat on column performance in UHPLC at the company Dionex-Softron in Germering. Since 2010 he has been working as Solutions Specialist at Dionex Softron GmbH, a subsidiary of Thermo Fisher Scientific.

Jens Trafkowski is global product manager for analytical HPLC at Agilent Technologies in Waldbronn, Germany, where he is responsible for the 1290 Infinity 2D-LC solution. Before his change to Agilent Technologies in 2011, he was for more than 6 years application specialist for LC-MS with responsibilities encompassing training courses, technical support, and application development. He studied food chemistry and received his PhD at the Institute of Forensic Medicine at the University of Bonn with a work on applications of PLLC-MS/MS in forensic and clinical toxicology.

Jochen Tuerk Jochen Tuerk studied chemistry and received his PhD at the University of Duisburg-Essen with a work on LC-MS/MS method development for the analysis of active pharmaceutical ingredients. He has been working since 2001 at the Institute of Energy and Environmental Technology e. V. (IUTA) in the area of employment protection, as well as environmental, residue, and pharmaceutical analysis. Since 2009 he has been divisional head at IUTA with focus on trace analysis, substance identification with LC-MS/MS and LC-HRMS, and the development of oxidative methods for the treatment of urban and industrial wastewater.

Christoph Portner Upon a professional training as laboratory assistant, Christoph Portner studied Water Science and received his PhD in analytical chemistry from the Carl von Ossietzky University in Oldenburg with a research work on the development of LC-MS methods for mycotoxin detection. Since 2006 he has been staff scientist at the Institute of Energy and Environmental Technology e. V. (IUTA) in Duisburg. His research is focused on identification and quantification of trace matter, transformation products, and metabolites with LC-MS/MS and LC-HRMS, effect-related analysis, and quality management.

Terence Hetzel Terence Hetzel finished his studies in "Instrumental Analytical Chemistry and Laboratory Management" in 2012. He started his PhD at the Institute of Energy and Environmental Technology e. V. (IUTA) in Duisburg, Germany. His research area is the development and characterization of miniaturized liquid chromatographic separation techniques in combination with mass spectrometric detection.

Steffen Wiese Steffen Wiese studied instrumental analytical chemistry and laboratory management and received his PhD from the University Duisburg-Essen, Institute of Analytical Chemistry. Since 2007 he has been staff scientist at the Institute of Energy and Environmental Technology e. V. (IUTA) in Duisburg. His research interests encompass computer-assisted method development in liquid chromatography and the development of chromatographic separation-, coupling-, and detection methods, such as high-temperature HPLC and capillary- and nano-HPLC.

Arno Simon Arno Simon is intensely engaged in chromatographic data systems since a number of years and works as consultant, trainer, and reference book author. As managing partner of beyontics GmbH, he is with his company partner of leading pharmaceutical companies and supports implementation and operation of CDS Waters Empower and Thermo Chromeleon. He was employed at Waters GmbH in the areas of service, support, and distribution until 1997.

Thorsten Teutenberg Thorsten Teutenberg studied chemistry at the Ruhr-University Bochum, Germany, and received his PhD at the Institute of Analytical Chemistry with a work on high-temperature HPLC. He moved to the Institute of Energy and Environmental Technology e. V. (IUTA) in Duisburg in 2004 as staff scientist. Since 2012 he has been leader of the division for research analytics. His research interests include the area of all aspects of high-temperature HPLC, miniaturized separation- and detection techniques, and multidimensional chromatographic methods.

Björn-Thoralf Erxleben Björn-Thoralf Erxleben studied biochemistry and received his PhD from the Institute of Forensic Medicine (Charité) at the Humboldt University in Berlin. Since 1993, he has been Product Specialist HPLC at Shimadzu Europe GmbH, and since end of 2006, he has been Product Manager HPLC/data processing.

Siegfried Chroustovsky After vocational training as laboratory assistant and training as chemical engineer, Siegfried Chroustovsky has been working for 20 years in a biotechnological company producing citric and gluconic acid. He developed a substantial part of electronics and programming for data acquisition for monitoring the fermentation reactions. For 25 years, he has been independent service technician and supervises mainly HP, Agilent, and Waters HPLC equipment.

Tobias Fehrenbach After his academic studies of chemistry, Tobias Fehrenbach received a doctorate on the analysis of amino acids and proteins in atmospheric aerosols at the chair of analytical chemistry of the Technical University of Munich. In 2006, he started his career as product specialist for LC systems at Dionex Softron GmbH, a part of Thermo Fisher Scientific since 2011, in Germering (Germany). In 2010, he became product manager for LC liquid handlers, column compartments, and fitting systems at Thermo Fisher Scientific. He has been self-employed since 2015.

Juri Leonhardt Juri Leonhardt finished his studies with the focus on instrumental analysis and laboratory management in 2011. Then he began his doctorate at the faculty of "Instrumental Analytical Chemistry" at the University Duisburg-Essen in Essen. His research is primarily focused on the development of miniaturized comprehensive multidimensional liquid chromatography systems based on nano- and microliquid chromatography and their hyphenation to different detection techniques. Since 2011 he has been research assistant at the department of Research Analysis at the Institut für Energie und Umwelttechnik e. V. (Institute of Energy and Environmental Technology) in Duisburg.

Index

The HPLC Expert II: Find and Optimize the Benefits of your HPLC/UHPLC, First Edition.
Edited by Stavros Kromidas.

f

g